T0212772

Lecture Notes in Artificial Intelligence　　9091

Subseries of Lecture Notes in Computer Science

LNAI Series Editors

Randy Goebel
  University of Alberta, Edmonton, Canada
Yuzuru Tanaka
  Hokkaido University, Sapporo, Japan
Wolfgang Wahlster
  DFKI and Saarland University, Saarbrücken, Germany

LNAI Founding Series Editor

Joerg Siekmann
  DFKI and Saarland University, Saarbrücken, Germany

More information about this series at http://www.springer.com/series/1244

Denilson Barbosa · Evangelos Milios (Eds.)

Advances in Artificial Intelligence

28th Canadian Conference on Artificial Intelligence,
Canadian AI 2015
Halifax, Nova Scotia, Canada, June 2–5, 2015
Proceedings

 Springer

Editors
Denilson Barbosa
University of Alberta
Edmonton
Canada

Evangelos Milios
Dalhousie University
Halifax
Canada

ISSN 0302-9743   ISSN 1611-3349 (electronic)
Lecture Notes in Artificial Intelligence
ISBN 978-3-319-18355-8   ISBN 978-3-319-18356-5 (eBook)
DOI 10.1007/978-3-319-18356-5

Library of Congress Control Number: 2015937368

LNCS Sublibrary: SL7 – Artificial Intelligence

Springer Cham Heidelberg New York Dordrecht London
© Springer International Publishing Switzerland 2015
This work is subject to copyright. All rights are reserved by the Publisher, whether the whole or part of the material is concerned, specifically the rights of translation, reprinting, reuse of illustrations, recitation, broadcasting, reproduction on microfilms or in any other physical way, and transmission or information storage and retrieval, electronic adaptation, computer software, or by similar or dissimilar methodology now known or hereafter developed.
The use of general descriptive names, registered names, trademarks, service marks, etc. in this publication does not imply, even in the absence of a specific statement, that such names are exempt from the relevant protective laws and regulations and therefore free for general use.
The publisher, the authors and the editors are safe to assume that the advice and information in this book are believed to be true and accurate at the date of publication. Neither the publisher nor the authors or the editors give a warranty, express or implied, with respect to the material contained herein or for any errors or omissions that may have been made.

Printed on acid-free paper

Springer International Publishing AG Switzerland is part of Springer Science+Business Media (www.springer.com)

Preface

The 28th Canadian Conference on Artificial Intelligence (AI 2015) built on a long sequence of successful conferences, bringing together Canadian and international researchers, presenting and discussing original research. The conference was held in Halifax, Nova Scotia, Canada, during June 2–2, 2015, and was collocated with the 41st Graphics Interface Conference (GI 2015), and the 12th Conference on Computer and Robot Vision (CRV 2015).

AI 2015 attracted 81 submissions from Canada and internationally. Each submission was reviewed in double-blind mode by at least three Program Committee members. For the conference and the proceedings 15 regular papers and 12 short papers were accepted, i.e., 18.5% and 15% of the total number of submissions, respectively. Regular papers were allocated 16 pages in the proceedings, while short papers were allocated 8 pages. The proceedings include eight additional papers from the Graduate Student Symposium.

The conference program was enriched by two keynote speakers, two invited speakers, and three tutorials. The academic keynote speaker was Janyce Wiebe, University of Pittsburgh, and the industry keynote speaker was Charles Elkan, Amazon.com and University of California, San Diego. The invited speakers were Csaba Szepesvári, University of Alberta, and Diana Inkpen, University of Ottawa. The tutorials were given by Axel Soto and Evangelos Milios, Paul Hollensen and Thomas Trappenberg, Dalhousie University, and Dominik Ślęzak, University of Warsaw and Infobright.com.

We want to extend our warm thanks to all the individuals who contributed to the success of the conference: Pawan Lingras, Saint Mary's University, and Stan Matwin, Dalhousie University, the General Chairs of the three collocated conferences, Artificial Intelligence, Graphics Interface, and Computer and Robot Vision (AI/GI/CRV); Sageev Oore and Jason Rhinelander, Saint Mary's University, the Local Arrangements Chairs for AI/GI/CRV; Andrew Valencik, the Registration Chair; Danny Silver, Acadia University, and Marina Sokolova, University of Ottawa, the Chairs of the Graduate Student Symposium; Marina Sokolova, the Publicity Chair; Armin Sajadi, AI 2015 Web Site Design. The Program Committee members and external reviewers provided timely and helpful reviews.

AI 2015 was sponsored by the Canadian Artificial Intelligence Association (CAIAC). Nathalie Japkowicz, the President of CAIAC, and the CAIAC Executive Committee, provided essential advice and guidance based on their experience from previous Canadian AI conferences.

June 2015

Denilson Barbosa
Evangelos Milios

Organization

The 28th Canadian Conference on Artificial Intelligence (AI 2015) was sponsored by the Canadian Artificial Intelligence Association (CAIAC), and held in conjunction with the 41st Graphics Interface Conference (GI 2015) and the 12th Conference on Computer and Robot Vision (CRV 2015).

General Chairs

Pawan Lingras   Saint Mary's University, Canada
Stan Matwin   Dalhousie University, Canada

Local Arrangements Chairs

Sageev Oore   Saint Mary's University, Canada
Jason Rhinelander   Saint Mary's University, Canada

Registration Chair

Andrew Valencik   Saint Mary's University, Canada

AI Conference Program Chairs

Denilson Barbosa   University of Alberta, Canada
Evangelos Milios   Dalhousie University, Canada

Graduate Symposium Chairs

Danny Silver   Acadia University, Canada
Marina Sokolova   University of Ottawa, Canada

Publicity Chair

Marina Sokolova   University of Ottawa, Canada

Program Committee

Esma Aimeur	Université de Montréal, Canada
Xiangdong An	York University, Canada
Aijun An	York University, Canada
Sophia Ananiadou	University of Manchester, UK
Ralitsa Angelova	Google Inc., Switzerland
Dirk Arnold	Dalhousie University, Canada
Fahiem Bacchus	University of Toronto, Canada
Ebrahim Bagheri	Ryerson University, Canada
Sotiris Batsakis	University of Huddersfield, UK
Virendra Bhavsar	University of New Brunswick, Canada
Scott Buffett	National Research Council Canada, IIT – e-Business, Canada
Cory Butz	University of Regina, Canada
Eric Charton	École Polytechnique de Montréal, Canada
Lyne Da Sylva	Université de Montréal, Canada
Gerard de Melo	Tsinghua University, China
Joerg Denzinger	University of Calgary, Canada
Angelo Di Iorio	University of Bologna, Italy
Chrysanne Dimarco	University of Waterloo, Canada
Carlotta Domeniconi	George Mason University, USA
Chris Drummond	NRC Institute for Information Technology, Canada
Jocelyne Faddoul	St. Francis Xavier University, Canada
Stefano Ferilli	Università degli Studi di Bari Aldo Moro, Italy
Paola Flocchini	University of Ottawa, Canada
Enrico Francesconi	ITTIG-CNR, Italy
Fred Freitas	Universidade Federal de Pernambuco (UFPE), Brazil
Ricardo José Gabrielli Barreto Campello	Universidade de São Paulo, Brazil
Yong Gao	University of British Columbia Okanagan, Canada
Ali Ghorbani	University of New Brunswick, Canada
Orland Hoeber	University of Regina, Canada
Jimmy Huang	York University, Canada
Frank Hutter	Albert-Ludwigs-Universität Freiburg, Germany
Diana Inkpen	University of Ottawa, Canada
Aminul Islam	Dalhousie University, Canada
Nathalie Japkowicz	University of Ottawa, Canada
Vangelis Karkaletsis	NCSR "Demokritos", Greece
Vlado Kešelj	Dalhousie University, Canada
Fazel Keshtkar	Southeast Missouri State University, USA
Svetlana Kiritchenko	National Research Council Canada, Canada
Ziad Kobti	University of Windsor, Canada
Jacob Kogan	University of Maryland, Baltimore County, USA
Grzegorz Kondrak	University of Alberta, Canada

Leila Kosseim	Concordia University, Canada
Anastasia Krithara	NCSR "Demokritos", Greece
Philippe Langlais	Université de Montréal, Canada
Guy Lapalme	RALI-DIRO, Université de Montréal, Canada
Simone Ludwig	North Dakota State University, USA
Yannick Marchand	Dalhousie University, Canada
Gordon McCalla	University of Saskatchewan, Canada
Wagner Meira Jr.	Universidade Federal de Minas Gerais, Brazil
Robert Mercer	University of Western Ontario, Canada
Marie-Jean Meurs	Concordia University, Canada
Shamima Mithun	Concordia University, Canada
Joseph Modayil	University of Alberta, Canada
Abidalrahman Moh'd	Dalhousie University, Canada
Malek Mouhoub	University of Regina, Canada
Gabriel Murray	University of the Fraser Valley, Canada
Alexandros Nanopoulos	University of Hildesheim, Germany
Raymond Ng	University of British Columbia, Canada
Jian-Yun Nie	Université de Montréal, Canada
Roger Nkambou	Université du Québec à Montréal (UQAM), Canada
Fred Popowich	Simon Fraser University, Canada
Doina Precup	McGill University, Canada
Sheela Ramanna	University of Winnipeg, Canada
Frank Rudzicz	University of Toronto, Canada
Samira Sadaoui	University of Regina, Canada
Fatiha Sadat	Université du Québec à Montréal (UQAM), Canada
Eugene Santos	Dartmouth College, USA
Mohak Shah	Bosch Research, USA
Weiming Shen	National Research Council Canada, Canada
Daniel L. Silver	Acadia University, Canada
Marina Sokolova	University of Ottawa and Institute for Big Data Analytics, Canada
Axel Soto	Dalhousie University, Canada
Bruce Spencer	University of New Brunswick, Canada
Luis Enrique Sucar	INAOE, Mexico
Csaba Szepesvári	University of Alberta, Canada
Stan Szpakowicz	University of Ottawa, Canada
Andrea Tagarelli	University of Calabria, Italy
Thomas Tran	University of Ottawa, Canada
Peter van Beek	University of Waterloo, Canada
Julita Vassileva	University of Saskatchewan, Canada
Julien Velcin	Université de Lyon 2, France
Pooja Viswanathan	University of Toronto, Canada

Xin Wang	University of Calgary, Canada
Leo Wanner	ICREA and Universitat Pompeu Fabra, Spain
Jacek Wolkowicz	Dalhousie University, Canada
Dan Wu	University of Windsor, Canada
Yang Xiang	University of Guelph, Canada
Jingtao Yao	University of Regina, Canada
Zhang Ying	Google Inc., USA
Fei Yu	Microsoft, USA
Harry Zhang	University of New Brunswick, Canada
Xinghui Zhao	Washington State University, USA
Xiaodan Zhu	National Research Council Canada, Canada
Nur Zincir-Heywood	Dalhousie University, Canada

Additional Reviewers

Gerald Penn	University of Toronto, Canada
Ricardo Amorim	Universidade Federal de Pernambuco (UFPE), Brazil
Elnaz Davoodi	Concordia University, Canada
Stephen Makonin	Simon Fraser University, Canada
Matthias Feurer	Albert-Ludwigs-Universität Freiburg, Germany
Katharina Eggensperger	Albert-Ludwigs-Universität Freiburg, Germany
Aaron Klein	Albert-Ludwigs-Universität Freiburg, Germany
Hilário Oliveira	Universidade Federal de Pernambuco (UFPE), Brazil
Mustafa Misir	Albert-Ludwigs-Universität Freiburg, Germany
Heidar Davoudi	York University, Canada
Andrea Pazienza	Università degli Studi di Bari Aldo Moro, Italy
Garrett Nicolai	University of Alberta, Canada
Ehsan Amjadian	Carleton University, Canada
Pablo Jaskowiak	Universidade de São Paulo, Brazil
Thomas Triplet	Centre de recherche informatique de Montréal (CRIM), Canada
Félix-Hervé Bachand	Concordia University, Canada
Leanne Wu	University of Calgary, Canada
Berardina Nadja De Carolis	Università degli Studi di Bari Aldo Moro, Italy
Yan Zhao	Dartmouth College, USA

Sponsoring Institutions

Canadian Artificial Intelligence Association (CAIAC) https://www.caiac.ca/
Saint Mary's University, http://www.smu.ca
Dalhousie University http://www.dal.ca

Theory Versus Practice in Data Science (Industry Keynote Talk)

Charles Elkan

Amazon Fellow, Amazon.com
Seattle, WA, USA &
Professor, Department of Computer Science and Engineering
University of California, San Diego, USA (on leave)

Abstract. In this talk I will discuss examples of how Amazon serves customers and improves efficiency using learning algorithms applied to large-scale datasets. I'll then discuss the Amazon approach to projects in data science, which is based on applying tenets that are beneficial to follow outside the company as well as inside it. Last but not least, I will discuss which learning algorithms tend to be most successful in practice, and I will explain some unsolved issues that arise repeatedly across applications and should be the topic of more research in the academic community. Note: All information in the talk will be already publicly available, and any opinions expressed will be strictly personal.

Biography

Charles Elkan is on leave from being a professor of computer science at the University of California, San Diego, working as Amazon Fellow and leader of the machine learning organization for Amazon in Seattle and Silicon Valley. In the past, he has been a visiting associate professor at Harvard and a researcher at MIT. His published research has been mainly in machine learning, data science, and computational biology. The MEME algorithm that he developed with Ph.D. students has been used in over 3000 published research projects in biology and computer science. He is fortunate to have had inspiring undergraduate and graduate students who are in leadership positions now such as vice president at Google.

An NLP Framework for Interpreting Implicit and Explicit Opinions in Text and Dialog (Academic Keynote Talk)

Jan Wiebe

Department of Computer Science
Co-director, Intelligent Systems Program
University of Pittsburgh, USA

Abstract. While previous sentiment analysis research has concentrated on the interpretation of explicitly stated opinions and attitudes, this work addresses the computational study of a type of opinion implicature (i.e., opinion-oriented inference) in text and dialog. This talk will describe a framework for representing and analyzing opinion implicatures which promises to contribute to deeper automatic interpretation of subjective language. In the course of understanding implicatures, the system recognizes implicit sentiments (and beliefs) toward various events and entities in the sentence, often attributed to different sources (holders) and of mixed polarities; thus, it produces a richer interpretation than is typical in opinion analysis.

Biography

Janyce Wiebe is Professor of Computer Science and Co-Director of the Intelligent Systems Program at the University of Pittsburgh. Her research with students and colleagues has been in discourse processing, pragmatics, and word-sense disambiguation. A major concentration of her research is subjectivity analysis, recognizing and interpretating expressions of opinions and sentiments in text, to support NLP applications such as question answering, information extraction, text categorization, and summarization. Her professional roles have included ACL Program Co-Chair, NAACL Program Chair, NAACL Executive Board member, Transactions of the ACL Action Editor, Computational Linguistics and Language Resources and Evaluation Editorial Board member, AAAI Workshop Co-Chair, ACM Special Interest Group on Artificial Intelligence (SIGART) Vice-Chair, and ACM-SIGART/AAAI Doctoral Consortium Chair. She received her PhD in Computer Science from the State University of New York at Buffalo, and later was a Post-Doctoral Fellow at the University of Toronto.

How to Explore to Maximize Future Return (Invited Talk)

Csaba Szepesvári

Department of Computing Science,
University of Alberta, Edmonton, Canada
csaba.szepesvari@ualberta.ca

Abstract. With access to huge-scale distributed systems and more data than ever before, learning systems that learn to make good predictions break yesterday's records on a daily basis. Although prediction problems are important, predicting what to do has its own challenges, which calls for specialized solution methods. In this talk, by means of some examples based on recent work on reinforcement learning, I will illustrate the unique opportunities and challenges that arise when a system must learn to make good decisions to maximize long-term return.

In particular, I will start by demonstrating that passive data collection inevitably leads to catastrophic data sparsity in sequential decision making problems (no amount of data is big enough!), while clever algorithms, tailored to this setting, can escape data sparsity, learning essentially arbitrarily faster than what is possible under passive data collection. I will also describe current attempts to scale up such clever algorithms to work on large-scale problems. Amongst the possible approaches, I will discuss the role of sparsity to address this challenge in the practical, yet mathematically elegant setting of "linear bandits". Interestingly, while in the related linear prediction problem, sparsity allows one to deal with huge dimensionality in a seamless fashion, the status of this question in the bandit setting is much less understood.

Biography

Csaba Szepesvári is a Professor at the the Department of Computing Science of the University of Alberta and a Principal Investigator of the Alberta Innovates Center for Machine Learning. He received his PhD from the University of Szeged, Hungary in 1999. The coauthor of a book on nonlinear approximate adaptive controllers and the author of a book on Reinforcement Learning, he published about 150 journal and conference papers. He is an Action Editor of the Journal of Machine Learning Research and the Machine Learning Journal. His research interests include reinforcement learning, statistical learning theory and online learning.

Detecting Locations from Twitter Messages (Invited Talk)

Diana Inkpen

School of Electrical Engineering and Computer Science,
University of Ottawa, Canada

Abstract. This talk will focus on machine learning methods for detecting locations from Twitter messages. There are two types of locations that we are interested in: location entities mentioned in the text of each message and the physical locations of the users. For the first type of locations, we detected expressions that denote locations and we classified them into names of cities, provinces/states, and countries. We approached the task in a novel way, consisting in two stages. In the first stage, we trained Conditional Random Field models with various sets of features. We collected and annotated our own dataset for training and testing. In the second stage, we resolved cases when more than one place with the same name exists, by applying a set of heuristics. For the second type of locations, we put together all the tweets written by a user, in order to predict his/her physical location. Only a few users declare their locations in their Twitter profiles, but this is sufficient to automatically produce training and test data for our classifiers. We experimented with two existing datasets collected from users located in the U.S. We propose a deep learning architecture for the solving the task, because deep learning was shown to work well for other natural language processing tasks, and because standard classifiers were already tested for the user location task. We designed a model that predicts the U.S. region of the user and his/her U.S. state, and another model that predicts the longitude and latitude of the user's location. We found that stacked denoising auto-encoders are well suited for this task, with results comparable to the state-of-the-art.

Biography

Diana Inkpen is a Professor at the University of Ottawa, in the School of Electrical Engineering and Computer Science. Her research is in applications of Computational Linguistics and Text Mining. She organized seven international workshops and she was a program co-chair for the AI 2012 conference. She is in the program committees of many conferences and an associate editor of the Computational Intelligence and the Natural Language Engineering journals. She published a book on Natural Language Processing for Social Media (Morgan and Claypool Publishers, Synthesis Lectures on Human Language Technologies), 8 book chapters, more than 25 journal articles and more than 90 conference papers. She received many research grants, including intensive industrial collaborations.

An Introduction to Deep Learning
(Invited Tutorial)

Paul Hollensen and Thomas P. Trappenberg

Faculty of Computer Science,
Dalhousie University, Halifax, Canada
{tt,hollensen}@cs.dal.ca

Abstract. Deep learning is a family of methods that have brought machine learning applications to new heights, winning many competitions in vision, speech, NLP, etc. The centre piece of this method is the learning of many layers of features to represent data at increasing levels of abstraction. While this has immediate consequences for the quality of classification and regression results, there is also some excitement in the AI community with the possibility to fuse such connectionist AI with symbolic AI.

This tutorial is an introduction to deep learning. We will motivate the excitement in this field with a survey of recent state-of-the-art results, and we will outline some of the theory behind representational learning. We will then discuss a small implementation of a convolutional network before discussing some of the software frameworks that make these powerful methods accessible and practical, including how these methods can be accelerated by inexpensive graphics hardware from high-level, productive languages. We will also highlight exemplary examples of published models with freely available implementations that can serve as a starting point for both advancing theory and application to new domains.

Biographies

Paul Hollensen studied Cognitive Science at the University of Toronto followed by Computer Science at Dalhousie University where he is a Ph.D. candidate. His principal research interests are computational neuroscience and hierarchical machine learning methods. He worked extensively with unsupervised and deep learning methods and several implementation platforms.

Thomas P. Trappenberg received a PhD in Physics from RWTH Aachen University, Germany, and he held research positions at Dalhousie University, Canada, the RIKEN Brain Science Institute, Japan, and Oxford University, England. He is a full professor in Computer Science and author of Fundamentals of Computational Neuroscience, now in its second edition. His research interests are computational neuroscience, machine learning, and neurocognitive robotics.

User-Centered Text Mining
(Invited Tutorial)

Axel J. Soto and Evangelos E. Milios

Faculty of Computer Science,
Dalhousie University, Halifax, Canada
{soto,eem}@cs.dal.ca

Abstract. Historically, text mining methods for extracting "knowledge" from text have increased in sophistication by the incorporation of both statistical learning and symbolic natural language processing. However, in scenarios where a domain user aims to make sense of document collections and derive insights from them, domain knowledge is necessary to inform the analysis, and text mining needs to be complemented by text visualization and user-driven interaction with the analytic process.

In this tutorial we introduce Visual Text Analytics as a multi-disciplinary field of research. We cover conceptual and practical methods and tools, review state-of-the-art systems that integrate text mining with visualization and user interaction, and identify promising research directions for making text mining more user-centered and accessible to users with an interest in domain-specific applications.

Biographies

Axel J. Soto is a Research Associate and Adjunct Professor with the Faculty of Computer Science at Dalhousie University (Canada). He received his B.Sc. in Computer Systems engineering and his PhD in Computer Science at Universidad Nacional del Sur (Argentina) in 2005 and 2010, respectively. He now conducts research in the area of Visual Text Analytics. Most of his research focuses on the area of multivariate statistics, machine learning and visualization.

Evangelos E. Milios received a diploma in Electrical Engineering from the NTUA, Athens, Greece, and Master's and Ph.D. degrees in Electrical Engineering and Computer Science from the Massachusetts Institute of Technology. Since July of 1998 he has been with the Faculty of Computer Science, Dalhousie University, Halifax, Nova Scotia, where he has been Associate Dean - Research since 2008. He currently works on modelling and mining of content and link structure of Networked Information Spaces, text mining and visual text analytics.

Rough Sets in KDD
(Invited Tutorial)

Dominik Ślęzak

[1] Institute of Mathematics, University of Warsaw, Poland
[2] Infobright Inc., Poland/Canada
`slezak@{mimuw.edu.pl,infobright.com}`
www.dominikslezak.org

Abstract. Rough sets provide foundations for a number of methods useful in data mining and knowledge discovery, at different stages of data processing, classification and representation. Thus, it is worth expanding rough set notions and algorithms towards real-world environments. We attempt to categorize some ideas of how to scale and apply rough set methods in practice. As case studies, we refer to research projects related to text processing, risk management and sensory measurements, conducted by the rough set team at University of Warsaw. We also give examples of commercial applications, which follow (or may follow) rough set principles in several areas of data analytics. Finally, we demonstrate functionality of some of existing rough set data mining packages, which can be helpful in academic research and for industry purposes.

Biography

Dominik Ślęzak received his Ph.D. in 2002 from University of Warsaw and D.Sc. in 2011 from Institute of Computer Science, Polish Academy of Sciences. In 2005 he co-founded Infobright, where he holds position of chief scientist. He is also associate professor at University of Warsaw. He worked as assistant professor at Polish-Japanese Academy of Information Technology and at University of Regina. He delivered invited talks at over 20 international conferences. He co-organized a number of scientific events, including a role of general program chair of Web Intelligence Congress 2014. He is in editorial board of Springer's Communications in Computer and Information Science and serves as associate editor in several journals. In 2014-2016 he is responsible for conference sponsorships in IEEE Technical Committee on Intelligent Informatics. In 2012-2014 he served as president of International Rough Set Society. His interests include databases, data mining and soft computing. He co-authored over 150 papers. He is also co-inventor in five granted US patents.

This work was partly supported by Polish National Centre for Research and Development (NCBiR) grants O ROB/0010/03/001 in frame of Defence and Security Programmes and Projects and PBS2/B9/20/2013 in frame of Applied Research Programmes, as well as by Polish National Science Centre (NCN) grants DEC-2012/05/B/ST6/03215 and DEC-2013/09/B/ST6/01568.

Contents

NLP, Text and Social Media Mining

Data Mining and Machine Learning

Graduate Student Symposium

Agents, Uncertainty and Games

Exploiting Semantics in Bayesian Network Inference Using Lazy Propagation

Anders L. Madsen[1,2]([⊠]) and Cory J. Butz[3]

[1] HUGIN EXPERT A/S, Aalborg, Denmark
anders@hugin.com
[2] Department of Computer Science, Aalborg University, Aalborg, Denmark
[3] Department of Computer Science, University of Regina, Regina, Canada
butz@cs.uregina.ca

Abstract. Semantics in Bayesian network inference has received an increasing level of interest in recent years. This paper considers the use of semantics in Bayesian network inference using Lazy Propagation. In particular, we describe how the semantics of potentials created during belief update can be determined using the Semantics in Inference algorithm. This includes a description of the necessary properties of Semantics in Inference to make the task feasible to be performed as part of belief update. The paper also reports on the results of an experimental analysis designed to determine the average number of potentials and distributions created during belief update on a set of real-world Bayesian networks.

Keywords: Bayesian network · Inference semantics · Lazy propagation

1 Introduction

A Bayesian network is an efficient and intuitive graphical representation of a joint probability distribution over a set of variables, see, e.g., [7,9,13,15,16,23,25]. As such Bayesian networks have been applied to solve many different types of real-world problems where reasoning with uncertainty is key. Unfortunately, both exact and approximate belief update in Bayesian networks are, in general, *NP-hard* [6,8]. This means that the use of exponential complexity algorithms is justified (unless P=NP). One of the main reasons for the popularity of Bayesian networks is that there exists tools implementing efficient (worst-case exponential complexity) algorithms for belief update.

The semantics in Bayesian network inference has for some years received an increasing level of interest. For instance, a differential semantics in inference in join trees is given by [24], whereas [18] gives a differential semantics in Lazy AR Propagation while [31] considers propagation in Conditional Linear Gaussian Bayesian networks based on semantic modelling. Related work includes a join tree propagation architecture for semantic modelling [5] and a join tree propagation architecture utilizing both Variable Elimination (VE) [29,30] and arc-reversal [2].

© Springer International Publishing Switzerland 2015
D. Barbosa and E. Milios (Eds.): Canadian AI 2015, LNAI 9091, pp. 3–15, 2015.
DOI: 10.1007/978-3-319-18356-5_1

The Semantics in Inference (SI) algorithm was introduced in [3] and refined in [4]. The SI algorithm uses d-separation to denote the semantics in every potential constructed during inference. A potential created during inference is a potential or a conditional probability distribution (CPD) with respect to the joint probability distribution of the Bayesian network. The SI algorithm can determine if the potential is a CPD with respect to the Bayesian network or not.

In this paper, we describe how the SI algorithm can be used to determine the semantics of potentials created by Lazy VE Propagation (LP) [19, 22] during belief update. The SI algorithm determines the semantics of every potential produced as a result of a variable elimination process. Inference using LP in a junction tree consists of three phases; (1) assignment of each CPD to a clique and insertion of evidence, (2) two rounds of message passing to make the junction tree consistent, and (3) calculation of posterior beliefs of each non-evidence variable given the evidence. We perform an experimental analysis on a set of real-world Bayesian networks to determine the average number of potentials and distributions created during message passing and calculation of posterior beliefs. We discuss how the SI algorithm can potentially be applied to improve the efficiency of calculating posterior beliefs.

Determining semantics of the potentials created during inference is of interest to the Bayesian network community [9, 16]. Understanding semantics is useful in practice because these partial results can be reused in subsequent computation [1]. While this paper focuses on establishing the semantics of inference in LP, future work will develop methods for exploiting P-distributions in subsequent LP inference.

This paper is organised as follows. In Section 2 preliminaries and the notation used in the paper are introduced. This section includes a description of LP. Section 3 presents the SI algorithm and provides an example illustrating the ideas presented in this paper and Section 4 describes the combination of LP and SI. Section 5 presents the results of a preliminary experimental analysis while Section 6 concludes the paper.

2   Preliminaries and Notation

Let $\mathcal{X} = \{X_1, \ldots, X_n\}$ be a set of discrete random variables such that $\mathrm{dom}(X)$ is the state space of X and $\|X\| = |\mathrm{dom}(X)|$. A (discrete) *Bayesian Network* $\mathcal{N} = (\mathcal{X}, G, \mathcal{P})$ consists of a set of random variables $\mathcal{X}$, an acyclic, directed graph (DAG) $G = (V, E)$, where $V \sim \mathcal{X}$ is the set of vertices and E is the set of edges, and a set of CPDs $\mathcal{P}$. It represents a factorization of a joint probability distribution into a set of conditional probability distributions:

$$P(\mathcal{X}) = \prod_{X \in \mathcal{X}} P(X \mid \mathrm{pa}(X)), \qquad (1)$$

where $\mathrm{pa}(X)$ denotes the parents of X in G and $\mathrm{fa}(X) = \mathrm{pa}(X) \cup \{X\}$. We let $An(Y)$ denote the ancestors of Y in G and let $De(Y)$ descendants of Y in G.

We use upper case letters, e.g., X_i and Y, to denote variables and lower case letters, e.g., x_j and y, to denote states of variables. Sets of variables are denoted using calligraphic letters, e.g., $\mathcal{X}$ and $\mathcal{Y}$. We only consider hard evidence, i.e., instantiations of variables, and a set of evidence is denoted ϵ, i.e., $\epsilon = \{X_{i_1} = x_{i_1 k_1}, \ldots, X_{i_m} = x_{i_m k_m}\}$. The set of variables instantiated by ϵ is denoted $\mathcal{X}_\epsilon$.

A probability *potential* $\psi(X_1, \ldots, X_n)$ is a non-negative and not-all-zero function over a set of variables while a probability distribution $P(X_1, \ldots, X_n)$ is a probability potential that sums to one [27,28]. A conditional probability potential $\psi(\mathcal{X} \mid \mathcal{Y})$ is a probability potential over $\mathcal{X}$ conditional on $\mathcal{Y}$. For probability potential $\psi(\mathcal{X} \mid \mathcal{Y})$ with *domain* $\mathrm{dom}(\psi) = \mathcal{X} \cup \mathcal{Y}$, we let $\mathrm{head}(\psi) = \mathcal{X}$ denote the *head* (i.e., the conditioned variables) and $\mathrm{tail}(\psi) = \mathcal{Y}$ denote the *tail* (i.e., the conditioning variables) of ψ.

Belief update is defined as the task of computing the posterior marginal distribution $P(X \mid \epsilon)$ for each non-observed variable $X \in \mathcal{X}$ given a set of evidence ϵ.

We define a *query* $Q = (\Psi, T, \epsilon)$ as follows [19]:

Definition 1 (Query). *A query Q is a triple $Q = (\Psi, T, \epsilon)$ where Ψ is a set of probability potentials, $T \subseteq \mathcal{X}$ is the set of variables of interest (the target set), and ϵ is the set of evidence.*

The solution to a query $Q = (\Psi, T, \epsilon)$ is the probability potential $\psi(T \mid \epsilon)$ of the target set T given the evidence ϵ. In LP, we will consider a factorization of $\psi(T, \epsilon)$ a solution to the query Q. Barren variables will have no impact on the solution of a query Q [26]. The notion of *barren variables* is defined as follows [26]:

Definition 2 (Barren Variable). *A variable X is a barren w.r.t. a set $T \subseteq \mathcal{X}$, evidence ϵ, and DAG G, if $X \notin T$, $X \notin \mathcal{X}_\epsilon$ and X only has barren descendants in G (if any).*

The notion of barren variables can be extended to graphs with both directed and undirected edges [19].

A junction tree (a.k.a., join tree) representation $T = (\mathcal{C}, \mathcal{S})$ with cliques $\mathcal{C}$ and separators $\mathcal{S}$ of $\mathcal{N}$ is created by moralisation and triangulation of G (see e.g. [11]). The compilation process of constructing a junction tree is as follows [7]:

1. Moralization: the moral graph G^m of G is formed.
2. Triangulation: the moral graph G^m is triangulated to obtain G^t. A graph is triangulated if every cycle of length greater than three has a chord.
3. Construction of the junction tree: a junction tree is constructed with nodes corresponding to the cliques of G^t (i.e., each clique C is a maximal complete subgraph of G^t).

The cliques of the junction tree are connected by separators such that the so-called junction tree property holds. The junction tree property ensures that whenever two cliques A and B are connected by a path, the intersection $S = A \cap B$ is a subset of every clique and separator on the path. The state space size of clique C is $s(C) = \prod_{X \in C} \|X\|$. The state space size of a separator is defined in the same way.

Belief update in $\mathcal{N}$ can be performed by message passing over T using different algorithms [12,17,22,28]. In most message passing algorithms each clique holds a probability potential and each separator holds two probability potentials: one for each message passed over it.

2.1   Lazy Propagation

LP combines a Shenoy-Shafer propagation scheme with VE for message computation and calculation of marginals [22]. The basic idea of LP is to decompose clique and separator potentials until combination becomes mandatory by a variable elimination operation and to exploit independence relations induced by evidence and barren variables. By maintaining factorizations of clique and separator potentials, LP aims at taking advantage of independence and irrelevance properties induced by evidence in a Shenoy-Shafer message passing scheme.

To simplify the presentation, we take advantage of a generalized notion of clique and separator potentials.

Definition 3 (Clique Potential). *A clique potential on $W \subseteq V$ is a singleton $\pi_W = (\Psi)$ where Ψ is a set of potentials on subsets of W.*

Separator potentials are defined in the same way. We call a potential π_W *vacuous* if $\pi_W = (\emptyset)$. We define the operation of potential *combination* next.

Definition 4 (Combination). *The combination of clique or separator potentials $\pi_{W_1} = (\Psi_1)$ and $\pi_{W_2} = (\Psi_2)$ denotes the potential on $W_1 \cup W_2$ given by $\pi_{W_1} \otimes \pi_{W_2} = (\Psi_1 \cup \Psi_2)$.*

LP adds the following steps [19,22] to the compilation process described above, i.e., after Step 3 perform:

4 Assign a vacuous potential to each clique $C \in \mathcal{C}$.
5 For each node X, assign $P(X \mid pa(X)) \in \mathcal{P}$ to a clique A, which can accommodate it, i.e., $fa(X) \subseteq A$.

After initialization each clique A holds a potential $\pi_A = (\Psi)$. The CPDs are instantiated to reflect the evidence ϵ. The set of clique potentials is invariant during propagation of evidence. The joint potential π_V on $T = (\mathcal{C}, \mathcal{S})$ is:

$$\pi_V = \bigotimes_{A \in \mathcal{C}} \pi_A = \left(\bigcup_{X \in V} \{P(X \mid pa(X))\} \right).$$

Message passing in LP proceeds in two rounds relative to a selected root R of T. First, a *collect* operation from the leaves to R is performed. This is followed by a *distribute* operation from R to the leaves. Messages are passed between cliques via the separators $\mathcal{S}$. The separator $S = A \cap B$ connecting two adjacent cliques A and B stores the two messages passed between A and B during a propagation.

In principle the message $\pi_{A \to B}$, which can be seen as a factorised solution to a query $Q = (\Psi, B, \epsilon)$, where $\Psi = \pi_A \otimes (\otimes_{C \in \text{adj}(A) \setminus \{B\}} \pi_{C \to A})$, is computed as:

$$\pi_{A \to B} = \left(\pi_A \otimes \left(\otimes_{C \in \text{adj}(A) \setminus \{B\}} \pi_{C \to A} \right) \right)^{\downarrow B}, \tag{2}$$

where $\pi_{C \to A}$ is the message passed from C to A. In the message passing process, VE [29,30] is used to eliminate variables. VE as used by LP eliminates X from a set of potentials Ψ to produce Ψ^* as a three-step process:

$$\Psi_X = \{\psi \in \Psi \mid X \in \text{dom}(\psi)\}$$
$$\psi_X = \sum_X \prod_{\psi \in \Psi_X} \psi$$
$$\Psi^* = \Psi \setminus \Psi_X \cup \{\psi_X\}.$$

We will denote $\prod_{\psi \in \Psi_X} \psi$ as ψ^X.

After a full round of message passing the posterior marginal $P(X \mid \epsilon)$ can be computed from any clique or separator containing X using VE on the (relevant) potentials. Assume $X \in C$, then $P(X \mid \epsilon)$ is computed as:

$$P(X \mid \epsilon) \propto \left(\pi_A \otimes \left(\otimes_{C \in \text{adj}(A)} \pi_{C \to A} \right) \right)^{\downarrow X}. \tag{3}$$

3 Semantics in Inference

To introduce and explain the notion of Semantics in Inference, we will use the Extended Student Bayesian Network (ESBN) example [3,16].

3.1 ESBN Example

The ESBN model $\mathcal{N} = (\mathcal{X}, G, \mathcal{P})$ from [16] is shown in Fig. 1 and its optimal junction tree (in total clique state space size) is shown in Fig. 2.

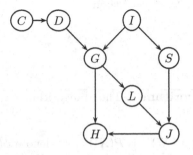

Fig. 1. The ESBN [16]

Fig. 2. A junction tree for the ESBN

3.2   The SI Algorithm

To decide the *semantics* of a CPD $\psi(X|Y)$ is to decide whether or not the probability potential $\psi(X|Y)$ equals $P(X|Y)$. The *Semantics of Inference* (SI) algorithm [3,4] decides the semantics of $\psi(X|Y)$ as follows. If $An(S \cup X) \cap De(S \cup X) = \emptyset$, then $\psi(X|Y) = P(X|Y)$, where S is the set of variables that have been marginalized away to build $\psi(X|Y)$. If $An(S \cup X) \cap De(S \cup X) \neq \emptyset$, then $\psi(X|Y) = \phi(X|Y)$. This condition is used in the SI algorithm shown in Algorithm 1.

For example [3], consider the following calculation on the ESBN:

$$\psi(G, H\,|\,I, J) = P(H\,|\,G, J) \sum_D P(G\,|\,D, I) \sum_C P(C)P(D\,|\,C).$$

Here, $X = \{G, H\}$ and $S = \{C, D\}$. From the ESBN, we obtain $An(X \cup S) = \{I, S, L, J\}$ and $De(X \cup S) = \{L, J\}$. Thereby, $An(X \cup S) \cap De(X \cup S) = \{L, J\}$. By the SI algorithm, $\psi(G, H|I, J)$ is not equal to $P(G, H|I, J)$ calculated as:

$$P(G, H\,|\,I, J) = \frac{P(G, H, I, J)}{P(I, J)}.$$

Data: Let ψ be a probability potential created during inference using VE.
Result: A classification of a probability potential ψ as a ϕ-potential or a
        P-distribution.
begin
    Let $\mathcal{H} = head(\psi)$
    Let $\mathcal{T} = tail(\psi)$
    Let S be the set of variables eliminated to create ψ
    if $An(S \cup \mathcal{H}) \cap De(S \cup \mathcal{H}) = \emptyset$ **then**
       | return $P(\mathcal{H}|\mathcal{T})$
    else
       | return $\phi(\mathcal{H}|\mathcal{T})$
    end
end

Algorithm 1. The SI algorithm.

If SI determines that $\psi(X|Y)$ is $P(X|Y)$, we denote $\psi(X|Y)$ as $P(X|Y)$ and call it a *P-distribution*. Otherwise, we denote $\psi(X|Y)$ as $\phi(X|Y)$ and call it a ϕ-potential. For example, the above $\psi(G, H|I, J)$ is denoted as $\phi(G, H|I, J)$ and called a ϕ-potential.

We need the proposition below in the following sections.

Proposition 1. *Let $\psi(\mathcal{X} \mid \mathcal{Y})$ be a P-distribution (a ϕ-potential), then $\psi(\mathcal{X}_1 \mid \mathcal{Y}) = \sum_{\mathcal{X}_2} \psi(\mathcal{X} \mid \mathcal{Y})$ where $\mathcal{X}_1 \cap \mathcal{X}_2 = \emptyset$ and $\mathcal{X} = \mathcal{X}_1 \cup \mathcal{X}_2$ is a P-distribution (a ϕ-potential).*

Proof. Since $\mathcal{S} \cup \mathcal{X} = \mathcal{S} \cup \mathcal{X}_1 \cup \mathcal{X}_2$ the result follows immediately.

From the proposition it follows that if ψ^X is a P-distribution (ϕ-potential), then ψ_X is a P-distribution (ϕ-potential).

3.3 Example Continued

For the ESBN example, we can compute the marginal $P(H)$ as follows

$$P(H) = \sum_{G,J} P(H \mid G, J) \sum_{S} \sum_{L} P(J \mid L, S) P(L \mid G) \sum_{I} P(I) P(S \mid I) \quad (4)$$

$$\sum_{D} P(G \mid D, I) \sum_{C} P(C) P(D \mid C), \quad (5)$$

where we have used the elimination order $\sigma = (C, D, I, L, S, J, G)$. This operation corresponds to a collect operation to the clique GHJ in Fig. 2, see e.g. [27]. Using this operation $P(H)$ is a P-distribution.

Let us consider the semantics in some of the potentials created during the calculation of $P(H)$. Consider the elimination of C, D, and I. The potentials ψ^X for each $X \in \{C, D, I\}$ are all P-distributions as $An(\mathcal{S} \cup \mathcal{X}) \cap De(\mathcal{S} \cup \mathcal{X}) = \emptyset$ for each potential. Next, for the elimination of L, we have $\psi^L = P(J, L \mid G, S)$ and $\psi_L = P(J \mid G, S)$. In this case, we have $\mathcal{S} = \{C, D, I\}$ and $\mathcal{X} = \{J, L\}$. As $An(\mathcal{S} \cup \mathcal{X}) \cap De(\mathcal{S} \cup \mathcal{X}) \neq \emptyset$, we conclude that ψ_L is a ϕ-potential.

4 Semantics in Inference with Lazy Propagation

Inference in a Bayesian network $\mathcal{N}$ using LP over a junction tree representation T of $\mathcal{N}$ proceeds in two steps as mentioned above. First a full round of message passing over T is performed and subsequently the marginal $P(X \mid \epsilon)$ is computed for each $X \in \mathcal{X} \backslash \mathcal{X}_\epsilon$. In LP, VE is used both for message computation using Eq. 2 and marginal computation using Eq. 3. The question we consider in this work is how often are the potentials produced by VE during LP a ϕ-potential and how often a P-distribution. The answer to this question is not only interesting from a theoretical point of view, but also from a practical point of view.

One option for exploiting the SI algorithm as part of belief update in LP is as follows. Assume we are in the process of computing the posterior marginal $P(X \mid \epsilon)$ from a set of potentials Ψ by eliminating variables in order σ where the length of σ is n. Consider the potential ψ from which the nth variable of σ, i.e., the last variable, is eliminated to produce $P(X \mid \epsilon)$. Let $Y = \sigma(n)$ such that $\psi(X, Y \mid \epsilon')$ where $\epsilon' \subseteq \epsilon$ is the potential from which Y is eliminated. If $\psi(X, Y \mid \epsilon')$ is a P-distribution, then we can compute both $P(X \mid \epsilon')$ and $P(Y \mid \epsilon')$ from it. Notice, however, that the evidence set ϵ' may be a proper subset of the evidence ϵ, i.e., we are not obtaining the posterior marginal given the entire set of evidence ϵ. For instance, consider a chain $X_1 \to X_2 \to X_3 \ldots \to X_n$ where $\epsilon = \{X_1 = x_{i_1}, X_n = x_{i_n}\}$. In the process of computing $P(X_{n-1} \mid \epsilon)$ we will eliminate X_{n-2} from $P(X_{n-1}, X_{n-2} \mid X_1 = x_{i_1})$.

For the analysis not to be dependent on implementation details, we consider the semantics in ψ^X and ψ_X, where $\psi_X = \sum_X \psi^X = \sum_X \prod_{i=1}^{n} \psi_i$, and where $\Psi_X = \{\psi \mid X \in \text{dom}(\psi)\} = \{\psi_1, \ldots, \psi_n\}$. Otherwise, we would have to consider the order in which potentials are combined when $|\Psi_X| > 2$. This could impact the results.

4.1   Example

Continuing Example 3.3, Fig. 3 shows the junction tree T (from Fig. 2) initialised for LP. That is, each CPD $P \in \mathcal{P}$ is assigned to a clique C that can accommodate it, i.e., $\text{dom}(P) \subseteq C$.

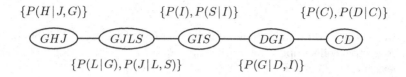

$$\{P(H \mid J, G)\} \qquad \{P(I), P(S \mid I)\} \qquad \{P(C), P(D \mid C)\}$$

$$\boxed{GHJ} - \boxed{GJLS} - \boxed{GIS} - \boxed{DGI} - \boxed{CD}$$

$$\{P(L \mid G), P(J \mid L, S)\} \qquad \{P(G \mid D, I)\}$$

Fig. 3. The ESBN junction tree initialized for LP

Assume clique CD is selected as root of T and $\epsilon = \emptyset$, i.e., no evidence is observed. Fig. 4 shows the messages passed over T using LP. During the collect operation, no potentials are created as all variables to eliminate are barren. Only the potential $\pi_{GIS \to DGI} = (\{P(I)\})$ is non-vacuous and it takes no calculations to create this message.

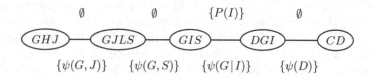

$$\emptyset \qquad \emptyset \qquad \{P(I)\} \qquad \emptyset$$

$$\boxed{GHJ} - \boxed{GJLS} - \boxed{GIS} - \boxed{DGI} - \boxed{CD}$$

$$\{\psi(G, J)\} \quad \{\psi(G, S)\} \quad \{\psi(G \mid I)\} \quad \{\psi(D)\}$$

Fig. 4. Message passing in ESBN junction tree using LP

During the distribute operation, four potentials are created. The potentials are computed as:

$$\psi(D) = \sum_C P(C) P(D \mid C), \tag{6}$$

$$\psi(G \mid I) = \sum_D \psi(D) P(G \mid D, I), \tag{7}$$

$$\psi(G, S) = \sum_I \psi(G \mid I) P(S \mid I) P(I), \tag{8}$$

$$\psi(G, J) = \sum_S \psi(G, S) \psi(J, L \mid G, S)$$

$$= \sum_S \psi(G, S) \sum_L P(J \mid L, S) P(L \mid G). \tag{9}$$

Table 1. Bayesian networks and their junction trees $\hat{T}$ where * means that the triangulation is optimal, ** means that the triangulation has been created using a maximum of 200,000 separators and no * means that the best known triangulation is used

$\mathcal{N}$	$|\mathcal{X}|$	$|\hat{\mathcal{C}}|$	$\max_{C \in \hat{\mathcal{C}}} s(C)$	$s(\hat{T})$
3nt*	58	41	3.5	4.1
Barley*	48	36	6.9	7.2
Diabetes	413	337	4.9	7.0
Hepar_II*	70	58	2.6	3.4
KK*	50	38	6.8	7.1
Mildew*	35	29	6.1	6.5
Munin1	189	162	7.6	7.9
Munin2	1,003	854	5.2	6.3
Munin3	1,044	904	5.2	6.5
Munin4	1,041	877	5.7	6.9
Water*	32	21	5.8	6.5
andes**	223	180	4.8	5.3
cc145*	145	140	3.0	3.6
cc245*	245	235	5.4	5.8
hailfinder*	56	43	3.5	4.0
medianus*	56	44	5.7	6.1
oow*	33	22	6.3	6.8
oow_bas*	33	19	5.7	6.3
oow_solo*	40	29	6.2	6.7
pathfinder*	109	91	4.5	5.3
sacso**	2,371	1,229	5.2	6.0
ship*	50	35	6.6	7.4
system_v57*	85	75	4.8	6.1
win95pts*	76	50	2.7	3.4

From Section 3.3, we know that $\psi(D)$, $\psi(G \mid I)$, and $\psi(G, S)$ are P-distributions while $\psi(J, L \mid G, S)$ is a ϕ-potential. Also, $\psi(G, J)$ is a ϕ-potential. Due to space limitations we do not include the analysis for ψ^X with $X \in \{C, D, I, L, S\}$. After a full round of message passing, marginals can be computed. For instance, the marginal $P(H)$ can be calculated from GHJ as:

$$P(H) = \sum_{J,G} P(H \mid J, G) \psi(J, H). \tag{10}$$

If we substitute recursively, Eq. 6 – 9 into Eq. 10, then we arrive at Eq. 4.

Table 2. The average number and variance of P-distributions and ϕ-potentials created during inference for both a full propagation (FP) and computing marginals only (MO)

	$FP\,\mu_\phi$	$FP\,\sigma^2_\phi$	$FP\,\mu_P$	$FP\,\sigma^2_P$	$MO\,\mu_\phi$	$MO\,\sigma^2_\phi$	$MO\,\mu_P$	$MO\,\sigma^2_P$
3nt	0.74	10.154	87.8	5978.55	0.38	3.369	33.64	886.576
Barley	35.62	1820.581	103.78	12132.8	15.98	590.909	46.34	3568.25
Diabetes	583.12	406584.6	578.76	311083	299.54	125289	150.12	83578.4
Hepar_II	23.88	1481.925	119.08	13559.3	12.22	552.88	65.64	5134.01
KK	35.72	2251.072	118.44	13907.3	15.84	727.166	54.14	4220.79
Mildew	20.22	580.759	63.48	4188.09	7.64	210.859	30.22	1351.83
Munin1	203.04	64697.25	309.5	105377	72.24	10228.1	97.52	16737.4
Munin2	747.86	934233.8	1277.48	1514712	284.54	160056	362.08	178051
Munin3	1064.22	1363521	1295.9	1160251	411.64	239721	337.94	154923
Munin4	754.46	1027148	1389.74	2363543	258.22	167574	410.38	294963
Water	13.84	475.206	108.78	8750.86	8.14	189.354	52.96	2896.85
andes*	462.4	149627.6	428.54	251688	160.46	35240.7	201.46	105046
cc145	7.9	470.495	133.68	16977.2	5.94	336.118	108.3	11802.9
cc245	89.62	15337.43	247.82	79488.8	62.2	9702.99	172.76	46766.5
hailfinder	21.6	589.98	105.5	7789.93	6.54	145.524	48.44	2041.91
medianus	23.38	974.925	118.86	12134.1	9.6	244.444	43.82	2295.28
oow	29.88	1671.743	76.44	4545.42	14.32	477.472	31.66	1246.87
oow_bas	9.54	224.069	52.52	2021.14	3.82	64.008	22.4	487.434
oow_solo	26.44	1530.875	93.7	8614.01	12.08	445.327	43.18	2629.83
pathfinder	4.64	157.324	153.56	19499.4	3.04	127.796	112.5	13648.5
sacso	1050.32	2211989	5182.88	1.8E+07	540.96	894532	2873.88	5822965
ship	36.86	2218.283	120.22	13796.5	20.14	813.112	54.02	3966.99
system_v57	73.32	9030.159	153.48	20889.5	33.8	2742.34	55.5	5594.46
win95pts	26.74	2542.76	193.9	21858.8	14.02	877.939	87	5788

5   Experimental Analysis

This section presents the results of an empirical evaluation of using semantics in Bayesian network inference using LP.

5.1   Setup

Table 1 shows statistics on the Bayesian networks used in the evaluation and their optimal (or believed to be near-optimal) junction tree $\hat{T} = (\hat{\mathcal{C}}, \hat{\mathcal{S}})$. In the table, $|\mathcal{X}|$ denotes the number of variables of $\mathcal{X}$, $|\hat{\mathcal{C}}|$ is the number of elements of $\hat{\mathcal{C}}$, i.e., the number of cliques in $\hat{T}$, $\max_{C \in \hat{\mathcal{C}}} s(C)$ is the state space size of the largest clique in $\hat{\mathcal{C}}$, and $s(\hat{T})$ is the total state space size of all cliques of $\hat{T}$. The state space sizes are on a log-scale in base 10. The junction trees have been generated using the *total weight* and *fill-in-weight* heuristics as implemented in the HUGIN tool [10, 21]. The test set consists of networks of different sizes and complexity in terms of the number of variables and size of the junction tree.

For each network, one hundred sets of evidence were generated at random. For each evidence set, LP computes the posterior marginal distribution of each non-evidence variable. This means that statistics were calculated over one hundred sets of evidence. In the experiments, the *fill-in-weight* heuristic [14] was applied to determine the on-line elimination order when computing messages and posterior marginals [20].

The experiments were performed using a Java implementation (Java (TM) SE Runtime Environment, Standard Edition (build 1.7.0_21-b11)) running on a Linux Ubuntu 14.04 (kernel 3.13.0-39-generic) PC with an Intel Core i7(TM) 920 Processor (2.67GHz) and 12 GB RAM.

5.2 Results

Table 2 shows the results of the experimental analysis performed. In the table, $^{FP}\mu_\phi$ is the average number of ϕ-potentials created during a full propagation (FP), i.e., both message passing and computing marginals, while $^{MO}\mu_\phi$ is the average number of ϕ-potentials created when computing marginals only (MO). Similarly, $^{FP}\mu_P$ is the average number of P-distributions created during a full propagation, while $^{MO}\mu_P$ is the average number of P-distributions created when computing marginals only. The variances are denoted using σ^2 in a similar manner.

The average number of ϕ-potentials (P-distributions) created during both message propagation and calculation of marginals is shown in the first (third) column while the variance is shown in the second (fourth) column. The average number of ϕ-potentials (P-distributions) created during calculation of marginals is shown in the fifth (seventh) column while the variance is shown in the sixth (eight) column.

In most cases, the average number of P-distributions is larger than the average number of ϕ-potentials. For instance, for the Mildew network, we have $^{FP}\mu_\phi = 20.22$ and $^{FP}\mu_P = 63.48$ for both message propagation and marginals while $^{MO}\mu_\phi = 7.64$ and $^{MO}\mu_P = 30.22$ for only marginals. This means that the average number of P-distributions for this network (and evidence) is three to four times higher than average number of ϕ-potentials for both message passing and marginals and marginals only.

It is only for the Diabetes and andes networks that the average number of ϕ-potentials is higher than the number of P-distributions when considering both message propagation and marginals. When considering marginals only, then it is only for the Diabetes and Munin3 networks that the average number of ϕ-potentials is higher than the average number of P-distributions.

6 Conclusion

In this paper, we have described how the SI algorithm can be used in LP to determine the semantics in potentials created during message passing and calculation of marginals. The results of a preliminary experimental analysis involving a set of real-world Bayesian networks of different complexity show that on average the number of P-distributions produced is often larger than the number of ϕ-potentials.

References

1. Butz, C.J., de S. Oliveira, J., Madsen, A.L.: Bayesian network inference using marginal trees. In: van der Gaag, L.C., Feelders, A.J. (eds.) PGM 2014. LNCS (LNAI), vol. 8754, pp. 81–96. Springer, Heidelberg (2014)
2. Butz, C.J., Konkel, K., Lingras, P.: Join tree propagation utilizing both arc reversal and variable elimination. Intl. J. Approx. Rea. **52**(7), 948–959 (2011)
3. Butz, C.J., Yan, W., Madsen, A.L.: d-separation:strong completeness of semantics in bayesian network inference. In: Zaïane, O.R., Zilles, S. (eds.) Canadian AI 2013. LNCS (LNAI), vol. 7884, pp. 13–24. Springer, Heidelberg (2013)
4. Butz, C.J., Yan, W., Madsen, A.L.: On semantics of inference in bayesian networks. In: van der Gaag, L.C. (ed.) ECSQARU 2013. LNCS (LNAI), vol. 7958, pp. 73–84. Springer, Heidelberg (2013)
5. Butz, C.J., Yao, H., Hua, S.: A join tree probability propagation architecture for semantic modelling. J. Int. Info. Sys. **33**(2), 145–178 (2009)
6. Cooper, G.F.: The computational complexity of probabilistic inference using Bayesian belief networks. Artificial Intelligence **42**(2–3), 393–405 (1990)
7. Cowell, R.G., Dawid, A.P., Lauritzen, S.L., Spiegelhalter, D.J.: Probabilistic Networks and Expert Systems. Springer (1999)
8. Dagum, P., Luby, M.: Approximating probabilistic inference in Bayesian belief netwoks is NP-hard. Artificial Intelligence **60**, 141–153 (1993)
9. Darwiche, A.: Modeling and Reasoning with Bayesian Networks. Cambridge University Press (2009)
10. Jensen, F.: HUGIN API Reference Manual, Version 8.1 (2014). www.hugin.com
11. Jensen, F.V., Jensen, F.: Optimal junction trees. In: Proc. of UAI, pp. 360–366 (1994)
12. Jensen, F.V., Lauritzen, S.L., Olesen, K.G.: Bayesian updating in causal probabilistic networks by local computations. Computational Statistics Quarterly **4**, 269–282 (1990)
13. Jensen, F.V., Nielsen, T.D.: Bayesian Networks and Decision Graphs, 2nd (edn.). Springer (2007)
14. Kjærulff, U.B.: Graph triangulation – algorithms giving small total state space. Technical Report R 90–09, University of Aalborg, Denmark (1990)
15. Kjærulff, U.B., Madsen, A.L.: Bayesian Networks and Influence Diagrams: A Guide to Construction and Analysis, 2nd (edn.). Springer (2013)
16. Koller, D., Friedman, N.: Probabilistic Graphical Models – Principles and Techniques. MITPress (2009)
17. Lauritzen, S.L., Spiegelhalter, D.J.: Local computations with probabilities on graphical structures and their application to expert systems. JRSS, B **50**(2), 157–224 (1988)
18. Madsen, A.L.: A differential semantics of lazy propagation. In: Proc. of UAI, pp. 364–371 (2005)
19. Madsen, A.L.: Variations Over the Message Computation Algorithm of Lazy Propagation. IEEE Transactions on Systems, Man. and Cybernetics Part B **36**(3), 636–648 (2006)
20. Madsen, A.L., Butz, C.J.: Ordering arc-reversal operations when eliminating variables in lazy ar propagation. International Journal of Approximate Reasoning **54**(8), 1182–1196 (2013)
21. Madsen, A.L., Jensen, F., Kjærulff, U.B., Lang, M.: HUGIN - The Tool for Bayesian Networks and Influence Diagrams. International Journal on Artificial Intelligence Tools **14**(3), 507–543 (2005)

22. Madsen, A.L., Jensen, F.V.: Lazy propagation: A junction tree inference algorithm based on lazy evaluation. Artificial Intelligence 113(1–2), 203–245 (1999)
23. Neapolitan, R.: Learning Bayesian Networks. Prentice Hall (2003)
24. Park, J.D., Darwiche, A.: A differential semantics for jointree algorithms. Artificial Intelligence 156(2), 197–216 (2004)
25. Pearl, J.: Probabilistic Reasoning in Intelligent Systems: Networks of Plausible Inference. Series in Representation and Reasoning. Morgan Kaufmann Publishers, San Mateo (1988)
26. Shachter, R.D.: Evaluating influence diagrams. Operations Research 34(6), 871–882 (1986)
27. Shafer, G.R.: Probabilistic Expert Systems. SIAM (1996)
28. Shafer, G.R., Shenoy, P.P.: Probability Propagation. Annals of Mathematics and Artificial Intelligence 2, 327–351 (1990)
29. Zhang, N.L., Poole, D.: A simple approach to bayesian network computations. In: Proc. Canadian Conference on AI, pp. 171–178 (1994)
30. Zhang, N.L., Poole, D.: Intercausal independence and heterogeneous factorization. In: Proc. of UAI, pp. 606–614 (1994)
31. Zhu, M., Liu, S., Yang, Y.: Propagation in CLG Bayesian networks based on semantic modeling. Artificial Intelligence Review 38(2), 149–162 (2012)

Darwinian Networks

Cory J. Butz$^{(\boxtimes)}$, Jhonatan S. Oliveira, and André E. dos Santos

Department of Computer Science, University of Regina, Regina, Canada
{butz,oliveira,dossantos}@cs.uregina.ca
http://www.darwiniannetworks.com

Abstract. We suggest *Darwinian networks* (DNs) as a simplification of
working with *Bayesian networks* (BNs). DNs adapt a handful of well-
known concepts in biology into a single framework that is surprisingly
simple, yet remarkably robust. With respect to modeling, on one hand,
DNs not only represent BNs, but also faithfully represent the testing
of independencies in a more straightforward fashion. On the other hand,
with respect to two exact inference algorithms in BNs, DNs simplify each
of them, while unifying both of them.

Keywords: Bayesian networks · d-separation · Variable elimination

1 Introduction

Many different platforms, techniques and concepts can be employed while modeling and reasoning with *Bayesian networks* (BNs) [25]. A problem domain is
modeled initially as a *directed acyclic graph* (DAG), denoted $\mathcal{B}$, and the strengths
of relationships are quantified by *conditional probability tables* (CPTs). Independencies are tested in $\mathcal{B}$ using *d-separation* [25] or *m-separation* [20,27]. Reasoning
with a BN can be done using $\mathcal{B}$, including inference algorithms such as *variable
elimination* (VE) [27] and *arc-reversal* (AR) [24,26]. Considering exact inference
in discrete BNs, a common task, called belief update, is to compute posterior
probabilities given evidence (observed values of variables). Before performing
number crunching, two kinds of variables can safely be removed, namely, *barren* variables [27] and what we call *independent given evidence* variables [23,27].
VE treats the removal of these variables as separate steps. Furthermore, VE
involves multiple platforms. VE first prunes barren variables from a DAG $\mathcal{B}$,
giving a sub-DAG $\mathcal{B}^s$, and then prunes independent by evidence variables from
the *moralization* [21] of $\mathcal{B}^s$, denoted $\mathcal{B}^s_m$. By adapting a few well-known concepts
in biology [10,11,13–15], all of the above can be unified into one platform to be
denoted $\mathcal{D}$.

Darwinian networks (DNs) are put forth to simplify working with BNs. A
CPT $P(X|Y)$ is viewed as a population $p(X,Y)$ with combative traits X and
docile traits Y. More generally, a BN is seen as a set of populations. In DNs, how
populations adapt to the deletion of other populations corresponds precisely with
testing independencies in BNs. Once abstract concepts like merge and replication
are used to represent multiplication, division, and addition, it readily follows

© Springer International Publishing Switzerland 2015
D. Barbosa and E. Milios (Eds.): Canadian AI 2015, LNAI 9091, pp. 16–29, 2015.
DOI: 10.1007/978-3-319-18356-5_2

that DNs can also represent VE and AR. Besides providing a single platform for testing independencies and performing inference, we show how DNs simplify d-separation, m-separation, VE, and AR.

Related works include [16,17]. Dechter has already given *bucket elimination* as an elegant framework for unifying inference. We extend this line of investigation to topics not yet unified, including testing independencies with d-separation and m-separation.

This paper is organized as follows. Section 2 reviews BNs. DNs are introduced in Section 3. Testing independencies is described in Section 4. Section 5 presents inference. Concluding remarks are made in Section 6.

2   Bayesian Networks

2.1   Modeling

Let $U = \{v_1, v_2, \ldots, v_n\}$ be a finite set of variables, each with a finite domain, and V be the domain of U. Let $\mathcal{B}$ denote a *directed acyclic graph* (DAG) on U. A *directed path* from v_1 to v_k is a sequence $v_1, v_2, \ldots, v_k$ with arcs (v_i, v_{i+1}) in $\mathcal{B}$, $i = 1, 2, \ldots, k - 1$. For each $v_i \in U$, the *ancestors* of v_i, denoted $An(v_i)$, are those variables having a directed path to v_i, while the *descendants* of v_i, denoted $De(v_i)$, are those variables to which v_i has a directed path. For a set $X \subseteq U$, we define $An(X)$ and $De(X)$ in the obvious way. The *children* $Ch(v_i)$ and *parents* $Pa(v_i)$ of v_i are those v_j such that $(v_i, v_j) \in \mathcal{B}$ and $(v_j, v_i) \in \mathcal{B}$, respectively. An *undirected path* in a DAG is a path ignoring directions. A *path* in an undirected graph is defined similarly. A singleton set $\{v\}$ may be written as v, $\{v_1, v_2, \ldots, v_n\}$ as $v_1 v_2 \cdots v_n$, and $X \cup Y$ as XY.

D-separation [25] tests independencies in DAGs and can be presented as follows [12]. Let X, Y, and Z be pairwise disjoint sets of variables in a DAG $\mathcal{B}$. We say X and Z are *d-separated* by Y, denoted $I_{\mathcal{B}}(X, Y, Z)$, if at least one valve on every undirected path between X and Z is closed. There are three kinds of valves v: (i) a *sequential* valve means v is a parent of one of its neighbours and a child of the other; (ii) a *divergent* valve is when v is a parent of both neighbours; and, (iii) a *convergent* valve is when v is a child of both neighbours. A valve v is either open or closed. A sequential or divergent valve is *closed*, if $v \in Y$. A convergent valve is *closed*, if $(v \cup De(v)) \cap Y = \emptyset$. For example, suppose $X = a$, $Y = c$, and $Z = f$ in DAG $\mathcal{B}$ depicted in Figure 1 (i). To test $I_{\mathcal{B}}(a, c, f)$ there are two undirected paths from a to f. On the path (a, c), (c, e), (e, f), valve c is closed, since c is a sequential valve and $c \in Y$. Valve d is closed on the other path, since d is a convergent valve and $\{d, h\} \cap Y = \emptyset$. As both paths from a to f have a closed valve, $I_{\mathcal{B}}(a, c, f)$ holds.

M-separation [20,27] is another method for testing independencies in DAGs, and is equivalent to d-separation. Let X, Y, and Z be pairwise disjoint sets of variables in a DAG $\mathcal{B}$. Then m-separation tests $I_{\mathcal{B}}(X, Y, Z)$ with four steps: (i) construct the sub-DAG of $\mathcal{B}$ onto $XYZ \cup An(XYZ)$, yielding $\mathcal{B}^s$; (ii) construct the *moral graph* [21] of $\mathcal{B}^s$, denoted $\mathcal{B}^s_m$, by adding an undirected edge between each pair of parents of a common child and then dropping directionality;

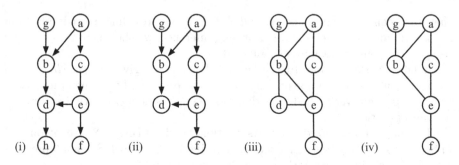

Fig. 1. (i) A DAG $\mathcal{B}$. (ii) Sub-DAG $\mathcal{B}^s$. (iii) Moralization $\mathcal{B}^s_m$. (iv) $\mathcal{B}^s_m$ with d and its edges deleted.

(iii) delete Y and its incident edges; and (iv) if there exists a path from any variable in X to any variable in Z, then $I_\mathcal{B}(X, Y, Z)$ does not hold; otherwise, $I_\mathcal{B}(X, Y, Z)$ holds. For example, in Figure 1, to test $I_\mathcal{B}(a, d, f)$ in $\mathcal{B}$ of (i), the sub-DAG $\mathcal{B}^s$ is in (ii). $\mathcal{B}^s_m$ is shown in (iii). Removing d and incident edges gives (iv). Since there exists a path from a to f, $I_\mathcal{B}(a, d, f)$ does not hold.

A *potential* on V is a function ϕ such that $\phi(v) \geq 0$ for each $v \in V$, and at least one $\phi(v) > 0$. A *uniform* potential on V is a function 1 that sets $1(v) = 1/k$, where $v \in V$, $k = |V|$ and $|\cdot|$ denotes set cardinality. Henceforth, we say ϕ is on U instead of V. A *joint probability distribution* is a potential P on U, denoted $P(U)$, that sums to one. For disjoint $X, Y \subseteq U$, a *conditional probability table* (CPT) $P(X|Y)$ is a potential over XY that sums to one for each value y of Y.

A *Bayesian network* (BN) [25] is a DAG $\mathcal{B}$ on U together with CPTs $P(v_1|Pa(v_1))$, $P(v_2|Pa(v_2))$, ..., $P(v_n|Pa(v_n))$. For example, Figure 2 (i) shows a BN, where CPTs $P(a), P(b|a), \ldots, P(g|e, f)$ are not shown.

We call $\mathcal{B}$ a BN, if no confusion arises. The product of the CPTs for $\mathcal{B}$ on U is a joint probability distribution $P(U)$. The *conditional independence* [25] of X and Z given Y holding in $P(U)$ is denoted $I(X, Y, Z)$. It is known that if $I_\mathcal{B}(X, Y, Z)$ holds by d-separation (or m-separation) in $\mathcal{B}$, then $I(X, Y, Z)$ holds in $P(U)$.

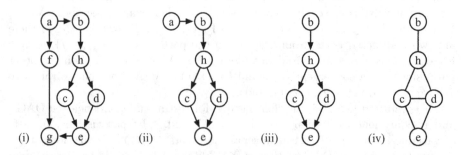

Fig. 2. [27] Given query $P(e|b = 0)$ posed to BN $\mathcal{B}$ in (i), pruning barren variables g and f in (ii) and independent by evidence variable a in (iii). (iv) is $\mathcal{B}^s_m$.

2.2   Inference

VE computes $P(X|Y = y)$ from a BN $\mathcal{B}$ as follows: (i) all barren variables are removed recursively, where v is *barren* [27], if $Ch(v) = \emptyset$ and $v \notin XY$; (ii) all independent by evidence variables are removed, giving $\mathcal{B}^s$, where v is an *independent by evidence* variable, if $I_{\mathcal{B}}(v, Y, X)$ holds in $\mathcal{B}$ by m-separation; (iii) build a uniform distribution $1(v)$ for any root of $\mathcal{B}^s$ that is not a root of $\mathcal{B}$; (iv) set Y to $Y = y$ in the CPTs of $\mathcal{B}^s$; (v) determine an *elimination ordering* [4,18], denoted σ, from the moral graph $\mathcal{B}^s_m$; (vi) following σ, eliminate variable v by multiplying together all potentials involving v, and then summing v out of the product; and, (vii) multiply together all remaining potentials and normalize to obtain $P(X|Y = y)$. For example, in Figure 2, given $P(e|b = 0)$ and BN $\mathcal{B}$ in (i), g and f are barren (ii) and a is independent by evidence (iii) for steps (i) and (ii). In steps (iii) and (iv), VE builds $1(b)$ and updates $P(h|b)$ as $P(h|b = 0)$. Step (v) can determine $\sigma = (c, d, h)$ from $\mathcal{B}^s_m$ shown in (iv). Step (vi) computes (step (vii) is discussed later):

$$P(c, e|d, h) = P(c|h) \cdot P(e|c, d), \tag{1}$$

$$P(e|d, h) = \sum_c P(c, e|d, h), \tag{2}$$

$$P(e|h) = \sum_d P(d|h) \cdot P(e|d, h), \tag{3}$$

$$P(e|b = 0) = \sum_h P(h|b = 0) \cdot P(e|h). \tag{4}$$

AR, unlike VE's step (vi), eliminates variable v_i by reversing the arc (v_i, v_j) between v_i and each child v_j in $Ch(v_i)$. For example, to eliminate variable a in Figure 3 (i), AR must reverse (a,d) in (ii), and then (a,c) in (iii), giving (iv), as:

$$P(a, d|b) = P(a) \cdot P(d|a, b), \tag{5}$$

$$P(d|b) = \sum_a P(a, d|b), \tag{6}$$

$$P(a|b, d) = P(a, d|b) \,/\, P(d|b), \tag{7}$$

$$P(a, c|b, d) = P(a|b, d) \cdot P(c|a, d), \tag{8}$$

$$P(c|b, d) = \sum_a P(a, c|b, d). \tag{9}$$

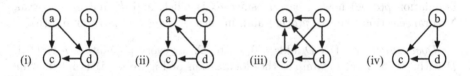

Fig. 3. Applying AR in (5)-(9) to eliminate a

3 Darwinian Networks

Two central tasks are adaptation and evolution.

3.1 Adaptation

A *trait* t can be combative or docile. A *combative* trait t_c is depicted by a clear (white) circle. A *docile* trait t_d is illustrated by a dark (black) circle. A *population* $p(C, D)$ contains a non-empty set CD of traits, where C and D are disjoint, C is exclusively combative, and D is exclusively docile. A population is depicted by a closed curve around its traits. For example, Figure 4 (i) shows eight populations, including $p(b, ag)$, short for $p(\{b\}, \{a, g\})$, illustrated with a closed curve around the (clear) combative trait b and two (dark) docile traits a and g.

Definition 1. *A* Darwinian network *(DN), denoted* $\mathcal{D}$, *is a finite, multiset of populations.*

A DN $\mathcal{D}$ is depicted by a dashed closed curve around its populations. For example, Figure 4 (i) depicts a DN $\mathcal{D} = \{p(a),\ p(b, ag),\ p(c, a),\ p(d, be),\ p(e, c),\ p(f, e),\ p(g),\ p(h, d)\}$, where $p(C, \emptyset)$ is succinctly written $p(C)$.

All combative traits in a given DN $\mathcal{D}$ are defined as $T_c(\mathcal{D}) = \{t_c \mid t_c \in C,$ for at least one $p(C, D) \in \mathcal{D}\}$. All docile traits in $\mathcal{D}$, denoted $T_d(\mathcal{D})$, are defined similarly. For example, considering DN $\mathcal{D}$ in Figure 4 (i), then $T_c(\mathcal{D}) = \{a_c, b_c, c_c, d_c, e_c, f_c, g_c, h_c\}$. In addition, $T_d(\mathcal{D}) = \{a_d, b_d, c_d, d_d, e_d, g_d\}$.

Populations are classified based upon characteristics of their traits. For adaptation, barren populations need only to be classified. Later, for evolution, we will extend the classification.

Given two DNs $\mathcal{D}$ and $\mathcal{D}'$, let t_c be a trait in $T_c(\mathcal{D})$. Trait t_c is *strong*, if $t_c \in T_c(\mathcal{D}')$; otherwise, t_c is *weak*. Trait t_c is *relict*, if $t_d \notin T_d(\mathcal{D})$. The notions of strong, weak, and relict are defined analogously for t_d.

Given DNs $\mathcal{D}$ and $\mathcal{D}'$, a population $p(t_c, D)$ is *barren*, if t_c is relict, and both t_c and t_d are weak. For example, consider the DNs $\mathcal{D}$ in Figure 4 (i) and $\mathcal{D}' = p(acf)$ in Figure 4 (v). Population $p(d, be)$ is not barren, since d_c is not relict. Population $p(h, d)$ is barren, since h_c is relict, and h_c and h_d are weak.

In adaptation, *natural selection* removes recursively all barren populations from a DN $\mathcal{D}$ with respect to another DN $\mathcal{D}'$.

Example 1. Referring to Figure 4, let us apply natural selection on the DN $\mathcal{D}$ in (i) with respect to DN $\mathcal{D}'$ in (v). First, barren population $p(h, d)$ is removed. Population $p(d, be)$ now is barren, since d_c is relict, and d_c and d_d are weak. Natural selection removes $p(d, be)$ and, in turn, $p(b, ag)$ and $p(g)$, giving (ii).

Docilization of a DN $\mathcal{D}$ adds $p(\emptyset, D)$ to $\mathcal{D}$, for every population $p(C, D)$ in $\mathcal{D}$ with $|D| > 1$. For example, the docilization of Figure 4 (ii) is itself, while the docilization of Figure 5 (ii) adds populations $p(\emptyset, ag)$ and $p(\emptyset, be)$, giving Figure 5 (iii).

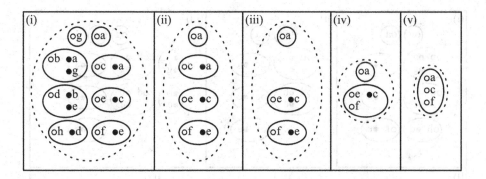

Fig. 4. Testing $I(a, c, f)$

To *delete* a population $p(C, D)$ from a DN $\mathcal{D}$ is to remove all occurrences of it from $\mathcal{D}$. For example, the deletion of $p(c, a)$ from Figure 4 (ii) gives Figure 4 (iii).

Two populations *merge* together as follows: for each trait t appearing in either population, if t is combative in exactly one of the two populations, then t is combative in the merged population; otherwise, t is docile. For example, the merge of populations $p(e, c)$ and $p(f, e)$ in Figure 4 (iii) is population $p(ef, c)$ in Figure 4 (iv).

Let $\mathcal{P}_X$, $\mathcal{P}_Y$, and $\mathcal{P}_Z$ be pairwise disjoint subsets of populations in a DN $\mathcal{D}$ and let DN $\mathcal{D}' = p(C)$, where $C = T_c(\mathcal{P}_X \mathcal{P}_Y \mathcal{P}_Z)$. We test the *adaptation* of $\mathcal{P}_X$ and $\mathcal{P}_Z$ given $\mathcal{P}_Y$, denoted $A(\mathcal{P}_X, \mathcal{P}_Y, \mathcal{P}_Z)$, in $\mathcal{D}$ with four simple steps: (i) let natural selection act on $\mathcal{D}$ with respect to $\mathcal{D}'$, giving $\mathcal{D}^s$; (ii) construct the docilization of $\mathcal{D}^s$, giving $\mathcal{D}_m^s$; (iii) delete $p(C, D)$ from $\mathcal{D}_m^s$, for each $p(C, D)$ in $\mathcal{P}_Y$; and, (iv) after recursively merging populations sharing a common trait, if there exists a population containing both a combative trait in $T_c(\mathcal{P}_X)$ and a combative trait in $T_c(\mathcal{P}_Z)$, then $A(\mathcal{P}_X, \mathcal{P}_Y, \mathcal{P}_Z)$ fails; otherwise, $A(\mathcal{P}_X, \mathcal{P}_Y, \mathcal{P}_Z)$ succeeds.

Example 2. Let us test $A(p(a), p(c, a), p(f, e))$ in the DN $\mathcal{D}$ of Figure 4 (i), where $\mathcal{P}_X = p(a)$, $\mathcal{P}_Y = p(c, a)$, and $\mathcal{P}_Z = p(f, e)$. As $T_c(\{p(a), p(c, a), p(f, e)\}) = \{a_c, c_c, f_c\}$, we obtain the DN $\mathcal{D}' = p(acf)$ in Figure 4 (v). In step (i), natural selection gives $\mathcal{D}^s$ in Figure 4 (ii). In step (ii), docilization of $\mathcal{D}^s$ gives $\mathcal{D}_m^s$ in Figure 4 (ii). In step (iii), the deletion of $p(c, a)$ from $\mathcal{D}_m^s$ gives Figure 4 (iii). Recursively merging populations in step (iv) yields Figure 4 (iv). As no population in Figure 4 (iv) contains a_c in $T_c(p(a))$ and f_c in $T_c(p(f, e))$, $A(p(a), p(c, a), p(f, e))$ succeeds.

Example 3. Let us now test $A(p(a), p(d, be), p(f, e))$ in the DN $\mathcal{D}$ of Figure 5 (i). In this case, DN $\mathcal{D}' = p(adf)$ is shown in Figure 5 (vi). In step (i), natural selection removes barren population $p(h, d)$ as shown in Figure 5 (ii). In step (ii), docilization of Figure 5 (ii) gives Figure 5 (iii). In step (iii), $p(d, be)$ is deleted as depicted in Figure 5 (iv). Recursively merging populations in step (iv) yields Figure 5 (v). By definition, $A(p(a), p(d, be), p(f, e))$ fails, since the population in Figure 5 (v) contains a_c and f_c.

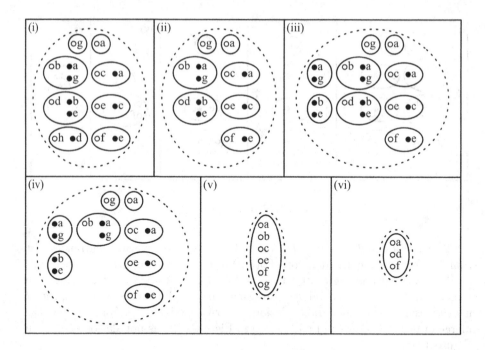

Fig. 5. Testing $I(a, d, f)$

3.2   Evolution

As promised, population classification is extended.

Let $\mathcal{P}_Y = \{p(t_c, D) \mid p(t_c, D) \in \mathcal{D} \text{ and } t_d \in D'\}$ and $\mathcal{P}_Z = \{p(t_c, D) \mid p(t_c, D) \in \mathcal{D} \text{ and } t_c \in C'\}$, given DNs $\mathcal{D}$ and $\mathcal{D}' = p(C', D')$. In $\mathcal{D}$, $p(t_c, D)$ is *independent*, if $A(p(t_c, D), \mathcal{P}_Y, \mathcal{P}_Z)$ succeeds, and is *evident*, if t_d is strong, and D is all relict. Population $p(C, D)$ in a DN $\mathcal{D}$ is *spent*, if there exists $p(C', D)$ in $\mathcal{D}$ such that $C' \subset C$ and $C - C'$ is all relict. In Figure 6, with $\mathcal{D}$ in (ii) and $D' = p(e, b)$ in (xiii), $p(a)$ is independent as $A(p(a), p(b, a), p(e, cd))$ succeeds, where $\mathcal{P}_Y = p(b, a)$ and $\mathcal{P}_Z = p(e, cd)$. In $\mathcal{D}$ of (iii) and D' of (xiii), $p(b, a)$ is evident as b_d is strong, and a_d is relict. In $\mathcal{D}$ of (vi), $p(ce, dh)$ is spent as $p(e, dh)$ is in $\mathcal{D}$ and c_c is relict.

New populations can be created in a DN as follows. *Replication* of a population $p(C, D)$ gives $p(C, D)$, as well as any set of populations $p(C', D)$, where $C' \subset C$. For instance, replication of $p(ce, dh)$ in Figure 6 (v) can yield $p(ce, dh)$ and $p(e, dh)$ in Figure 6 (vi).

The *evolution* of a DN $\mathcal{D}$ into a DN $\mathcal{D}'$ occurs by natural selection removing recursively all barren, independent, and spent populations, merging existing populations, and replicating to form new populations.

Example 4. In Figure 6, consider one explanation of the evolution of $\mathcal{D}$ in (i) into $\mathcal{D}' = p(e, b)$ in (xiii). Natural selection removes barren populations $p(g, ef)$ and $p(f, a)$, yielding (ii). Next, natural selection removes independent population

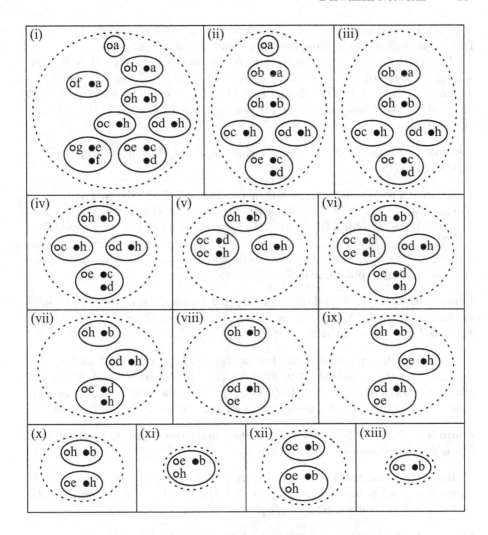

Fig. 6. Graphically representing VE's computation

$p(a)$, giving (iii), and evident population $p(b, a)$, yielding (iv). Then, $p(c, h)$ and $p(e, cd)$ merge to form $p(ce, dh)$ in (v). Replication gives (vi). The rest of the example involves natural selection (vii), merge (viii), replication (ix), natural selection (x), merge (xi), replication (xii), and natural selection (xiii), leaving $\mathcal{D}'$ with population $p(e, b)$.

4   Testing Independencies

Testing independencies in BNs can be seen as testing adaptation in DNs. D-separation can use specialized terminology not referenced in inference such as open sequential valves and closed divergent valves. In contrast, no specialized concepts are used in adaptation. And whereas m-separation requires DAGs, sub-DAGs, and moral graphs, adaptation uses but one platform.

$\mathcal{D} = \{p(v, Pa(v)) \mid P(v|Pa(v)) \text{ is in } \mathcal{B}\}$ is the DN for a given BN $\mathcal{B}$. Conversely, the *directed graph* (or simply *graph*) $\mathcal{G}(\mathcal{D})$ of a DN $\mathcal{D}$ has variables $T_c(\mathcal{D})$ and arcs $\{(v_i, v_j) \mid p(C, D) \in \mathcal{D} \text{ and } v_i \in D \text{ and } v_j \in C\}$. The *undirected graph* $U(\mathcal{D})$ of a DN $\mathcal{D}$ has variables $T_c(\mathcal{D})$ and edges $\{(v_i, v_j) \mid p(C, D) \in \mathcal{D}$ and $v_i, v_j \in CD\}$.

Lemma 1. *Every BN $\mathcal{B}$ can be represented as a DN $\mathcal{D}$, and the graph of $\mathcal{D}$ is $\mathcal{B}$, that is, $\mathcal{G}(\mathcal{D}) = \mathcal{B}$.*

Omitted proofs will be provided in [5]. The BN $\mathcal{B}$ in Figure 1 (i) can be represented as the DN $\mathcal{D}$ in Figure 4 (i). The graph of $\mathcal{D}$ is $\mathcal{B}$, i.e., $\mathcal{G}(\mathcal{D}) = \mathcal{B}$.

Let $\mathcal{D}$ be the DN for a BN $\mathcal{B}$ on U. The populations for $W \subseteq U$, denoted $\mathcal{P}_W$, are $\mathcal{P}_W = \{p(C, D) \mid p(C, D) \in \mathcal{D} \text{ and } C \subseteq W\}$. Thus, given pairwise disjoint subsets X, Y, and Z in $\mathcal{B}$, it is necessarily the case that $\mathcal{P}_X$, $\mathcal{P}_Y$, and $\mathcal{P}_Z$ are pairwise disjoint populations in $\mathcal{D}$.

Lemma 2. *Let $\mathcal{B}^s$ be the sub-DAG constructed from a BN $\mathcal{B}$ in step (i) of testing the independence $I_\mathcal{B}(X, Y, Z)$ using m-separation. Then $\mathcal{B}^s = \mathcal{G}(\mathcal{D}^s)$, where $\mathcal{D}^s$ is the DN constructed in step (i) of testing $A(\mathcal{P}_X, \mathcal{P}_Y, \mathcal{P}_Z)$ in the DN $\mathcal{D}$ for $\mathcal{B}$.*

Step (i) of m-separation when testing $I_\mathcal{B}(a, d, f)$ in the BN $\mathcal{B}$ of Figure 1 (i) constructs the sub-DAG $\mathcal{B}^s$ in Figure 1 (ii). On the other hand, step (i) of adaptation when testing $A(p(a), p(d, be), p(f, e))$ in the DN $\mathcal{D}$ in Figure 5 (i) constructs the DN $\mathcal{D}^s$ in Figure 5 (ii). As guaranteed by Lemma 2, $\mathcal{B}^s = \mathcal{G}(\mathcal{D}^s)$.

Lemma 3. $\mathcal{B}^s_m = U(\mathcal{D}^s_m)$, *where $\mathcal{B}^s_m$ is the moralization of $\mathcal{B}^s$ in Lemma 2, and $\mathcal{D}^s_m$ is the docilization of $\mathcal{D}^s$ in Lemma 2.*

Recall the moralization $\mathcal{B}^s_m$ in Figure 1 (iii) and the docilization $\mathcal{D}^s_m$ in Figure 5 (iii), when testing $I_\mathcal{B}(a, d, f)$ and $A(p(a), p(d, be), p(f, e))$, respectively. As Lemma 3 guarantees, $\mathcal{B}^s_m = U(\mathcal{D}^s_m)$.

Lemma 4. *The undirected graph of the DN obtained by deleting the populations in $\mathcal{P}_Y$ from $\mathcal{D}^s_m$ is the same graph obtained by deleting Y and its incident edges from $\mathcal{B}^s_m$, where $\mathcal{D}^s_m$ and $\mathcal{B}^s_m$ are in Lemma 3.*

When testing $A(p(a), p(d, be), p(f, e))$, deleting population $p(d, be)$ in $\mathcal{P}_Y$ from Figure 5 (iii) gives Figure 5 (iv). The undirected graph of the DN in Figure 5 (iv) is Figure 1 (iv). This is the same graph obtained by deleting variable d and incident edges from $\mathcal{B}^s_m$ in Figure 1 (iii) in testing $I_\mathcal{B}(a, d, f)$ using m-separation.

Theorem 1. $I_\mathcal{B}(X, Y, Z)$ *holds in a BN $\mathcal{B}$ if and only if $A(\mathcal{P}_X, \mathcal{P}_Y, \mathcal{P}_Z)$ succeeds in the DN $\mathcal{D}$ for $\mathcal{B}$.*

Proof. ($\Rightarrow$) Suppose $I_\mathcal{B}(X, Y, Z)$ holds in a BN $\mathcal{B}$. Let $\mathcal{B}^s$ be constructed in step (i) of m-separation when testing $I_\mathcal{B}(X, Y, Z)$. Let $\mathcal{B}^s_m$ be the moralization of $\mathcal{B}^s$ in step (ii) of m-separation. Let $\mathcal{D}$ be the DN for $\mathcal{B}$. Let $\mathcal{D}^s$ be constructed in step (i) of adaptation when testing $A(\mathcal{P}_X, \mathcal{P}_Y, \mathcal{P}_Z)$ in $\mathcal{D}$. Let $\mathcal{D}_m^s$ be the docilization of $\mathcal{D}^s$ in step (iii) of adaptation. By Lemma 1, $\mathcal{G}(\mathcal{D}) = \mathcal{B}$. By Lemma 2, $\mathcal{G}(\mathcal{D}^s) = \mathcal{B}^s$. By Lemma 3, $U(\mathcal{D}_m^s) = \mathcal{B}_m^s$. By Lemma 4, the undirected graph of the DN obtained by deleting the populations in $\mathcal{P}_Y$ from $\mathcal{D}_m^s$ is the same graph obtained by deleting Y and its incident edges from $\mathcal{B}_m^s$. By assumption, there is no path from a variable in X to a variable in Z after deleting Y and its incidents edges from $\mathcal{B}_m^s$. By construction, the recursive merging process in step (iv) of adaptation corresponds precisely with path existence in step (iv) of m-separation. Hence, there is no population containing both a combative trait in $T_c(\mathcal{P}_X)$ and a combative trait in $T_c(\mathcal{P}_Z)$. By definition, $A(\mathcal{P}_X, \mathcal{P}_Y, \mathcal{P}_Z)$ succeeds. ($\Leftarrow$) Follows in a similar fashion as for ($\Rightarrow$).

Theorem 1 indicates that testing adaptation in DNs can be used to test independencies in a BN $\mathcal{B}$ replacing d-separation and m-separation. $I_\mathcal{B}(a, c, f)$ holds by d-separation in Figure 1 (i) and $A(p(a), p(c, a), p(f, e))$ succeeds in Example 2. Similarly, $I_\mathcal{B}(a, d, f)$ does not hold in Figure 1 (i) by m-separation and $A(p(a), p(d, be), p(f, e))$ fails as shown in Example 3.

5   Performing Inference

BN inference using VE and AR is unified and simplified using DN evolution.

Recall how VE computes query $P(e|b = 0)$ posed to the BN $\mathcal{B}$ in Figure 2 (i). Referring to Figure 6, $\mathcal{B}$ is $\mathcal{D}$ in (i), while $P(e|b = 0)$ is DN $\mathcal{D}'$ in (xiii). The removal of barren populations $p(g, ef)$ and $p(f, a)$ in (ii) corresponds to VE pruning barren variables g and f in Figure 2 (ii). Natural selection removes independent population $p(a)$ in (iii) and VE removes independent by evidence variable a in Figure 2 (iii). VE then builds $1(b)$ for the evidence variable b, while natural selection removes evident population $p(b, a)$ in (iv). As for the elimination of c, d, and h in (1) - (4): the multiplication in (1) is the merge of $p(c, h)$ and $p(e, cd)$ in (iv), yielding $p(ce, dh)$ in (v); the marginalization in (2) is the replication $p(ce, dh)$ and $p(e, dh)$ in (vi), followed by the removal of spent population $p(ce, dh)$ in (vii); (3) is shown in (vii) - (x); and, (4) is in (x) - (xiii).

The robustness of DNs only is partially revealed in this example in which DNs detect and remove barren variables, detect and remove an independent by evidence variable, and represent multiplication and marginalization to eliminate variables c, d, and h. Below, we show how DNs represent division and normalization akin to VE's step (vii). In [4], we show how DNs can determine the elimination ordering $\sigma = (c, d, h)$ used above by VE. And, VE can impose superfluous restrictions on the order in which variables are removed by using separate steps, i.e., barren in (i), independent by evidence in (ii), and evidence in (iii). For instance, if $P(f|c)$ is posed to $\mathcal{B}$ in Figure 1 (i), then VE necessarily removes variables h, d, b, g, a, c in this order. In DNs, natural selection can remove populations following this order, or a, c, h, d, b, g, or even a, h, c, d, b, g, for instance.

Note that $p(e, b)$ above represents $P(e|b)$ and not $P(e|b = 0)$. Additional notation, explicitly denoting evidence values such as $p(e, b = 0)$ could be introduced,

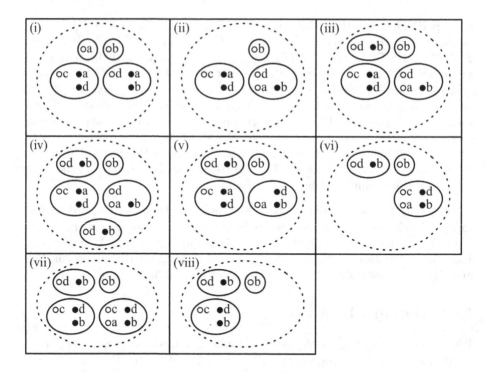

Fig. 7. Graphically representing AR's computation

but will not be, since trait characteristics and population classification would remain unchanged.

Now recall the AR example in (5) - (9) and consider the DN evolution in Figure 7. The BN $\mathcal{B}$ in Figure 3 (i) is the DN $\mathcal{D}$ in Figure 7 (i). The multiplication $P(a, d|b) = P(a) \cdot P(d|a, b)$ in (5) is the merge of populations $p(a)$ and $p(d, ab)$ in Figure 7 (i), yielding $p(ad, b)$ in Figure 7 (ii). The marginalization in (6) is the replication in Figure 7 (iii). And, the division $P(a|b, d) = P(a, d|b)/P(d|b)$ in (7) can be the replication $p(ad, b)$ and $p(d, b)$ in Figure 7 (iv), followed by the merge of $p(ad, b)$ and $p(d|b)$, giving $p(a, bd)$, in Figure 7 (v). The multiplication in (8) is the merge in Figure 7 (vi). Lastly, the marginalization in (9) is the replication in Figure 7 (vii), followed by natural selection removing spent population $p(ac, bd)$, giving Figure 7 (viii). Observe that DN evolution simplifies AR in that merge can represent both multiplication and division. Thus, DNs can represent VE's step (vii), which multiplies to obtain $P(X, Y = y)$, marginalizes to yield $P(Y = y)$, and divides to obtain $P(X|Y = y)$.

6   Concluding Remarks

While many works have generalized BNs, we seek to simplify reasoning with BNs. DNs, a biological perspective of BNs, are surprisingly simple, yet remarkably robust. DNs can represent the testing of independencies using d-separation and m-separation, as well as belief update using VE and AR. DNs simplify each of these separate techniques, while unifying them into one network. The DN keystone is the novel representation of a population $p(C, D)$ using both combative traits C (coloured clear) and docile traits D (coloured dark).

DNs offer both theoretical and practical benefits. Recall how the testing of independence $I(a, d, f)$ using m-separation in Figure 1 is represented as testing adaptation in Example 3. The moralization step adds undirected edges (a, g) and (b, e) in Figure 1 (iii), while docilization adds populations $p(\emptyset, ag)$ and $p(\emptyset, be)$ in Figure 5 (iii). The important point is that adding $p(\emptyset, ag)$ is wasteful, since the subsequent merge in step (iv) of adaptation of population $p(b, ag)$ and $p(\emptyset, ag)$ yields $p(b, ag)$ itself. As docilization is designed to perfectly mimic moralization, this implies that adding edge (a, g) is wasteful. Clearly this is the case, since analysis of Figure 1 (iii) reveals that the addition of (a, g) does not change the connectivity of the undirected graph. Returning once more to DNs, a population $p(\emptyset, D)$ need only be added when the corresponding population $p(C, D)$ will be deleted in step (iii) of adaptation. For m-separation, this means that undirected edges need only be added between parents for a common child that is in the conditioning set (Y in $I_B(X, Y, Z)$). For example, edge (b, e) needs to be added in Figure 1 (iii), since common child d is in Y for $I_B(a, d, f)$. In stark contrast, edge (a, g) does not need to be added in Figure 1 (iii), since common child b is not in Y for $I_B(a, d, f)$. The above discussion reveals that m-separation can be wasteful and thereby can be simplified. In [7], we simplify both d-separation and m-separation. For our discussion here, the salient feature is that DN analysis provides a deeper understanding of BNs.

An important practical consideration in BN inference is determining good *elimination orderings* [18]. Empirical results show that *min-neighbours* (MN), *min-weight* (MW), *min-fill* (MF), and *weighted-min-fill* (WMF) are four heuristics that perform well in practice [19]. Given a query $P(X|Y)$ posed to a BN $\mathcal{B}$, all variables except XY are recursively eliminated from the moralization $\mathcal{B}_m^s$ based upon a minimum score $s(v)$. In [4], we show how these four heuristics can be represented in DNs. More importantly, we propose a new heuristic, called *potential energy* (PE), based on DNs themselves. Our analysis of PE shows that it can: (i) better score a variable; (ii) better model the multiplication of the probability tables for the chosen variable; (iii) more clearly model the marginalization of the chosen variable; and (iv) maintain a one-to-one correspondence between the remaining variables and probability tables.

DNs themselves are a unification of our past work. The graphical depiction of DNs is based upon [1]. The structure of a population $p(C, D)$ corresponds to CPT structure $P(C|D)$, first established for join tree propagation [9] and then extended with semantics for VE [2,8]. As *lazy propagation* (LP) [23] can be represented in DNs [6], our future work will look at exploiting DNs to improve AR orderings in LP [22] and combining AR and VE in propagation [3].

References

1. Butz, C.J., Hua, S., Chen, J., Yao, H.: A simple graphical approach for understanding probabilistic inference in bayesian networks. Information Sciences **179**, 699–716 (2009)
2. Butz, C.J., Yan, W., Madsen, A.L.: d-separation: strong completeness of semantics in bayesian network inference. In: Zaïane, O.R., Zilles, S. (eds.) Canadian AI 2013. LNCS (LNAI), vol. 7884, pp. 13–24. Springer, Heidelberg (2013)
3. Butz, C.J., Konkel, K., Lingras, P.: Join tree propagation utilizing both arc reversal and variable elimination. International Journal of Approximate Reasoning **52**(7), 948–959 (2011)
4. Butz, C.J., Oliveira, J.S., dos Santos, A.E.: Determining good elimination orderings with darwinian networks. In: Twenty-Eighth International FLAIRS Conference (2015, to appear)
5. Butz, C.J., Oliveira, J.S., dos Santos, A.E.: On testing independencies in darwinian networks (2015, unpublished)
6. Butz, C.J., Oliveira, J.S., dos Santos, A.E.: Representing lazy propagation in darwinian networks (2015, unpublished)
7. Butz, C.J., Oliveira, J.S., dos Santos, A.E.: Simplifying d-separation in bayesian networks (2015, unpublished)
8. Butz, C.J., Yan, W.: The semantics of intermediate cpts in variable elimination. In: Fifth European Workshop on Probabilistic Graphical Models, pp. 41–49 (2010)
9. Butz, C.J., Yao, H., Hua, S.: A join tree probability propagation architecture for semantic modeling. Journal of Intelligent Information Systems **33**(2), 145–178 (2009)
10. Campbell, N.A., Reece, J.B.: Biology. Pearson Benjamin Cummings (2009)
11. Coyne, J.A.: Why Evolution is True. Oxford University Press (2009)
12. Darwiche, A.: Modeling and Reasoning with Bayesian Networks. Cambridge University Press (2009)
13. Darwin, C.: On the Origin of Species. John Murray (1859)
14. Dawkins, R.: The Selfish Gene. Oxford University Press (1976)
15. Dawkins, R.: The Greatest Show on Earth: The Evidence for Evolution. Free Press (2009)
16. Dechter, R.: Bucket elimination: A unifying framework for reasoning. In: Twelfth Conference on Uncertainty in Artificial Intelligence, pp. 211–219 (1996)
17. Dechter, R.: Bucket elimination: A unifying framework for reasoning. Artificial Intelligence **113**(1), 41–85 (1999)
18. Kjærulff, U.: Triangulation of graphs - algorithms giving small total state space. Tech. Rep. R90–09, Aalborg University, Denmark, March 1990
19. Koller, D., Friedman, N.: Probabilistic Graphical Models: Principles and Techniques. MIT Press (2009)
20. Lauritzen, S.L., Dawid, A.P., Larsen, B.N., Leimer, H.G.: Independence properties of directed markov fields. Networks **20**, 491–505 (1990)
21. Lauritzen, S.L., Spiegelhalter, D.J.: Local computation with probabilities on graphical structures and their application to expert systems. Journal of the Royal Statistical Society **50**, 157–244 (1988)
22. Madsen, A.L., Butz, C.J.: Ordering arc-reversal operations when eliminating variables in lazy ar propagation. International Journal of Approximate Reasoning **54**(8), 1182–1196 (2013)

23. Madsen, A.L., Jensen, F.V.: Lazy propagation: A junction tree inference algorithm based on lazy evaluation. Artificial Intelligence **113**(1–2), 203–245 (1999)
24. Olmsted, S.: On Representing and Solving Decision Problems. Ph.D. thesis, Stanford University (1983)
25. Pearl, J.: Probabilistic Reasoning in Intelligent Systems: Networks of Plausible Inference. Morgan Kaufmann (1988)
26. Shachter, R.D.: Evaluating influence diagrams. Operations Research **34**(6), 871–882 (1986)
27. Zhang, N.L., Poole, D.: A simple approach to bayesian network computations. In: Tenth Canadian Conference on Artificial Intelligence, pp. 171–178 (1994)

Improvements to the Linear Optimization Models of Patrol Scheduling for Mobile Targets

Jean-Francois Landry[1], Jean-Pierre Dussault[2], and Éric Beaudry[1(✉)]

[1] Université du Québec à Montréal (UQAM), Montreal, Canada
{landry.jean-Francois,beaudry.eric}@uqam.ca
[2] Université de Sherbrooke, Sherbrooke, Canada
dussault.jean-pierre@usherbrooke.ca

Abstract. Recent work in the field of security for the generation of patrol schedules has provided many solutions to real-world situations by using Stackelberg models and solving them with linear programming software. More specifically, some approaches have addressed the difficulties of defining patrol schedules for moving targets such as ferries to minimize their vulnerabilities to terrorist threats. However, one important aspect of these types of problems that hasn't been explored yet concerns the concept of time-windows. In this work, we show the relevance of considering time-windows when generating solutions such as to attend to a broader class of problems and generate sound solutions. We propose some improvements to the model for the generation of patrol schedules for attacks on mobile targets with adjustable time durations while keeping the constraints linear to take advantage of linear programming solvers. To address the scalability issues raised by this new model, we propose a general column-generation approach composed of a master and slave problem, which can also be used on the original problem of patrol generation without time-windows. Finally, we discuss and propose a new two-phase equilibrium refinement approach to improve the robustness of the solutions found.

1 Introduction

In the last few years, there has been a growing interest in the field of security to improve and automate some of the complex tasks which would normally need to be done by a field expert. The tasks we are referring to relate to the definition of optimal patrol schedules to protect specific locations or targets. Traditionally, these tasks have been done by hand using expert knowledge of the problem to find the optimal schedule which would minimize the chances for a hostile agent to successfully perform an attack or other form of single action.

Without specific metrics to objectively quantify the quality of the schedules generated, a great risk is always present that a small window of vulnerability may have been overlooked, leaving some targets exposed to potential attacks.

J.-F. Landry–This work was funded by MITACS, Menya Solutions inc. and the Natural Sciences and Engineering Research Council of Canada (NSERC).

© Springer International Publishing Switzerland 2015
D. Barbosa and E. Milios (Eds.): Canadian AI 2015, LNAI 9091, pp. 30–41, 2015.
DOI: 10.1007/978-3-319-18356-5_3

A natural approach to solve these types of problems is to model them using game theory. The problem is formulated in such a way that the goal is to find an optimal set of schedules that minimize the probability of success for hostile agents. More specifically, Stackelberg models allow for a representation of this class of problems such that they can often be solved using standard linear programming software very efficiently. Such a model is based on the premise that a defender will select a mixed-strategy, knowing that a potential attacker has a full knowledge of this strategy. It becomes therefore mandatory to minimize the worst-case scenario when searching for a solution. An optimum solution found is called a Strong Stackelberg Equilibrium (see [1] for more details). Various flavors of security problems have been explored in [2], [3], [4], [5] where aspects such as the number of patrollers present, the topology of the graphs or the type of actions of potential attackers. Many real-life examples can be also be found in [6] describing how many of these challenges have been confronted.

The specific problem addressed in this paper relates to the problem of moving targets, previously studied in [7], for which a linear program model was proposed with discretized values for time and distances to generate the patrol schedules. The solution provided was very clever but had some scalability issues for larger problems (as studied for the discrete case in [8]).

In this work, we provide a scalable model for the continuous version of this problem. Our contribution consists in adding the notion of a minimal time window required by the attacker to successfully complete an attack. In order to do so, we work on a full representation of the problem rather than a compact one. This different representation allows for a column generation approach to be used where a master problem will work on a subset of all the paths possible in the problem, and a slave problem will generate new paths for the master problem by solving a shortest path problem.

We then propose a two-phase equilibrium refinement approach where some of the constraints of the problems are relaxed in a second phase of the algorithm to minimize the utilities subject to a fixed bound. Finally we present some results showing a great improvement for the utility values for the refinement approach.

2   Domain Definition

The problem we are trying to solve concerns moving targets, in this case ferries, which have a fixed regular schedule moving between point A and point B. The schedule of a target is defined as a piecewise linear function normalized between 0 and 1. More precisely, $F_q : q = 1..Q$ is a set of Q mobile targets with their corresponding schedules. These targets can be attacked anywhere on their trajectories, and we are trying to define the schedules of patrols to cover the targets and prevent the chances of successful attacks, weighted by a utility function. An example with three different targets can be seen in Figure 1, each traveling on a different schedule. We define the utility gained by an attacker is defined as a piecewise linear function with a value between 0 and 10. The time is discretized in M points, or $M-1$ regular intervals and the distance is discretized in N points at intervals (also $N-1$ regular intervals), both normalized between 0 and 1. This discrete representation results in a a transition graph as displayed in Figure 1. Each node in the graph corresponds to a variable $p(i,k)$, for which we are trying to define the optimal value. It is the probability that the node i will be covered at time-step k by a patrol. To keep graph consistency, it is also necessary to have arc variables $f(i,j,k)$ on which constraints will be applied. Instead of associating probabilities to a list of a full trajectories, probabilities are associated to each $f(i,j,k)$ where a patrol goes from node i to node j at time-step k. This compact representation helps to reduce the number of variables needed to represent the problem in this formulation. From these probabilities, a full trajectory can be extracted as a Markov strategy when needed. Once solved, the solution of this problem will have two things: a set of trajectories for patrols to follow, and a set of probabilities associated with them. This will allow a patrol to decide on the trajectory to follow for the day given a set of probabilities (i.e. rolling the dice for choosing one path over another).

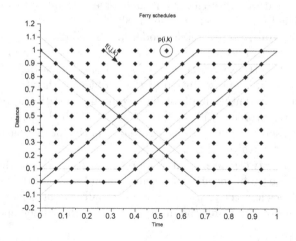

Fig. 1. Discretized transition graph in compact representation

3   Minimum Time-Windows

The problem of patrol schedule assignment to protect moving targets travelling in a straight line between two points can be seen as the assignment of position and time to a patrol, in regards to maximum speed constraints.

The ideal formulation of the problem would take into account the continuous aspects of the patrols' trajectories in time. However, such a formulation is not possible given the complexity it would entail to find a solution which generates continuous trajectories maintaining a Stackelberg equilibrium.

An easier and faster way is to approach this problem through a linear programming model to take advantage of the many very efficient and fast existing solvers. The solution proposed in [7] was to discretize the time and distance variables to reduce the problem to a linear one. What we propose in this paper is related to that model but with the added complexity of minimum time-windows for attacks to occur.

Using a model without time-windows as done in [7] can be very efficient but lacks credibility for some applications. It is not unrealistic to expect an attack on a moving target to occur with a certain duration. The hypothesis we make for this work is that an attacker's utility will be greatly decreased if a patrol covers the target at least once during this interval, diminishing chances of a successful attack. Figure 2 shows the utilities for an example scenario where patrol schedules were generated for 2 patrols and 3 targets over a 30 minutes time-span (normalized between 0 and 1). The thin black line shows the utility as computed for each time-steps and intersections. It is possible to see that some spikes in the utility only occur at a very small time-interval. More precisely, the utility at time 0.26 is equal to 3.57. However, this spike has a very short duration, and if we were to suppose that an attack has a minimum time-window of the size of one time-step, the real utility value would be reduced to 1.6, which is showed by the bold blue curve in the graph. The bold blue curve was generated by iteratively going through all time-steps and intersections, and taking the minimum value of the utility for the size of the time-window (1/16).

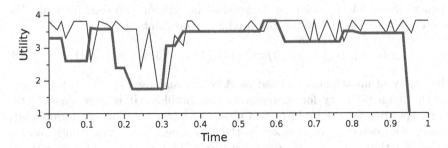

Fig. 2. Shows in bold blue the reduced attacker's utility when taking into account a time window of width $\frac{1}{16}$ as compared to the pointwize utility in thin black

Although we could intuitively use such a heuristic (taking the max of the values computed over an interval) to compute the utility in regards to a time-window for a solution generated using the previous model, the resulting values will not be accurate if the values from a compact representation are used for reasons detailed in the next section.

3.1   Model

To properly take into account minimum time-windows for the attacker utility
the following model needs to be solved:

$$\min_{f(i,j,k),p(i,k)} z \tag{1}$$

$$f(i,j,k) \in [0,1] \qquad \forall i,j,k \tag{2}$$

$$f(i,j,k) = 0 \;\forall i,j,k : |d_j - d_i| > v_m \delta t \tag{3}$$

$$p(i,k) = \sum_{j=1}^{N} f(j,i,k-1) \qquad \forall i,k > 1 \tag{4}$$

$$p(i,k) = \sum_{j=1}^{N} f(i,j,k) \qquad \forall i,k < M \tag{5}$$

$$\sum_{i=1}^{N} p(i,k) = 1 \qquad \forall k \tag{6}$$

$$z \geq AttEU(F_q,t) \quad \forall q,t : (t + \alpha_t) < t_M \tag{7}$$

The constraints (2),(4),(5),(6) are defined such as to keep graph consistency.
More precisely, to insure that probabilities inherent to a patrol's schedule remains
possible such that there are as many patrols that enter a node as there are that
leave it. Constraint (3) insures that a ship's maximum speed is taken into account
when verifying which nodes can be reached in-between two time intervals. For
future reference, we define the graph related constraints as

$$G = \{f(i,j,k), p(i,k) : (2), (3), (4), (5), (6) \text{ are satisfied}\} \tag{8}$$

The utility of an attacker, defined as $AttEU()$ and used on line (7), is where
lies the main difficulty for representing this problem. It is dependent on the
utility specified for the problem as a piecewise linear function, on a probability
of successful detection of attacker by the patrol, and more importantly, on the
probability that a patrol's planned trajectory will intersect at least once with the
target it is protecting. The computation of this utility relies on the hypothesis
that we can directly extract the chances of a successful attack from the current
graph representation. However, using a compact graph representation, crucial
information is lost in regards to the probabilities of a successful attack over a
time-window.

3.2   Probability of a Successful Attack

One of the difficulties in implementing a minimum time-window for potential
attacks relates to the computation of the probabilities of success of the attack.
Using a compact model with discretized time-steps as previously described has
the advantage of significantly reducing the number of variables necessary to

represent the problem from $O(N^M)$ to $O(MN^2)$. The problem with this representation, however, is that for attacks with time windows some of the information necessary to compute the probabilities of an attack being detected is lost.

To evaluate the probabilities of an attacker being detected in a time-window, it becomes necessary to evaluate the full trajectory of patrols in this time window. This can be seen in Figure 3 where in a model without time-windows and compact representation, probabilities are associated to edges $E_{1..6}$ time-steps t_k and intersections θ. In this figure, to compute the probability that a patroller will cross in the window of attack, a naïve and wrong approach would be to add the probabilities of all edges E crossing with the attack window. Doing so will result in possibly overestimating the probabilities. In the example (Figure 3), four trajectories may be followed by a patroller. E_1-E_6, E_2-E_4, E_2-E_5 and E_3-E_5. If all intersections are added together, then one would add the probabilities of taking edges E_2, E_5, E_6, even though a patrol taking the path E_2-E_5 should only count once. The correct way of computing the probability that a patrol crosses at least once with the attack-window corresponds to $E_1 * E_6 + E_2 * E_5 + E_3 * E_5$, which introduces a nonlinear expression to the problem in the compact form.

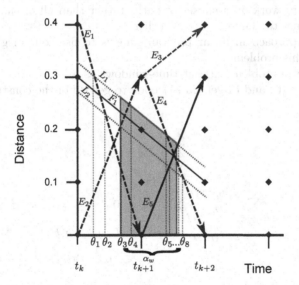

Fig. 3. L_1 and L_2 represent the distance from which the target F_1 is considered as covered by the patrol. $E_{1..6}$ is the compact representation of trajectories to be followed by the patrols. The darkened area shows a minimum attack time-window (of size α_n).

One way of representing the probability of detection of an attack for a time-window combined with the utility gained by the attacker is to use a full representation in the following way:

$$\min \quad z$$
$$\text{s.t.}$$

$$\sum_{p \in P} \lambda_p = 1$$

$$z \geq \left(1 - \sum_{p \in P} I_p^{qt} * \lambda_p\right) * U(t) \qquad \forall q, t \tag{9}$$

$$\lambda_p \geq 0$$

where P is the set of all possible paths in the graph and λ_p is the variable associated with the probability of following path p. Using the same trick as in [7] to take into account the continuity of the problem, I_p^{qt} is a vector indicating whether path p intersects at least once with ferry q between t and $t + \alpha_w$.

$U(t)$ is the utility gained by the attacker on average between time t and time $t + \alpha_w$. It is assumed that $t \in T$ is the set of all time steps as well as time-intersections shown as θ in Figure 3.

3.3   Column Generation for Mobile Targets

To solve the optimization model with paths instead of arcs, it becomes necessary to find a way to work on a subset of paths rather than all of them, since the problem becomes too large as stated earlier.

A classic approach in linear programming is to use column generation to work around this problem.

Rewriting the problem without time-windows as a list of arcs $x_{ij} \in A$, and adding a source (0) and target (M+1) node to get rid of the constraints $p(i, k)$ we get:

$$\min_x z$$
$$\text{s.t.} \qquad \sum_{j:(0,j) \in A} x_{ij} = 1$$

$$\sum_{j:(i,M+1) \in A} x_{iM+1} = 1$$

$$\sum_{j:(i,j) \in A} x_{ij} - \sum_{j:(j,i) \in A} x_{ji} = 0 \qquad i > 1, i < M+1 \tag{10}$$

$$z + \sum_{ij} I_{ij}^{qt} x_{ij} * U(q, t)) \geq U(q, t) \quad \forall q, t$$

$$x_{ij} \in [0, 1] \quad \forall i, j$$

where I_{ij}^{qt} is a pre-calculated vector for all patrols intersecting with F_q at time t equal to 1 (0 otherwise).

$$I_{ij}^{qt} = \left.\begin{bmatrix} I_{1,1}^1 & \cdots & I_{2,2}^1 \\ \vdots & & \\ I_{1,1}^{q*t} & \cdots & I_{2,2}^{q*t} \end{bmatrix}\right\} q*t \qquad X = \left.\begin{bmatrix} x_{1,1,1} & \cdots & x_{1,1,3} \\ \vdots & & \\ x_{2,2,1} & \cdots & x_{2,2,3} \end{bmatrix}\right\} ij$$

$$\underbrace{}_{ij} \qquad \qquad \underbrace{}_{p}$$

We can replace the x_{ij} by $\sum_{p \in P} x_{ijp} \lambda_p$ where the constant $x_{ijp} \in \{0, 1\}$ represents path p and λ_p the flow associated with this path, and divide the problem in a master and slave problem. The master problem will work on a subset of paths $P_s \in P$, while the slave problem will generate new paths to be added to the master problem.

We get the problem:

$$\min \quad z$$
$$\text{s.t.}$$
$$\sum_{p \in P_s} \lambda_p = 1 \tag{11}$$
$$-z - \sum_{ij} I_{ij}^{qt} \sum_{p \in P} x_{ijp} \lambda_p * U(q, t)) \leq -U(q, t) \; \forall q, t$$
$$\lambda_p \geq 0$$

The reduced cost can be computed in relation to the earlier representation, which we will call the Master Problem (MP):

$$\min_x \quad z$$
$$\text{s.t.}$$
$$\sum_{p \in P} \lambda_p = 1 \qquad\qquad : \pi_0 \tag{12}$$
$$z + \sum_{p \in P} (\sum_{ij} I_{ij}^{qt} x_{ijp}) * \lambda_p \geq \sum_{ij} I_{ij}^{qt} \quad \forall q, t \qquad : \pi_1$$
$$\lambda_p \geq 0 \qquad \forall p$$

Taking π_0 and π_1 as the dual variables, the reduced cost of one column (path) can be computed as:

$$c_p = 1 - (1 + \sum_{ij} I_{ij}^{qt} x_{ij}) * \pi_1 + \pi_0 \quad \forall q, t.$$

The column-generation sub-problem, which we will define as Slave Problem (SP), will be to find a path which minimizes the reduced cost:

$$\min_{x_{ij}} \quad 1 - \pi_1 + \sum_{ij} I_{ij}^{qt} x_{ij} * \pi_1 + \pi_0$$
$$\text{s.t.}$$
$$\sum_{j:(0,j) \in A} x_{ij} = 1$$
$$\sum_{j:(i,M+1) \in A} x_{iM+1} = 1 \tag{13}$$
$$\sum_{j:(i,j) \in A} x_{ij} - \sum_{j:(j,i) \in A} x_{ji} = 0 \quad i > 1, i < M + 1$$
$$x_{ij} \in [0, 1] \; \forall i, j$$

One issue of this model concerns the computation of the full matrix I for all possible intersections. It is impossible to only generate the intersections for the relevant paths in the master problem since when generating new columns (paths) to add in the slave problems, the reduced cost has to take into account the constraints in the master problem. In the case of a single resource, this means there will be as many constraints as there are variables. However when more resources are available, the number of variables will expand exponentially, making this approach much more interesting in terms of scalability.

4   Equilibrium Refinement

The notion of equilibrium refinement is not a new one and has been explored in few papers such as [7], [9] and [10]. When doing a min-max approach in the presented model, we aim to minimize the worst-case scenario. This means we generate solutions under the assumption that a defender will want to minimize his maximum potential lost (attacker utility) at any single point in time.

Operating under such an assumption is logical for this security problem since the attacker has no reason to attack a target for which his gain (utility) will be lower than if he attacks at another time.

However, as was pointed out by [7], this may leave other points of attack needlessly vulnerable if the attacker doesn't attack to the critical point (we consider a critical point as a maximum value which cannot be further decreased). This weakness in the model becomes most obvious when using a LP solver based on the simplex method. The solutions which are generated by this method will normally be on extreme points, which will often mean most of the utilities at all time steps will be at the maximum value.

If we slightly relax the initial hypothesis on the rationality of the attacker, the problem may now shift significantly and have an important impact on the results. If, for example, the attacker doesn't possess all of the information concerning the schedules of the patrols, or is unable to recreate the model used in this paper to generate the probabilities associated with each patrol, he may decide to attack at a sub-optimal point in time. There is no reason why we shouldn't take advantage of this, and this is where the notion of equilibrium refinement comes into play. We can keep our original solution, where we minimized the worst-case scenario for the attack, but then also see if we can't improve and refine the solution for other time-steps.

The problem can be seen as a desire to minimize the time as well as the utility to be gained for performing an attack, or more precisely, the integral of the function defining the attacker's utility.

Obviously, the approach we propose here is just one of many possible tools which should allow an expert to better select a solution approach, and its value should be defined in regards to the type of problem to be resolved.

We define a critical time-step c_t as a point in time for which the utility of the attacker will be the highest, given a selection of discretized time and distance points.

In the following model, these critical time-steps will be a subset of the intersections of the position of the patrols in time with the targets they are covering. Recall the definition of the graph constraints (8).

Phase 1:

$$\min_{f(i,j,k),p(i,k)} z \tag{14}$$

$$f(i,j,k), p(i,k) \in G \tag{15}$$

$$z \geq AttEU(F_q, t_k) \ \forall q, k \tag{16}$$

Phase 2:

$$\min_{f(i,j,k),p(i,k)} \sum_{i=1}^{P} v_i \tag{17}$$

$$f(i,j,k), p(i,k) \in G \tag{18}$$

$$v_i \geq d_k * AttEU(F_q, t_k) \ \forall q, k, i \tag{19}$$

$$v_i \leq z \quad \forall i \tag{20}$$

In *Phase*1, a solution will be generated for the initial problem to compute the solution to the Stackelberg model. Once this solution is available, another model will be solved in a second phase where the objective function will now be modified to be the summation of the values at each critical time-points, rather than the max as was done in phase 1. By setting an upper bound on the values of each critical time-point, we insure that the solution stays valid as a Strong Stackelberg Equilibrium, but we also improve the other time-points where possible attacks could occur, in the case of a sub-optimal choice by the attacker.

Figure 1 shows the trajectories of three ferries, for which two patrols are to be assigned to protect. Using similar parameters as in [7], we can see in Figure 4 an initial patrol schedule generated for the problem. It is possible to see that the patrols sometime actually take an arc that doesn't protect any of the ferries while a better choice would be possible. This sort of behavior is a result of doing a min-max approach. On Figure 5 we can see a net improvement over the trajectories that are generated, meaning that fewer unnecessary moves will be done by the patrol when they could be monitoring a ferry. Figures 6 and 7 show the utilities which have been computed at each phase, and we can see the difference between those two sets of utilities in Figure 8. It is obvious from this last figure that the approach we propose has a great impact on the improvement of the final solution for the patrol schedules since only at one point does the

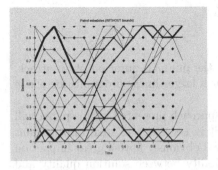

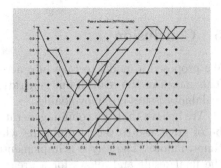

Fig. 4. Trajectories generated in phase 1 of base model

Fig. 5. Trajectories generated in phase 2 with bounded utility value for all time steps

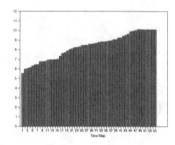

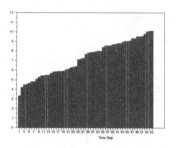

Fig. 6. Utilities ordered by magnitude with base model

Fig. 7. Utilities ordered by magnitude after second pass

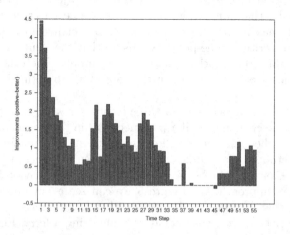

Fig. 8. Difference of utility values between first and second pass

value slightly worsens (but is still bounded by the maximum utility computed in the first phase).

5   Conclusion

We proposed a model to take into account the minimal duration of an attack. This increases the realism but generates much larger linear programs for which scalability issues are unavoidable.

We believe that the column generation approach presented in this paper will be of great use to tackle problems where it is necessary to take into account time-windows. The way is is implemented will obviously have a great impact on the final result and allows for a great flexibility between solution quality and memory/cpu usage. Refinements can also be done to the model, possibly for the generation of the paths for the slave problem, by using heuristics such as to generate promising trajectories first and possibly converge earlier to a solution.

Equilibrium refinement is a heuristic mean to reduce the time windows of high vulnerability and the two-phase approach proposed yields great improvements. The results we showed were only applied to the case where targets travel between two points, but as was demonstrated in [7], it is straightforward to generalize to more complex cases.

References

1. Korzhyk, D., Yin, Z., Kiekintveld, C., Conitzer, V., Tambe, M.: Stackelberg vs. nash in security games: An extended investigation of interchangeability, equivalence, and uniqueness. CoRR abs/1401.3888 (2014)
2. Basilico, N., Gatti, N., Amigoni, F.: Patrolling security games: Definition and algorithms for solving large instances with single patroller and single intruder. Artificial Intelligence **184–185**, 78–123 (2012)
3. Basilico, N., Gatti, N., Amigoni, F.: Leader-follower strategies for robotic patrolling in environments with arbitrary topologies. In: Proceedings of the 8th International Conference on Autonomous Agents and Multiagent Systems, AAMAS 2009, vol. 1. International Foundation for Autonomous Agents and Multiagent Systems, Richland, pp. 57–64 (2009)
4. Bošanský, B., Lisý, V., Jakob, M., Pěchouček, M.: Computing time-dependent policies for patrolling games with mobile targets. In: The 10th International Conference on Autonomous Agents and Multiagent Systems, AAMAS 2011, vol 3. International Foundation for Autonomous Agents and Multiagent Systems, Richland, pp. 989–996 (2011)
5. Lisý, V., Píbil, R., Stiborek, J., Bosanský, B., Pechoucek, M.: Game-theoretic approach to adversarial plan recognition. In: Raedt, L.D., Bessiére, C., Dubois, D., Doherty, P., Frasconi, P., Heintz, F., Lucas, P.J.F., (eds.) ECAI. Frontiers in Artificial Intelligence and Applications, vol. 242. IOS Press, pp. 546–551 (2012)
6. Tambe, M.: Security and Game Theory: Algorithms, Deployed Systems, Lessons Learned, 1st edn. Cambridge University Press, New York (2011)
7. Fang, F., Jiang, A., Tambe, M.: Protecting Moving Targets with Multiple Mobile Resources. Journal of Artificial Intelligence **48**, 583–634 (2013)
8. Xu, H., Fang, F., Jiang, A.X., Conitzer, V., Dughmi, S., Tambe, M.: Solving zero-sum security games in discretized spatio-temporal domains. In: Proceedings of the 28th Conference on Artificial Intelligence (AAAI 2014), Québec, Canada (2014)
9. Miltersen, P.B., Sørensen, T.B.: Computing proper equilibria of zero-sum games. In: van den Herik, H.J., Ciancarini, P., Donkers, H.H.L.M.J. (eds.) CG 2006. LNCS, vol. 4630, pp. 200–211. Springer, Heidelberg (2007)
10. Fudenberg, D., Tirole, J.: Game Theory, 1st edn. The MIT Press, Cambridge (1991)

Trajectory Generation with Player Modeling

Jason M. Bindewald$^{(\boxtimes)}$, Gilbert L. Peterson, and Michael E. Miller

Air Force Institute of Technology, Wright-Patterson AFB, OH, USA
{jason.bindewald,gilbert.peterson,michael.miller}@afit.edu

Abstract. The ability to perform tasks similarly to how a specific human would perform them is valuable in future automation efforts across several areas. This paper presents a k-nearest neighbor trajectory generation methodology that creates trajectories similar to those of a given user in the *Space Navigator* environment using cluster-based player modeling. This method improves on past efforts by generating trajectories as whole entities rather than creating them point-by-point. Additionally, the player modeling approach improves on past human trajectory modeling efforts by achieving similarity to specific human players rather than general human-like game-play. Results demonstrate that player modeling significantly improves the ability of a trajectory generation system to imitate a given user's actual performance.

Keywords: Human-computer interaction · Player modeling · Clustering · Trajectory generation

1 Introduction

The ability to perform tasks similarly to how a human would perform them is valuable to future automation efforts. For example, in adaptive automation (AA), where a human-machine team (HMT) achieves a goal by varying the allocation of tasks between human and machine entities during system operation, the actions of the releasing entity up to that point in time can positively or negatively impact the HMT's future performance [3]. Therefore, the similarity of the machine's task performance to that of the human affects the overall human-machine team's performance.

One specific arena that can benefit from similar action performance is automated trajectory generation. Given a specific state, the ability to generate a trajectory that is similar to a trajectory generated by a unique user would provide opportunities in many areas. In object motion tracking, generating similar

---

The views expressed in this document are those of the author and do not reflect the official policy or position of the United States Air Force, the United States Department of Defense, or the United States Government.

This work was supported in part through the Air Force Office of Scientific Research, Computational Cognition & Robust Decision Making Program (FA9550), James Lawton Program Manager.

© Springer International Publishing Switzerland 2015
D. Barbosa and E. Milios (Eds.): Canadian AI 2015, LNAI 9091, pp. 42–49, 2015.
DOI: 10.1007/978-3-319-18356-5_4

trajectories to those of a specific performer allows for prediction of future movements. In Air Traffic Control environments, generating trajectories similar to those of a specific person might enable collision avoidance in crowded areas.

This paper contributes an automated "full maneuver" trajectory generation system that creates trajectory responses similar to those of a given user in the *Space Navigator* tablet computer game [3]. Through player modeling it improves on past efforts in two ways. First, the trajectory generator produces trajectories as whole entities rather than creating them point-by-point. Second, the player modeling approach achieves significantly better similarity to specific human players over past human trajectory modeling efforts that mimic general human-like game-play.

This paper proceeds as follows. Section 2 reviews related work in the fields of trajectory generation and player modeling. Section 3 presents the specific user trajectory generation system methodology through the *Space Navigator* testbed, a k-nearest neighbor (k-NN) trajectory generation process, and a clustering-based technique for player modeling. Section 4 gives experimental results showing the clustering-based player modeling method's improvements. Section 5 summarizes the information presented and proposes potential future work.

2 Related Work

Most existing trajectory generation and prediction methods create trajectories in a piecemeal fashion using various methods such as trajectory libraries [11] or Gaussian mixture models [15]. As states are recognized, the trajectory generator predicts the next point on the trajectory. It then either returns this single point as the prediction or recursively finds further points until completing a full trajectory. Generating a trajectory one point at a time is unlike the "full maneuver"-based manner humans apply when generating trajectories for flight tasks [9]. Additionally, there is little research that imitates the behavior of a *specific* operator generating trajectories in a dynamic environment.

Within other application domains, player modeling has provided the ability to distinguish a single user from among several others. There are two main types of player models: model-based and model-free [16]. In a model-based player model the player types are pre-defined according to some set of features identified by the practitioner. Examples include the use of supervised neural networks [5], trait theory to pre-determine player types [2], and defining strategy groupings based on game design features [14].

Model-free player modeling learns player types that arise from the collected data and then defines player types from the groupings. A model-free method finds frequently occurring player groupings within the feature data. Previous research relies on clustering methods such as hierarchical clustering [6], evolutionary algorithms [12], and self-organizing maps [6]. Lazy learning methods, such as k-NN [11], provide model-free player modeling methods that are useful when a large number of training examples are available [1]. Within a database of example state-action pairs, the system is presented a state and a k-NN search returns the

k states that most closely resemble it. The system generates a response by using the k responses associated with each state in some way. Methods of this type have been used successfully in imitation learning in robotics [1,7] and trajectory generation [11].

3 Trajectory Generator with Player Modeling

To generate "full maneuver" trajectories similar to the ones generated by specific users, a cluster-based player modeling method was developed. The trajectory generator performs the steps in Fig. 1 in two sections: trajectory generation and player modeling. Operating in the *Space Navigator* environment, both sections rely on a consistent representation technique for dynamic states and a method for comparing trajectories of differing length. These techniques are used first by a k-NN based trajectory generation algorithm and then a clustering-based player modeling method allows the trajectory generator to create trajectories similar to those of a specific user.

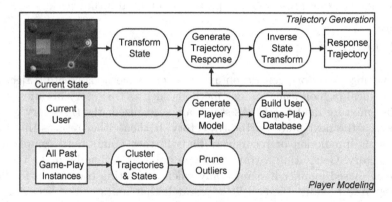

Fig. 1. A player modeling trajectory generator

3.1 Application Environment Considerations

Space Navigator [3] is a tablet computer-based trajectory generation game, in which spaceships appear from the sides of the screen. The player draws lines on the screen directing each ship to a destination planet. Points accumulate when ships encounter their destination planets or bonuses that appear in the play area. Points decrement when ships collide and when a ship traverses one of several "no-fly zones" (NFZs) that move to random locations at set intervals. The game ends after five minutes, a time that provided a high level of user engagement in a pilot study. *Space Navigator* has only one action available to the player, but maintains enough dynamism that an automation cannot achieve a "best" input. Game-play data was collected from 32 participants. The tablet presented each of four difficulty settings in random order to each participant four times, resulting in 16 games per person. When the player draws a trajectory, the game captures data associated with the game state (e.g. the drawn trajectory and bonus locations).

To address the enormous state-space of *Space Navigator*, state representations contain only elements that affect a user's score (other ships, bonuses, and NFZs) scaled to a uniform size. The feature vectors transform the state-space by rotating and translating so the selected ship is at the origin with the X-axis as the straight-line trajectory between the ship and destination planet, and scaling it so the ship to planet trajectory is unit length. The feature vectors account for differing element locations by dividing the state-space into the six zones delineated in Fig. 2. To compare inconsistent numbers of objects, each zone accumulates a score in a manner modeled after [7] over each object within the zone. Scores use a Gaussian weighting function on the shortest Euclidean distance from the object to the straight-line trajectory, as shown in Fig. 2. The resulting feature vector consists of 18 positive real-valued scores that can be adjusted using different weighting methods.

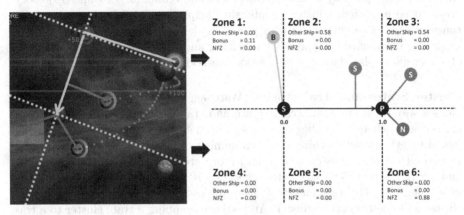

Fig. 2. A *Space Navigator* state and its transformed feature vector representation

Several steps in the trajectory generator shown in Fig. 1 require trajectory comparison. However, *Space Navigator* trajectories contain differing numbers and locations of points. Trajectory comparison requires both transformation and re-sampling. Trajectories are geometrically transformed in the same manner as their paired states. Then the trajectories are re-sampled to 50 points (approximately the average number of points in a trajectory found during experiments) using trajectory smoothing and linear interpolation [10]. Trajectory comparison uses $\underline{A}$verage $\underline{C}$oordinate $\underline{D}$istance (ACD) [8], which is a measure of distance between same-length trajectories based on ordinally comparing pairs of points.

3.2   Trajectory Generation

The trajectory generator portion of Fig. 1 leverages the state representation and trajectory re-sampling methods to create new trajectory responses from a game-play database. A k-NN search finds the k closest states, according to Euclidean distance, to a new state out of the user's game-play database. The trajectory generator weights each of the k user response trajectories based on Euclidean distance from the query using a Gaussian Radial Basis Function (RBF) [4]. The RBF gives more weight to states that are closer to and less weight to those that are farther away from the new state's representation. An influence variable can be used to determine how much more weight is given to closer states than distant ones. The response trajectory returned by the generator is a weighted average of the k-NN response trajectories given. Finally, the combined trajectory is normalized over the total weight of all k trajectories.

3.3   Player Modeling

To imitate a specific user's game-play, the trajectory generator implements the clustering based player modeling technique in the bottom of Fig. 1, by modifying the game-play databases that feed the trajectory response generator. Clustering state-trajectory pairs by state and trajectory allows common state and trajectory types to be analyzed. Cluster membership helps prune outlier instances from consideration. The clusters form a player model of state to trajectory cluster mappings representing a player's tendencies, and player models form the basis of a user game-play database population method.

Cluster States and Trajectories. Ward agglomerative clustering [13] provides a way to group instances by state and trajectory type. In agglomerative clustering, each instance begins as its own cluster. Clusters are progressively joined together until reaching a chosen number of clusters. Ward's method was proven effective for clustering in a trajectory creation environment in [9]. The number of clusters, 500, was determined heuristically and has not been tested experimentally. The agglomerative clustering assigns each instance to one state cluster and one trajectory cluster. An instance mapping a state cluster to a trajectory cluster demonstrates a user's proclivity to react with a given maneuver in a specific game situation. The frequency of state to trajectory cluster mappings reveals common responses and outlier actions.

Prune Outliers. A trajectory response from only one user in one instance is unlikely to be replicated by future users, and proves counterproductive in predicting future user action in preliminary experiments. So, outlier instances are pruned from the set of game-play instances available for use in building a specific user's game-play database. Instances with outlier trajectory responses are removed by removing all instances assigned to the least populated trajectory clusters. Instances falling in the bottom 25% of all trajectory clusters according to cluster size are removed. Outlier state clusters are removed in two ways. First, instances that fall in the bottom 25% of all state clusters according to cluster size are removed. Second, instances where the state cluster was encountered by an extremely small subset of users are removed. The number of instances to prune was determined through visual inspection of the resulting clusters.

Generate Player Model. The player model creation method begins with a user and the set of all state to trajectory cluster mappings left after pruning. It first creates a set of counters to help determine how many of the user's state-trajectory pairs belong to each cluster mapping and then processes each of the state to trajectory cluster mappings. For each cluster mapping, the number of instances provided by the user that belong to it is recorded. The player's model is a set of likelihoods that a given state-trajectory pair chosen at random from the pruned game-play database belongs to a specific cluster mapping, normalized by the total number of instances within the pruned game-play database.

Build User Game-Play Database. The trajectory generator then uses the player model to create a new game-play database. The database builder method adds state-trajectory pairs until the user game-play database has reached a desired size. At each iteration, a state to trajectory cluster mapping is selected according to player model probabilities. Then the database builder method checks whether the given cluster mapping contains any more instances. If so, a random instance from the cluster is added to the new game-play database. The resulting user game-play database is returned.

4   Results

An experiment using 10-fold cross-validation tests the effectiveness of the player modeling trajectory generator. Four 10-fold, 1,500 instance user game-play databases are created for each user: (1) a straight-line database where each response trajectory is the line from the ship to its destination planet; (2) 1,500 random instances from the user's original database; (3) a generic player database created with the player modeling method across all user game-play instances; and (4) a specific player database created with the player modeling method on only one user's game-play instances. For each game-play database type, the experiment sets aside one fold (150 instances) as the test set and the remaining (1,350) instances as the game-play database. One trajectory response is generated from each of the 32 single-user game-play databases, for each of the 150 test instances' state representations. For k-NN based methods, a value of $k = 5$ is set based on preliminary experimental results. The experiment records the ACD from the generated response trajectory to the user's actual trajectory. This is repeated over each of the current user's ten folds.

Comparison testing of the game-play databases shows that the trajectories generated using the specific player model database improved specific user imitation results when compared to those generated by the other three game-play databases. Table 1 contains results comparing trajectories generated using each database with the actual trajectory provided by the user, showing the mean ACD and 99% confidence interval across all 1,500 instances and 32 users.

The worst performing user game-play database was the straight line database. The mean ACD of 0.2319 and its 99% confidence interval (CI) are both the largest values for any database. The original user database slightly outperformed

Table 1. Mean and 99% CI of the average coordinate distances (ACD) across all users

Database	Mean ACD	99% CI
Straight-Line	0.2319	$[0.2299, 0.2339]$
Original User DB	0.2311	$[0.2292, 0.2331]$
Generic Player Model	0.2186	$[0.2169, 0.2204]$
Specific Player Model	0.2036	$[0.2018, 0.2055]$

the straight-line database. The straight-line database performs very well in many instances where the best possible trajectory response is extremely close to a straight line, while the original user database contains all of a user's outlier instances. The generic player model database improves on the straight-line and original user databases, because it maintains the instances where straight-line response trajectories are useful while avoiding the outlier instances of the original user's database. The specific player model provides a significant improvement over the player model at the 99% confidence level. This database improves on the generic database by ensuring a representation of common instances based on a user's game-play habits and strategies in that user's game-play database. The specific player model outperformed all others for 30 of 32 users.

Although there is improvement using the specific player model, further improvements could be gleaned by addressing three issues. First, a more detailed state representation scheme could improve the accuracy of generated trajectories, but would hinder the ability of the computer to generate real-time trajectories. In its current form, there is information present within each state that is not represented, such as the pre-drawn trajectories of other ships. Secondly, a user will not always make a similar decision when presented with same situation. Some players will operate within the game inconsistently, and over time all players' game-play patterns will change. Lastly, user actions are often unpredictable in situations when learning is occurring, which was likely the case over the course of this experiment. As a result, the players' tendencies likely changed over time.

5   Summary and Future Work

The results show that the clustering-based player modeling method for user game-play database creation allows a k-NN based trajectory generator to imitate specific user game-play patterns. This research demonstrates that by eliminating outlier game-play instances and ensuring that the game-play database represents a user's demonstrated tendencies, the ability of the trajectory generation system to imitate the user's actual trajectories improves in efficiency. These results show an improvement over incremental trajectory generation techniques by using a "full maneuver" approach, which allows for modeling specific users.

Future work in this research effort will include improvements to the real-time performance of the player modeling method and a follow-up *Space Navigator* data collection. Real-time performance improvement will focus on reducing "state-to-response" speed and enabling real-time incremental player model learning. To ensure that ACD captures the perceptions of human player, a further data collection will compare objective measures for trajectory similarity (e.g. Euclidean distance, ACD) with subjective measures of trajectory similarity provided by human game players. The resulting similar trajectory generator could allow insights into what makes specific users unique (both good and bad) in a trajectory generation environment. These insights can then be used to aid in areas such as adaptive automation, automated training, and adversary modeling.

References

1. Argall, B.D., Chernova, S., Veloso, M., Browning, B.: A survey of robot learning from demonstration. Robotics and Autonomous Systems **57**(5), 469–483 (2009)
2. Bateman, C., Lowenhaupt, R., Nacke, L.E.: Player typology in theory and practice. In: Proceedings of DiGRA (2011)
3. Bindewald, J.M., Miller, M.E., Peterson, G.L.: A function-to-task process model for adaptive automation system design. International Journal of Human-Computer Studies (2014)
4. Boyd, J.P., Bridge, L.R.: Sensitivity of RBF interpolation on an otherwise uniform grid with a point omitted or slightly shifted. Applied Numerical Mathematics **60**(7), 659–672 (2010)
5. Charles, D., Black, M.: Dynamic player modeling: A framework for player-centered digital games. In: Proc. of the International Conference on Computer Games: Artificial Intelligence, Design and Education, pp. 29–35 (2004)
6. Drachen, A., Canossa, A., Yannakakis, G.: Player modeling using self-organization in tomb raider: Underworld. In: IEEE CIG 2009, pp. 1–8, September 2009
7. Floyd, M.W., Esfandiari, B., Lam, K.: A case-based reasoning approach to imitating robocup players. In: FLAIRS Conference, pp. 251–256 (2008)
8. Hu, W., Xie, D., Fu, Z., Zeng, W., Maybank, S.: Semantic-based surveillance video retrieval. IEEE Transactions on Image Processing **16**(4), 1168–1181 (2007)
9. Huang, V., Huang, H., Thatipamala, S., Tomlin, C.J.: Contrails: Crowd-sourced learning of human models in an aircraft landing game. In: Proceedings of the AIAA GNC Conference (2013)
10. Li, X., Hu, W., Hu, W.: A coarse-to-fine strategy for vehicle motion trajectory clustering. In: ICPR 2006, vol. 1, pp. 591–594. IEEE (2006)
11. Stolle, M., Atkeson, C.G.: Policies based on trajectory libraries. In: ICRA 2006, pp. 3344–3349. IEEE (2006)
12. Togelius, J., De Nardi, R., Lucas, S.M.: Towards automatic personalised content creation for racing games. In: IEEE CIG 2007, pp. 252–259. IEEE (2007)
13. Ward Jr., J.H.: Hierarchical grouping to optimize an objective function. Journal of the American Statistical Association **58**(301), 236–244 (1963)
14. Weber, B., Mateas, M.: A data mining approach to strategy prediction. In: IEEE CIG 2009, pp. 140–147, September 2009
15. Wiest, J., Hoffken, M., Kresel, U., Dietmayer, K.: Probabilistic trajectory prediction with gaussian mixture models. In: IEEE Intelligent Vehicles Symposium, pp. 141–146. IEEE (2012)
16. Yannakakis, G.N., Spronck, P., Loiacono, D., André, E.: Player modeling. Artificial and Computational Intelligence in Games **6**, 45–59 (2013)

Policies, Conversations, and Conversation Composition

Robert Kremer[1][(✉)] and Roberto Flores[2]

[1] University of Calgary, Calgary, Canada
rkremer@ucalgary.ca
[2] Christopher NewPort University, Newport News, USA
roberto.flores@cnu.edu

Abstract. This paper motivates the use of inter-agent conversations in agent-based systems and proposes a general design for agent conversations. It also describes an instance of that design implemented in the CASA agent-based framework. While the agent community is generally familiar with the concept of a conversation, this paper's contribution is a particular model in which flexible composition of conversations from smaller conversations is possible.

Keywords: Agents · Conversations · Multi-agent aystems · Communication protocols

1 Introduction

Through the years it has become understood that inter-agent messages are not just isolated tokens of data but that meaning and purpose is also attached to a sequence of messages exchanged between participants. These sequences, called conversations, have properties that pertain to sequencing (which places in the conversation when messages can take place) and turn taking (the set of participants that could send allowed messages to advance a conversation).

Conversations are normally defined by developers at design time and can be encoded programmatically or through rules executed at runtime. An advantage of rules over program instructions is they could entail flexible (malleable, learnable, and blended) conversations given existing agent state. A flexible conversation would be one that is able to handle unexpected messages (which could either be erroneously sent or sent with the intent to refine the context of interaction) by leading to commonly understood conversation states from which agents could recover and reconvene to an originally intended conversation state.

In this paper we present our approach to encode flexible conversations using rules, and in particular policies as a higher abstraction to rules, in which policies can exist in the context of a conversation template. To this end, Section 2 presents our conceptualization of an agent as a software program that holds conversation policies reacting to message events, has the ability to send and receive messages, and contains an event queue in which events are scheduled and acted upon. Sections 3 presents conversations and conversation composition,

© Springer International Publishing Switzerland 2015
D. Barbosa and E. Milios (Eds.): Canadian AI 2015, LNAI 9091, pp. 50–58, 2015.
DOI: 10.1007/978-3-319-18356-5_5

and Section 4 describes how they are implemented in the CASA project. Lastly, Section 5 presents related work, and Section 6 our conclusions.

2 Policies

While agents may be anywhere on the continuum between reactive to contemplative, we focus on the reactive part because all useful agents need to be reactive to some extent. A *purely* contemplative agent would never produce any output, so would have no useful purpose from our perspective. While we don't rule out contemplation, we model an agent as reacting to events, including, importantly, communicative events. Thus, we will assume an agent has an *event queue* where it will look when it's ready to process the next event. Following many existing systems, we use an event queue rather than react immediately to incoming events because it is not always possible to react immediately to every incoming event. In order to specify how an agent is to process an event, we choose to use a rule-based system, where a set of rules will dictate one or more behaviours based on the event at hand. We call these rules *policies* because, as the reader will see, they differ from typical rules seen in the literature.

Since agents are computational entities, we assume all perceivable events come to it through an event queue. An agent observing an event (really, dequeueing it from its event queue) will attempt to apply policies to determine what to do. In this paper, we won't go into details on how the agent deals with multiple applicable policies since this issue is independent of policy structure and application, and can be usefully implemented in a variety of different ways. Most rule implementations have an antecedent consisting of a boolean expression; however since the event is the key element which actually stimulates policy application, we feature the matching of an *event descriptor* and an event as the antecedent of a policy. However, we also include a *precondition* (which *is* a boolean expression) which must evaluate to *true* in order for the policy to be applied. We use the precondition to filter on the bases of history, agent state, environmental state, or some combination thereof in the same way as more traditional rules do. Thus, we can compare the antecedent of a tradition rule to $match(eventDescriptor, event) \land precondition$. This use is similar to COOL [1].

Traditional rules also have consequents, and policies do too. We model the consequent as a list of *actions* which the agent will execute if a policy is fired. Note that we do not assume that these actions are primitive actions, but may include conditional actions, angelic choice among actions, repeated actions, parameterized actions, etc. Actions may include anything the agent can do, such as sending messages, changing state, etc. As well as at the consequent, we include in policies a *postcondition* which resembles the precondition in that it is a boolean expression describing some state of the agent or the environment, but it is interpreted as the *expected* state immediately after the policy is fired. The postcondition is useful for planning and analysis (not described in this paper), but there is no guarantee the postcondition holds after a policy fires, as unanticipated outcomes (such as errors and failures) can always happen. The postcondition is similar to FIPA's *rational effect* [2].

Formally, a policy consists of an antecedent, precondition, postcondition, and consequent:

$$Policy \equiv (\ antecedent : Antecedent,$$
$$precondition : Precondition,$$
$$postcondition : Postcondition,$$
$$consequent : Consequent)$$

The next subsections will describe these four components of a policy in detail. However, we first give some general definitions of properties:

$$Property \equiv Identifier \mapsto \top$$
$$Properties \equiv Identifier \nrightarrow \top$$

That is, we use *Property* to describe a single attribute/value pair, and *Properties* to describe a (possibly empty) set of attribute/value pairs.

Antecedents. An antecedent to a policy is just a description of an event that may occur. If the event description matches the event that just occurred, then the policy *may*[1] be eligible to be fired.

An event is an abstract occurrence that is observable by the agent. We take events to have abstract types arranged in a type lattice. For example, in some application, both *horseRace* and *race* could be events with a subsumption relation between them which we describe as *horseRace $\prec$ race*. Formally, the subsumption relation is partially ordered set (poset) which we will call the event ontology:

$$EventOnt \equiv (id : Identifier, \prec)$$

such that EvenOnt is a lattice. That is, an event ontology is some set of identifiers and a subsumption relation among these identifiers.

An individual event is slightly more than that, as an individual event might have properties specific to the individual. Thus, the type of an event is *Event*:

$$Event \equiv (type : EventOnt, props : Properties)$$

That is, we describe an event as a type identifier (from an event ontology) and a set of attribute/value pairs. Unlike in most object-oriented programming languages, we choose *not* to constrain particular types to hold specific properties, but let the concept of type and the concept of properties float independently of one another.

We want to fire policies when we match the antecedent of the policy with an event that has occurred. Thus, we need an object to describe an event. We call that object an event descriptor, and it differs only slightly from an event itself. The event descriptor looks like an event, but for each property, it holds not only a value, but a comparator operator to use in comparing the values of the property to that of an event. Thus, we describe *event descriptors* as:

$$EventDescriptor \equiv (type : EventOnt, props : Identifier \nrightarrow (\top, op : Operator))$$
$$where\ Operator \equiv \top \times \top$$

---

[1] The policy *will* be eligible to be fired if the precondition is also true (see §2).

That is, an Event Descriptor looks just like an event, except that its properties map identifiers not only to values, but also to operators, which are comparator predicates describing how the values are to be compared to corresponding values in an event. In practice, useful operators we have encountered are subsumption (via the ontology), equality, the simple math inequalities, and regular expression comparators [4].

An antecedent of a policy in nothing more than an event descriptor:

$$Antecedent \equiv EventDescriptor$$

More complex antecedents are possible, such as disjuncts of event descriptors, but we have not yet observed the need in practice.

The policy passes the antecedent if the event descriptor *matches*. An event descriptor matches an event if its type subsumes the type of the event and each of its properties' values is a match with the corresponding property value of the event according to operator associated with that property:

$$matches(event, desc) \equiv event.type \prec desc.type \land$$
$$\forall t : dom\, desc.props \bullet$$
$$desc.props(t).op(event.props(t),\, first\, desc.props(t))$$

It may seem odd that an antecedent is no more than an event descriptor because agents typically take into account the environment or history when making decisions. We choose to separate the matching of the event from decisions about the environment or history by using *both* the antecedent and requiring the *precondition* to hold as well. The precondition is described in the next section.

Preconditions. A *precondition* describes the state (of the agent or the environment) that must be satisfied in order for the policy to be eligible to fire.

$$Precondition \equiv boolean\,expression$$

We interpret *state* broadly (to include history), and it is therefore no more constrained than any arbitrary boolean expression.

Consequents. A *consequent* is a list of actions that will be executed if the policy fires. We do not describe actions in detail in this paper. Actions are application dependent, although there obviously include speech acts which are used by almost all muliti-agent systems, but vary in detail between systems. Actions may be more than atomic or primitive, but may include composition of actions including conditional actions, angelic choice between actions, repeated actions, parameterized actions, etc. All we can say about actions is that they are of type *Action*. A consequent then, is merely a list of actions:

$$Consequent \equiv seq\,Action$$

Postconditions. A postcondition is a partial description of the *expected* state of the environment immediately after the policy is executed. The word *expected* here is important as there is no guarantee that the state will actually conform to that state: errors may have occurred, or some other unforeseen process may have influenced the state. However, the postcondition is useful for planning and analysis (which we don't describe in this paper). Since agents are purely computation entities, a partial state can be described by an arbitrary boolean expression, and therefore the type of the postcondition is merely a boolean expression:

$$Postcondition \equiv \text{boolean expression}$$

Applicability of Policies. When an event happens, an agent must determine which of its policies is *applicable*. In a simple rule-based system, this is accomplished by merely evaluating the antecedents of the rules, and all those who's antecedents evaluate to *true* are applicable. This situation is only slightly more complex in policies in that we evaluate the precondition *and* whether the antecedent matches the current event:

$$applicable(policy, event) \equiv matches(policy.antecedent, event)$$
$$\wedge\ policy.precondition$$

As already explained, an event may have multiple applicable policies, and choice of which policies to fire is application dependent. However, our default approach is to fire almost all applicable policies, basing the order on a measure of specialization of the antecedent match. Space limitations preclude a detailed discussion on this topic.

3   Conversations

Naively, a designer could write a policy-based agent in which the agent has a set of policies that it applies to each event it sees. However, the designer would quickly see the agent becoming increasingly complex as the agent has to account for multiple concurrent conversations, erroneous messages, maliciously construed messages from other agents, and data associated with individual conversations.

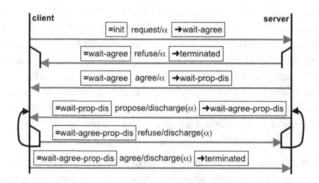

Fig. 1. A simple request conversation. Arrows are messages between agents. Messages are labeled with the performative of the message (e.g.: "request", followed by a slash and the content of the message (e.g.: "α"). Boxes with "=" represent preconditions of the state property of the policy handling the message. Boxes with "→" represent setting the state property in the consequent of policies handling the message.

The problem is one of history-state, similar to the states in a state diagram. For example, a simple conversation protocol could be that buyer may *request* an item from a supplier at a certain price, to which the supplier could *agree* or *refuse* (see Figure 1). Given this simple protocol, a devious supplier might unilaterally send an *agree* to a buyer to sell a certain item at a certain (high) price. A naive buyer could be tricked into purchasing the unwanted item. The problem can be easily handled with state: in its *initial* state, the protocol would only entertain a *request*, after which it would only entertain an *agree* or a *refuse*. An out-of-the-blue *agree* would be handled as if the message was not understood (because, out of context, it isn't). The reader may have the impression this design is a finite state machine, however, conversation state is not *required*, merely a helpful concept that we have found convenient for conversational design. The problem stated here can also be solved by referring explicitly to past events, but modeling these with state tends to be easier for the conversation designer.

In addition to the simple history-based state described above, conversations usually have other state information associated with them such as "who is the debtor and and who is the creditor in the conversation?" Furthermore, an agent may carry on several conversations with several different agents concurrently. All this suggests that, although agents certainly need policies at the top level, conversations should be objects that contain conversation-specific policies and represent their own state (where that state might contain history).

With respect to events, policies, and conversations, we model an agent as containing an *event queue*, a set of *global policies*, a set of *conversation templates*, and a set of *current conversations*. When an agent dequeues an event from the event queue, it checks to see if it's a message event (sent, received, or just observed), and if it is, there should be a conversation id associated with it. The agent then checks to see if the conversation id matches the id of any of its current conversations. If it does, it checks to see if any of the conversation's policies are applicable, and if there are, it fires these policies. On the other hand, if either there is no matching conversation or the matching conversation has no applicable policies, the agent consults its global policies to determine a course of action. (If nothing matches, there is usually a catch-all policy which handles the error.)

Among the global policies are policies that will *create* a conversation. Obviously, in situations like the receipt of a *request* message, there is no current conversation, so there must be a global policy that recognizes a *request* message whose consequent includes creating a conversation to handle the request. Such policies create new conversations by cloning a new conversation from the conversation templates.

Under this model, a conversation is a set of policies, a state variable, and a set of properties:

$$SimpleConversation \equiv (\ policies : (\mathbb{P}\ Policy),$$
$$state : Identifier \mapsto Identifier$$
$$props : Properties\ )$$
$$\text{s.t. } state \in props \wedge first\ state = \text{``}state\text{''}$$

Here, we have chosen to include state as just-another-property, however we distinguish it for reasons that will become apparent later.

Policies are no different from the global policies described in §2, except that they will be evaluated and fired within the context of the conversation. That is, the conversation object and its properties will be available to any code in the antecedent, precondition, consequent, or postcondition.

As already mentioned, the conversation state is modeled as just another conversation property who's value is a simple identifier. State is distinguished among the other properties only because it is so commonly used in conversation modeling. We use state to model the over-all state of the conversation, similar to the "current" state in a state diagram. While more complex models of state, such as those entailed in Petri nets, are possible, we choose to keep the state model simple.

Conversation properties are simple attribute/value pairs. Since we endeavor to keep our model as general as possible, we impose no restriction on properties.

Conversation Composition. The scheme described above allows for composition of conversations: conversations may be easily nested to create more complex conversations. While space limitations preclude a full description, suffice to say that conversation composition is complicated by synchronization among the conversations, which cannot necessarily be handled by control transfer as is done in programming languages. Instead, our approach is to share state information. However, state variable names from different conversations may not always correspond, so our solution is to provide a mapping from the outer conversation namespace to the contained conversation's namespace. It turns out that a mapping between values (as opposed to variable names) is also sometimes required. One such example is that of the state variable `state`, where the *terminated* value for the `state` variable in a sub-conversation signals the sub-conversation's termination. But in the larger context of the super-conversation, this *terminated* value is just an intermediate step where the sub-conversation has completed but the over-all conversation continues.

4   Implementation

Policies and conversations have been implemented as described here in CASA (Collaborative Agent System Architecture) [4]. CASA is written in Java and uses Common Lisp as a run-time scripting language. CASA's main goal is to provide flexibility in agent development by using plug-in and scriptable components to implement all major agent functionality. For example, policies and conversation protocols are never hard-coded, but are read in at agent startup from script files. Since the scripting language is Lisp, and Lisp is a full programming language on its own, these scripts can be as simple or as complex as desired. For example, a definition of a policy to ignore outgoing messages with a performative of "not-understood" can be written as follows:

```
(policy
  '(msgevent-descriptor event_messageSent :=performative not-understood)
  '(nil))
```

Here, `policy` is a lisp function call which defines a policy object. It has two required parameters: an antecedent which is an *EventDescrptor* object (see §2) and a consequent which is executable code (a list of actions). Both these parameters are quoted (single backquote before the open parenthesis) so that they may be executed and instantiated in the context of the event at runtime, and not in the context of agent load time.

5   Relation to Other Work

The current work is most closely related to COOL [1] which Barbuceanu and Fox consider addressing the "conventions" (coordination) level of agent interaction. Their model includes a numerically-labeled *state* in a finite state machine (analogous to our state property), messages with *performatives* (almost identical to ours), *conversation rules* (analogous to our conversation policies), *error-recovery rules* (the concept is captured within our global policies), *continuation rules* (analogous to our global policies), *conversation classes* (analogous to our conversation templates), and *actual conversations* (analogous to our instantiated conversations). Furthermore, COOL's rules are similar to our policies in that they have a *received* field (analogous to our antecedent), a *such-that* field (analogous to our precondition), and a collection of fields such as *next-state* and *transmit* describing actions (analogous to our consequent). COOL even uses conversation-scope variables similar to ours. What COOL does not have, however, is any mechanism to handle composition of conversations, as does CASA. Of less import, we also consider our work to have greater flexibility in terms of actions: COOL handles a limited set of possible actions as part of its definition, whereas our work leaves actions open to the full expressibility of the scripting language (Lisp). Also somewhat related to our work is JADE [5], and Cougaar [3], which form somewhat similar agent frameworks, but implement specific protocols, in contrast to CASA's flexibility.

6   Conclusion

This work extends previous work in explaining one way to specify inter-agent conversations. In particular, these conversations are easily *composable*: complex conversations can be built up from simpler conversations. To this end, policies feature a very flexible event matching scheme as part of their antecedents and conversations feature run-time local variables and a powerful mapping scheme for both variable *names* and variable *values* between super- and sub-conversations.

References

1. Barbuceanu, M., Fox, M.S.: Cool: a language for describing coordination in multi agent systems. In: Proceedings of the First International Conference on Multi-Agent Systems. pp. 17–24. AAA Press/The MIT Press, June 1995

2. Foundation for Intelligent Physical Agents (FIPA): FIPA communicative act library specification. document number SC00037J, FIPA TC communication, Dec 2003. http://www.fipa.org/specs/fipa00037/SC00037J.html
3. Helsinger, A., Thome, M., Wright, T.: Cougaar: a scalable, distributed multi-agent architecture. In: 2004 IEEE International Conference on Systems, Man and Cybernetics, vol. 2, pp. 1910–1917 vol. 2, Oct 2004
4. Kremer, Rob, Flores, Roberto A.: Flexible conversations using social commitments and a performatives hierarchy. In: Dignum, Frank P.M., van Eijk, Rogier M., Flores, Roberto (eds.) AC 2005. LNCS (LNAI), vol. 3859, pp. 93–108. Springer, Heidelberg (2006)
5. Telecom Italia Lab: Jade (java agent development environment), May 2008. http:// jade.cselt.it/

Task Allocation in Robotic Swarms: Explicit Communication Based Approaches

Aryo Jamshidpey[1](✉) and Mohsen Afsharchi[2](✉)

[1] Institute for Advanced Studies in Basic Sciences, Zanjan, Iran
aryo.jamshidpey@iasbs.ac.ir
[2] Department of Computer Science, University of Zanjan, Zanjan, Iran
afsharchim@znu.ac.ir

Abstract. In this paper we study multi robot cooperative task allocation issue in a situation where a swarm of robots is deployed in a confined
unknown environment where the number of colored spots which represent
tasks and the ratios of them are unknown. The robots should discover
the spots cooperatively and spread proportional to the spots area. We
proposed 4 self-organized distributed methods for coping with this scenario. In two different experiments the performance of the methods is
analyzed.

1 Introduction

Swarm robotics is inspired by social insects and other nature colonies that show
complex behavior although they have simple members. The action of assigning
tasks to agents for performing is called task allocation. From a control architectural perspective Burger [3] distinguishes between Heteronomous, Autonomous
and Hybrid methods in task allocation. In this paper, we introduce a practical
scenario for the issue of task allocation in swarm robotics and 4 hybrid methods
for solving it in unknown environments which, the number, locations and ratios
of tasks are unknown to robots.

Market-based mechanism is one of the main approaches that tackle the task
allocation problem. TraderBots is one of the works in this subject that is presented by Dias [6]. A comprehensive study of market-based multi-robot coordination can be found in [7]. Most solutions in self-organized task allocation is
threshold-based that are inspired by models initially proposed to describe the
behavior of insect societies [1]. In this case we can mention Krieger and Billeter
work [9] which benefits from a simple threshold-based model for task allocation
in a foraging scenario. Labella et al. [10] and Lui et al. [11] proposed two probabilistic task allocation approaches which use adaptive thresholds. Brutschy et al.
in[2] presented a task allocation strategy in which robots specialize to perform
tasks in the environment in a self-organized manner. Jones and Mataric [8] introduced an adaptive distributed autonomous task allocation method for identical
robots. Dahl et al. [4] proposed a method that controls the group dynamics of
task allocation. Dasgupta [5] presented a communication-based method for task

© Springer International Publishing Switzerland 2015
D. Barbosa and E. Milios (Eds.): Canadian AI 2015, LNAI 9091, pp. 59–67, 2015.
DOI: 10.1007/978-3-319-18356-5_6

allocation. Robots can only partially complete tasks and one after the other contribute to progressing them. Our proposed practical scenario like Dasgupta's one is about unknown environments.

2   Problem Definition

This scenario involves a colony of identical robots with limited energy levels that are rechargeable and an environment full of obstacles and colored spots which represent types of tasks. The individuals are unaware of the size of the population and the distribution of the other robots. At any moment, each robot is able to do only one of the *forage for green* or *forage for black* subtasks. Depending on the area of the spots, the number of robots to do cleaning and sampling actions in them varies. Obviously it is not necessary to fill the whole capacity of spots with robots. Even a robot is able to do cleaning and sampling actions on its own but it takes much time and is not desirable. In abstract the scenario can be defined in terms of finding more colored spots with minimal energy waste (One of the causes of this waste is unnecessary robot turns in the environment), maximum spreading of robots in different spots proportional to their area, avoiding robots from collision with obstacles and finally preventing robots from remaining idle.

3   Methods

In all proposed methods the robots are initially in random places. In each method the energy of each robot is divided into three parts. First part is for foraging. Second part is to do cleaning and sampling actions and the third part is dedicated to returning to the charging station. All methods are organized based on transferring messages and all messages have the same structure. Each message may have several rows that each of them presents a distinct spot's information. Information of each spot consists of 10 features. Message updating procedure is the same in all of proposed methods. Every robot has a message that contains information about the spots which is called private message. This message is empty at first and then updated via occurrence of two different events. First when a robot finds a non-observed spot for the first time, and second when it receives a message from one of its neighbors.

3.1   Static Communication Based Method

At the beginning, one of the two subtasks of exploring for green or black spot is assigned to each robot which is called worker statically (With a probability of 0.5). Each robot moves in the environment by random walk and avoids collision with other robots, obstacles and walls by the help of its distance computing sensors. Each robot updates its private message periodically and broadcasts it to every other robot in the coverage area of its radio frequency transmitter. By this process, messages will propagate among the robots. If a robot could not

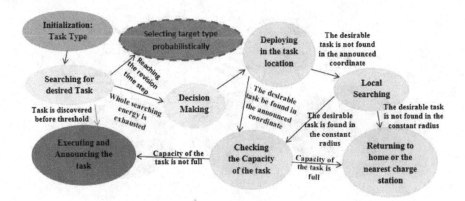

Fig. 1. Method1 and Method 2 state diagram

find a match before running out of its searching energy, it would go to one of
the spots indicated in its private message. By having this policy, our goals for
preventing idle robots existence, maximum spot coverage and also preventing
spot starvation will be guaranteed in a desirable level. The decision step is as
follows: The robot should sort its private message based on three fields; color,
hop count and its current distance from the spots. At first this sort is done based
on color field, so that spots with the same color as the robot current state will be
placed at the top. Then these lists are sorted based on the hop count field. Each
sorted list then will be categorized according to sets of 5 hops. Robots prefer to
go to spots with low hop count number to prevent starvation. In the third step
the sorted list will be further sorted based on Euclidean distance. So in each
hop category, spots are sorted based on their distance from the robot. The state
diagram of the robot's controller while using this method is shown in Figure 1.

3.2 Dynamic Communication Based Method

This method (figure 1) is similar to the first method in a way that it supports
dynamic task allocation in such a way that the robot will change its target color
probabilistically after a time step. For example consider that a robot has 4 spots
in its private list after 100 iterations containing 1 black and 3 green. The robot
will set its target color to black by probability 1/4 and also will set it to green
by probability 3/4.

3.3 Decentralized Chapar Method

In this method (figure 2 (a)) we have some radio turrets called Chapar stations which are used for radio communications and also we assume two groups
of robots; workers and Chapars. Chapars transfer messages between workers
and Chapar stations. High speed robots with simple structure are considered

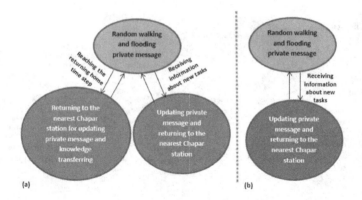

Fig. 2. (a): Method 3 State Diagram, (b): Method 4 State dDiagram

as Chapars. Chapars broadcast their updated private messages which are modified by some workers to the area of their radio frequency coverage. Once a Chapar realizes a new row in its private message it quickly goes to the nearest Chapar station and sends its private message to it and also updates the message based on the content of the messages sent by the Chapar stations. In addition each Chapar goes to the nearest Chapar station periodically to update its private message. Chapar stations are also in the coverage of each other and so they replicate messages to keep the whole system up to date.

3.4   Centralized Chapar Method

This method (Figure 2 (b)) is similar to the decentralized one in which there is only one Chapar station that covers the entire environment. In this method, Chapars have a single task which is to transfer messages from robots to the Chapar station and they do it once they realize a new row in their private messages.In spite of the centralized method, each Chapars only goes to the Chapar station when it encounters a new spot in its private message.

4   Simulations

We have used e-puck robots in simulations. Since the purpose of this article is to involve a wide range of robots and the use of simpler hardware, we considered them without any camera. All experiments have been implemented in a 3m x 3m square environment enclosed with walls by using of Webots as robotic simulation software. In our general experiments, performance of the four proposed methods is evaluated individually before and after energy consumption and in both of them 10 robots which are called workers with IDs from 1 to 10 are used. Initially foraging mode of the robots with IDs from 1 to 5 is adjusted to green and the robots with IDs from 6 to 10 is adjusted to black. For simplicity, the details of cleaning and sampling operations are ignored, the colored spots are considered as 30cm x 30cm squares and finally the length of each robot's communication radius is considered larger than the diameter of each colored square. In both experiments, for each 300 square centimeters of each spot one worker is sufficient for covering it desirably.

4.1   First Experiment: Performance Evaluation of Proposed Methods Before the Threshold Energy

Since the third and fourth methods use the worker robot which its controller is that of the first or second method, it is not necessary to compare the performance of all methods before foraging energy consumption (before the threshold). So we have compared only the performance of the first and second methods before the threshold. For this purpose four environments covered by green and black spots with obstacles are considered. The first environment has 3 green spots and 3 black ones, the second has 4 green spots and 2 black ones, the third includes 5 green spots and 1 black ones and finally the fourth contains only 6 green spots. In each of the four areas, both the first and second methods are tested 10 times separately. The average number of successful robots before the energy threshold for the fist and second methods is shown in Figure 3 (b).

It can be seen that, except for the first environment in which the average number of successful robots before the threshold are equal for both methods, in other environments, the average number of successful robots in the second method is higher than the first one. This disparity grows by moving from the first environment to the fourth and its reason is the changing attitude mechanism which is used by second method's robots during their foraging operation. As a result this leads to increasing in the number of robots which their attitude changes when the number of green spots rise and the number of black ones falls respectively. Subsequently this process will result in forming approximate stability in the number of successful robots before the threshold. As it is shown in Figure 3 (b), we can conclude that in unknown environments where the number of colored spots and the ratios of them are unknown, the second method is more successful than the first one in terms of the robots' attempts in finding spots by themselves before energy threshold.

In the next step the average number of green spots that have been found by workers before the threshold is shown in Figure 3 (a). In this figure the obtained curve from the second method is steeper than the first method's one. The reason that the second method's curve is more steeper, is increasing of the number of workers which search for green spots during the search time. As mentioned before this is because of changing the robot's attitudes during foraging operations. Figure 3 (a) shows that in unknown environments where the number of colored spots and their ratios are unknown, the second method is more successful than the first one in spot finding before the energy threshold.

Figure 4 shows the average number of robots deployed in the green spots before the end of the foraging energy in both the first and second methods. It can be observed that the steepness of the second method's curve is ascending linear but the steepness of the first method's curve is sub linear. In the first method

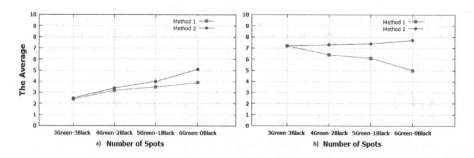

Fig. 3. (a): Average number of discovered spots. (b): Average number of successful robots before the threshold.

the maximum number of robots in green spots is 5 because only 5 robots have green initial foraging modes and there is no any changing attitude mechanism before threshold. But in the second method there is changing attitude and so the maximum number of robots in green spots might be 10. We can conclude that in unknown environments where the number of colored spots and the ratios of them are unknown, adaptability has a significant positive impact on the performance of robots. To sum up, from the above results in unknown environments with the features which are mentioned above, the second method is more efficient than the first one before the energy threshold.

4.2   Second Experiment: Performance Evaluation of Proposed Methods After the Threshold Energy

We use an environment consisting of 3 green spots and 3 black ones for evaluating the performance of the methods after foraging energy consumption. In this area, all proposed methods are separately tested 10 times with random initial distribution of robots. In the third and fourth methods, in addition to 10 workers, another 3 Chapar robots that are faster than the workers are used too. It should be mentioned about Chapar station that the third method is equipped with 3 Chapar stations which their communication radius cover each other sequentially and the fourth method is equipped with one of them, which is omniscient. Table 1 shows the results of the second experiment. The absorption percentage is the percentage of successfulness of finding spots by robots after the threshold by applying the decision making mechanism.

As expected, the fourth method has the highest absorption percentage which means 100%. This is due to the use of global message transferring system. The third method is in second place with 87.09% and after it the first and the second methods with approximate absorption 76% are both in third place.

5   Discussion

Most studies in swarm robotic define their own test scenario, which is then used to build a concrete swarm robotic system capable of solving the problem. This leads

Table 1. Results for the second experiment

Method Number	Meth. 1	Meth. 2	Met. 3	Meth. 4
The Average Number of Successful Robots Before the Threshold.	7.2	7.3	6.9	7.3
The Average Number of Unsuccessful Robots Before the Threshold.	2.8	2.7	3.1	2.7
Absorption Percentage.	75	77.78	87.09	100
Absorption Percentage of Robots Appropriate to Their Initial Foraging State.	46.42	37.03	70.96	83
The Average Number of Absorbed Robots.	2.1	2.1	2.7	2.7
The Average Number of Absorbed Robots Appropriate to Their Initial Foraging State.	1.3	1	2.2	2.2

to a huge amount of differently designed global missions and as a result to many different solutions which are hard to compare[3]. Thus in most of the proposed methods in this area, researchers have only introduced their own methods and refrained from comparing with other methods. Our scenario also possesses different features, goals and finally distinct global foraging mission compared to previous scenarios in task allocation field. Accordingly applying current approaches in our scenario and consequently comparing them with each other is so difficult and in most of the times is impossible, except for changing the scenario which in turn leads to changing both the problem and the solutions. As pointed out before, Dasgupta's method is practical for unknown environments with this difference that its global mission is different from what have been proposed here. However our proposed methods have distinct advantageous over Dasgupta's.

The proposed methods have the least waste time for robots as apposed to that of Dasgupta in which, robots might be idle in environment for a long time. These methods have been designed for the energy constrained scenario compared to Dasgupta's. Consequently the priority of discovering new spots is higher than accomplishing a task and then continuing unlimited foraging. Further more in contrast with Dasgupta's method(which robots may be in idle mode for a long time) our methods are close to optimal in the sense of idle mode. On the other hand, our approaches are very practical for scenarios which detecting a task should be done in a limited time. However in Dasgupta's approach, since after detecting a task other neighbor robots quit foraging and wait for accomplishing the task sequentially, the probability of identifying other tasks in a limited time will decrease. Moreover the proposed methods in this paper are practical for blind robots while Dasgupta's method is practical for robots with cameras which ignoring use of them obviously reduces cost in swarm robotics. Also there is no more need to apply sophisticated techniques of machine vision in order to determine starvation (which may have some errors).

5.1   Scalability and Robustness in Term of Single Point of Failure

If by adding or removing robots the performance of a method drops off, we will call it unscalable. Accordingly, both the first and second methods are scalable

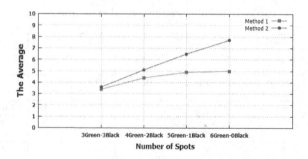

Fig. 4. Average number of robots deployed in green spots

because both are distributed, autonomous and based on the local communication.So Adding or removing robots will maintain their effectiveness. The decentralized Chapar method also has scalability property. This is due to using of the limited range Chapar station, the worker robots and the Chapar robots that are distributed and behave locally. But the centralized Chapar method, due to using of omniscient Chapar station with respect to its expected efficiency does not have scalability property.

With respect to robustness, the first and second methods are completely robust against single point of failure because they behave in a distributed and autonomous manner. If a robot fails outside a spot, others will consider it as an obstacle. If it fails inside a spot and before announcing the center of it, there is no problem because it looks like the situation in which the spot is not discovered yet. But if it fails after the center announcement, it causes only decreasing in speed of the operations in the spot, because they can be carried out by other active robots on the spot yet.

The third method is less robust. The strength of the method for worker robots is such as the first and second methods. In the case of Chapar robots failures, there will be not any problem because in the worst case the speed of communication will be reduced and other robots largely compensate this loss. But about Chapar station as we have mentioned previously in describing the method to benefit the high-speed informing relative to the cost, it is necessary that their communication radius cover each other sequentially. Thus, if one of Chapar stations fails, the connection between two parts of the environment will be lost. It can be said that this method is robust against single point of failure. But this robustness is less than the first two methods. Centralized Chapar method has the lowest robustness than the previous three methods because this method is only have a wide range Chapar station.

6   Conclusions

In this paper a new practical scenario in swarm robotics is presented which is about task allocation in unknown environments. Here we consider that there is limited source of energy for each Robot. It is pointed out that energy management in the form of a 3 level structure is essential and four self-organized threshold-based methods are proposed for solving the scenario. Moreover, in two general experiments, the performance of them is analyzed.

References

1. Bonabeau, E., Dorigo, M., Theraulaz, G.: Swarm intelligence: From natural to artificial systems. Oxford University Press (1999)
2. Brutschy, A., Tran, N.L., Baiboun, N., Frison, M., Pini, G., Roli, A., Birattari, M.: Costs and benefits of behavioral specialization. Robotics and Autonomous Systems **60**(11), 1408–1420 (2012)
3. Burger, M.: Mechanisms for Task Allocation in Swarm Robotics (Doctoral dissertation, Diploma thesis, Ludwig-Maximilians-Universitt Mnchen) (2012)
4. Dahl, T.S., Matari, M., Sukhatme, G.S.: Multi-robot task allocation through vacancy chain scheduling. Robotics and Autonomous Systems **57**(6), 674–687 (2009)
5. Dasgupta, P.: Multi-Robot Task Allocation for Performing Cooperative Foraging Tasks in an Initially Unknown Environment. In: Jain, L.C., Aidman, E.V., Abeynayake, C. (eds.) Innovations in Defence Support Systems -2. SCI, vol. 338, pp. 5–20. Springer, Heidelberg (2011)
6. Dias, M.B.: Traderbots: A new paradigm for robust and efficient multirobot coordination in dynamic environments (Doctoral dissertation, Carnegie Mellon University) (2004)
7. Dias, M.B., Zlot, R., Kalra, N., Stentz, A.: Market-based multirobot coordination: A survey and analysis. Proceedings of the IEEE **94**(7), 1257–1270 (2006)
8. Jones, C., Mataric, M.J.: Adaptive division of labor in large-scale minimalist multi-robot systems. In: Proceedings of the 2003 IEEE/RSJ International Conference on Intelligent Robots and Systems, IROS 2003, vol. 2, pp. 1969–1974. IEEE. Integration. Technical report, Global Grid Forum (2002)
9. Krieger, M.J., Billeter, J.B.: The call of duty: Self-organised task allocation in a population of up to twelve mobile robots. Robotics and Autonomous Systems **30**(1), 65–84 (2000)
10. Labella, T.H.: Division of labour in groups of robots (Doctoral dissertation, Ph. D. thesis, Universite Libre de Bruxelles) (2007)
11. Liu, W., Winfield, A.F.T., Sa, J., Chen, J., Dou, L.: Strategies for Energy Optimisation in a Swarm of Foraging Robots. In: Şahin, E., Spears, W.M., Winfield, A.F.T. (eds.) SAB 2006 Ws 2007. LNCS, vol. 4433, pp. 14–26. Springer, Heidelberg (2007)

AI Applications

M Applications

Budget-Driven Big Data Classification

Yiming Qian, Hao Yuan, and Minglun Gong[✉]

Department of Computer Science, Memorial University of Newfoundland,
St. John's, Newfoundland, NL A1B 3X5, Canada
{yq4048,hy5761}@mun.ca, gong@cs.mun.ca

Abstract. A practical large-scale data classification approach is presented in this paper. By exploiting online learning framework, our approach learns a set of competing one-class Support Vector Machine models, one for each data class. The presented approach enjoys three budget-driven features: 1) it is capable of handling classification when data cannot fit in memory; 2) both training and labeling process is user controllable; 3) the classifiers can easily adapt to changes in dynamic data with minimal computational cost. Compared with the most popular big data classification tool, LibLinear, our approach is shown to be competent at processing extreme large data, while consuming a fractional of memory and time.

Keywords: Big data · Budget-driven classification · Support Vector Machine (SVM) · One-class SVM (1SVM)

1 Introduction

We are in the era of big data - data that is both large and complex. As one of the main big data analytic tools, big data classification is used to support many applications in different scientific disciplines, e.g. economics, social networks, climate prediction, etc. There are three main computational challenges [28] in big data classification: (1) *volume*, which corresponds to the ever increasing amount of data. For example, Facebook reports about 6 billion new photos every month and 72 hours of video are uploaded to YouTube every minute [21]. It leads to the problem that the excessive data volume cannot fit in computer memory, especially for commodity machines, whereas most of existing methods assume data can be stored in memory. (2) *velocity*, which means data is streaming in at unprecedented speed. To cope with that, online learning model is proposed to provide immediate response to the newly generated data, e.g. [7,18,23,26]. (3) *variety*, which refers to data diversity. Real data is often heterogeneous, coming in different types of formats and from different sources.

Motivated by the above challenges, we propose a practical One-Class Support Vector Machine (1SVM) approach, which is specifically designed for big data classification under a limited-budget environment. Grown out of the kernel classifier, i.e., SVM, the proposed approach inherits the flexibility of using kernels, which helps to address the issue of *variety*. The proposed approach incorporates online learning framework into 1SVM training to address the issue of *volume*

© Springer International Publishing Switzerland 2015
D. Barbosa and E. Milios (Eds.): Canadian AI 2015, LNAI 9091, pp. 71–83, 2015.
DOI: 10.1007/978-3-319-18356-5_7

and *velocity*. Training examples are read into memory in an online sequence, allowing data reading and training to happen at the same time. Hence, the proposed approach only needs to allocate buffer for the current training example and the already-selected support vectors, rather than storing all data volume. Moreover, only dominant support vectors which influence decision boundaries are used, further saving memory. The training and labeling process are both user controllable. When a new example is observed during training, only a simple score function computation is required to update the 1SVM models, allowing our approach to react quickly to deal with data velocity. Users may suspend the training process anytime by stop feeding the data, and the partial trained model can be used to perform classification while achieving reasonable accuracy.

Our experiments are performed on a real dataset used in the well-known Yahoo! Large-scale Flickr-tag Image Classification Grand Challenge [1]. The dataset exhibits the three challenges of big data classification. Quantitative evaluations are performed on different problem size levels. The proposed approach achieves superior classification performance on extremely large data, compared with two state-of-the-art methods.

The rest of the paper is organized as follows. Related works are reviewed in Section 2. Section 3 presents the details of the proposed approach. Experimental results and comparisons to existing methods are provided in Section 4. Finally, Section 5 concludes the paper.

2   Related Work

There is a vast body of existing work in the areas of big data classification. The most related ones are discussed in the subsections below.

2.1   Budget-Driven Classification

Linear classification models have been shown to handle big data classification well. Existing work on training linear models under a limited-budget environment can be categorized into memory-driven [4,6,29,31] and time-driven [17,21,26] approaches. Memory-driven approaches focus on solving the problem when data cannot fit in memory. LibLinear [11] is the most popular solver for linear SVM. Yu *et al.* [29] proposed a block optimization framework to handle the memory limitations of LibLinear. They split data into blocks and store them in compressed files. By training on one block of examples at a time, the required training memory is reduced. Following [29], various block-optimization based approaches are also proposed [4,6,31].

Time-driven approaches try to solve the problem in linear time. Joachims [17] proposed SVMPerf by reformulate the original SVM function to a structural SVM, achieving linear time complexity for sparse training features. Unlike SVMPerf and many other methods, which formulate SVM as a constrained optimization problem, Shalev-Shwartz *et al.* [26] relied on an unconstrained optimization formulation of linear SVM and achieved liner training time. Very recently,

Nie *et al.* [21] proposed a new primal SVM solver with linear computational cost for big data classification.

Another way to scale up big data classification is through parallelism, such as [8,13,16,25]. However, the development of parallel classification is limited by not only parallel program implementation difficulties but also by data synchronization/communication overheads on distributed systems. More recent advances of big data classification is surveyed in [28,30].

Different from most existing approaches that focus on solving linear SVM, our approach relies on a competing 1SVMs model, and is designed for running on a single commodity PC with limited memory and processing power. In addition, our approach is not constrained to linear kernel, making it easily customizable for handling a variety of data.

2.2   Online vs. Batch Learning

Batch learning has been the standard methods for data classification [9,10,12, 24]. When employing batch learning for large-scale datasets, one often has to fight with a number of bottlenecks such as memory issue and computational costs. One notable exception is [5], where a scalable batch learning methods is proposed. Nevertheless, it requires complex speeding-up techniques such as disk swapping and chunking, which unfortunately also introduce quite a few tuning parameters. This is in sharp contrast to online learning methods, such as [18,19,23,26], that are usually very simple and efficient. Online learning model also allows users to interact with the training process by controlling the training time.

Online 1SVM [7,14,15] is a newly proposed classification approach and is shown to work well with large-scale and dynamic video data under online learning setting. Qian *et al.* [22] further developed a parameter self-tuning algorithm for online 1SVM and achieved competitive performance than multiclass SVM in terms of both accuracy and speed. These observations motivate us to utilize online 1SVM as a basis of our approach. However, different from the existing approaches that only address small/median-scale problem, our approach focuses on big data classification under a limited budget environment.

3   Proposed Approach

The budget-driven approach is presented in this section. Grown out of 1SVM training model, the proposed approach possesses two distinct features for limited-budget classification: controllable training time and low memory requirement.

3.1   Competing 1SVM

Given a set of examples that can either belong to or not belong to a given class, 1SVM uses a hypersphere to surround some of the examples. Examples inside the hypersphere are considered as inliers and otherwise outliers. Unlike the existing

1SVM training that is designed for handling one class classification problem, our approach builds upon the Competing 1SVM (C-1SVM) model proposed in [15], where two 1SVMs are trained to solve a binary classification problem. We here extend the C-1SVM for multiclass classification and apply it to big data. The key idea is to train and maintain multiple 1SVMs, each models the data distribution of a given class. Each 1SVM may label a testing example as inlier or outlier independently, and hence *competes* with each other. A testing example is *jointly* labeled by those 1SVMs. Compared with the SVM model using hyperplanes to split data, modeling different classes separately using multiple 1SVMs produces larger margins among different classes as shown in Figure 1.

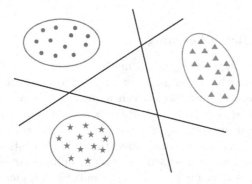

Fig. 1. Comparison between multiclass SVMs under the one-vs-rest training and C-1SVMs. Red dots, green triangles and blue stars represent training instances from three different classes, respectively. The straight lines indicate the decision boundaries of the multiclass SVMs, whereas the ellipsoids show the boundaries of the three C-1SVMs. Different from multiclass SVMs which use hyperplanes, C-1SVMs uses hyperspheres to capture the structures of different classes.

3.2 Budget-Driven Online Learning Model

Given a large set of examples, the conventional 1SVM training minimizes the volume of the hypersphere while at the same time includes as many "in-class" examples as possible. This is a classical batch learning problem. When dealing with big data problems, optimization in batch 1SVM learning often has to address memory issue and expensive computational costs. Since the C-1SVMs used in this paper need to train multiple models for different classes, it requires *even more* memory. To alleviate these bottlenecks on big data, we propose to employ online learning in C-1SVMs training. Our approach inherits the following distinct advantages of online learning for budget-driven problems. Different from batch learning that considers examples in a batch mode, examples are observed by the learner one by one in a time sequence under online learning. The online learner is gradually refined based on its current partial model and the newly observed example. The training process terminates when the user stops presenting examples, and the partially trained model can be used to per-form classification. As a result, training time is user controllable under online learning, which is promising when trying to classify big data under limited train-ing time while still aiming for good performance. Furthermore, our approach can easily handle dynamic data such as video stream and online user generated web-data, etc. When a new training example (e.g. a new image uploaded to social media) is observed, the minimization function of the batch 1SVM is changed, and thus it should be solved again to obtain the new solution. Hence, batch learning needs to start over, which is time-consuming. In comparison, by incorporating online learning into C-1SVM model, we need to refine the current model using the newly observed example, which only involves a simple computation of score function, as shown in Algorithm 1.

Grown out of the online learner in [15], our approach differs from it in several ways to fit our demand: budget-driven big data classification. First, all examples are shown repetitively to the online learner in [15] and it requires a large number of iterations to converge. In contrast, we believe redundancies exist in big data, and competitive classification accuracy can be obtained by considering only a portion of training examples. In each iteration, a randomly selected example is read, based on which the C-1SVM is updated. Reading data and training happen at the same time in our approach, allowing to only store one training example and the already-selected support vectors of each class during training. Notice that the algorithm in [15] reads all data into memory since examples are used in multiple rounds of training. Hence, our approach is particularly useful on big data classification when data cannot fit in memory. Experimental results in Figure 3 also show that the presented approach can converge in only one-round data reading when the problem size is large enough, which further confirms our assumption of redundancies in big data.

Second, as the training set becomes larger, the number of support vectors increase significantly, especially when training multiple C-1SVMs for a big data problem. Storing all these support vectors would require vast memory. To fur-ther reduce memory requirement, our approach assumes the hyperspheres of C-1SVMs are primarily determined by the *dominant* support vectors, i.e. those with high supporting weights. In practice, we set the number of support vectors of each class as a fixed value and store only the most dominant support vec-tors in memory. Compared with the conventional 1SVM algorithm that keeps all support vectors, our approach achieves competitive performance as reported in Figure 2. Moreover, controlling the number of support vectors helps reduce training time in each iteration since the computational cost for evaluating the score function is linearly dependent on the number of support vectors; see Equa-tion 1. By only keeping dominant support vectors, the model refinement in each iteration would be more efficient.

3.3   The Training Algorithm

The training algorithm of our approach is presented in Algorithm 1, which is very easy to implement. When a new example and its label $\{x_t, y_t\}$ is observed/read

---

Algorithm 1. Budget-driven C-1SVMs

Require: training examples with corresponding labels $\{x_t, y_t\}$, kernel function $K(\cdot, \cdot)$,
cut-off value C, support vector buffer size n
Ensure: support vector set S
1: **Initialize** each S_i as an empty set for each i_{th} class
2: **repeat**
3:     randomly read an example (x_t, y_t) at time t
4:     compute score $f_t(x_t) \leftarrow \sum_{j=1}^{n} w_j \chi \left(S_{y_t}(j) \neq x_t \right) K(S_{y_t}(j), x_t)$
5:     $w_t \leftarrow \max \left(0, \min \left(\frac{\gamma - f_t(x_t)}{K(x_t, x_t)}, C \right) \right)$
6:     **if** x_t already exists in S_{y_t} **then**
7:         update the weight of x_t with w_t
8:     **else if** $w_t >$ the minimal weight in S_{y_t} **then**
9:         replace the SV with minimal weight in S_{y_t} with $\{x_t, w_t\}$
10:    **end if**
11: **until** User Termination or S keep unchanged for T times

---

at time t, our approach first evaluates the score of x_t based on the existing support vectors from the same class and computes the weight of x_t. Following the decision function of SVM, the score function is defined as:

$$f_t(x_t) = \sum_{j=1}^{n} w_j \chi \left(S_{y_t}(j) \neq x_t \right) K(S_{y_t}(j), x_t), \tag{1}$$

and its supporting weight:

$$w_t = \max \left(0, \min \left(\frac{\gamma - f_t(x_t)}{K(x_t, x_t)}, C \right) \right), \tag{2}$$

where S_{y_t} is the support vector set of the class y_t, w_j is the weight of the j_{th} support vector in S_{y_t}. $\gamma := 1$ is the margin, and $\chi(\cdot)$ is an indicator function with $\chi(true) = 1$ and $\chi(false) = 0$. $K(\cdot, \cdot)$ is the kernel function, and different kernels can be used for different applications. It is worth noting that duplication is common in big data. Users may upload the same image to social media. In the conventional online learning, all duplicates are added into support vector sets as long as their supporting weights are large enough. These duplicate support vectors come from the same example but have different weights. In contrast, when observing a duplicate example x_t that is already in the corresponding support vector set, the proposed approach computes the score with the duplicate x_t excluded and the original weight of x_t is substituted by w_t. By summing the supports of the remaining support vectors, $f_t(x_t)$ can better shows how well the current model can predict x_t. The supporting weight w_t is computed by comparing the margin and the score. In addition, training examples can be corrupted by different label noises. Assigning a large weight on the corrupted examples would increase the chance of adding them into support vector sets, thus distorting the decision boundary. To address this, a cut-off value C is used to bound w_t, thus limiting the effects of label noise. $C = 0.5$ is used in our implementation. Finally, if the weight w_t of the newly observed x_t turns out larger than the minimal weight in S_{y_t}, the support vector with the minimal weight is replaced by $\{x_t, w_t\}$.

With the benefits of online learning, the training process can be terminated by stopping feeding training examples. Users can also selectively evaluate the partially trained model using a validation dataset to check its effectiveness. Besides user controllable termination, our algorithm itself can converge fast and achieve competitive accuracy at the same time. Following the convergence condition used in the conventional 1SVM training, the proposed approach considers the training algorithm converges until support vector sets keep unchanged for at least $T = 100$ times while different new examples are observed. The labeling process is straightforward in our approach, where an example is labeled as the class with the highest score based on Equation 1.

4   Experiments

In this section, we evaluate performance of the proposed approach for classifying a large-scale social media dataset under limited memory space and processing time. Quantitative comparisons with previous approaches in the literature are also performed. Besides, detailed analyses on the proposed budget-driven features are presented. We implemented the proposed approach in C++ and all experiments were run on a Intel Core i5 (3.20GHz) machine with 4GB RAM. Histogram intersection kernel [3] is used in our implementation and the adaptive kernel method proposed in [22] is applied.

4.1   Datasets

The Flickr-tag image dataset summarized in Table 1 is extracted from the one provided by Yahoo! Large-scale Flickr-tag Image Classification Grand Challenge [1,2] in ACM Multimedia 2013, It is also the data we found that is challenging enough to evaluate a classification algorithm towards the three aforementioned properties of big data. The main challenge of this dataset is that it consists of 10 classes (e.g. food, people) with $200K$ images for each class. A large intra-class visual diversity also exists, resulting some images from the same class not sharing any visual similarities. Furthermore, all tags are annotated by Flickr users, which are quite noisy. Following previous approaches [20,27], instances tagged with "2012" are removed in the dataset used in this paper since the tag "2012" is assigned based on the image capture time and is unrelated to visual information. Unlike previous works on the Flickr-tag dataset which concentrate

Table 1. Summary of the Flickr-tag dataset

# Train Instances	# Test Instances	# Classes	# Features
1,350,000	450,000	9	400

on exploring multiple image features, we are interested in the classification technique itself. Hence, the provided bag-of-words feature representation is used in
our implementation so that we can evaluate different classification techniques on
the same ground.

4.2   Impact of the Number of Support Vectors

We first evaluate the influence of parameter n, the support vector buffer size.
Without explicitly setting a common n for different classes, we observe that the
optimal buffer size is related to the number of examples in each class. Denote
the ratio of examples selected as support vectors as α. For the i_{th} class, we set
the support vector buffer $n_i = \alpha N_i$ in our experiments, where N_i is the number
of the observed examples belonging to the i_{th} class.

Figure 2 plots the classification accuracy of our approach under different
α values on the Flickr-tag dataset. Keeping only a small number of support
vector in the buffer produces lower accuracy, since the limited number of support
vectors cannot capture complex example distributions. As the α value increases,
the classification accuracy improves rapidly at first and then levels off. It is
worth noting that when using a large α value, e.g. 0.05, our approach becomes
the conventional 1SVM training, because the support vector buffers are large
enough to store all support vectors. Our approach with a smaller α value can
achieve similar performance to the conventional 1SVM training, while only uses
a fraction of support vectors and saves time and memory, as shown in Figure 2.
It confirms our assumption that the decision boundaries of C-1SVMs are mainly
influenced by dominant support vectors, while competitive performance can still
be achieved no matter how large the training size is. While users can set any
α value based on their memory limitation, we observe that setting $\alpha = 0.01$
provides a good trade-off regardless the training size. Hence, we use $\alpha = 0.01$
for the following experiments.

4.3   Comparisons with Existing Approaches

In this subsection, we compare the performance of our approach with LibLinear
and the KNN classifier. LibLinear [11] is the most popular solver for large-scale
data classification, and its extension [29] can handle data that cannot fit in
memory using a block optimization method. The KNN classifier is used in the
recent work [27] and is shown to be efficient on the Flick-tag prediction task[1].

To evaluate budget-driven features of our approach, performances on different
problem sizes are evaluated. The comparisons among our approach, LibLinear,
and KNN classifier are shown in Figure 3. We use the original LibLinear for
$< 500K$ dataset. When data cannot fit in memory, the extension of LibLinear is

---

[1] Note that our results of KNN method are different from the ones reported in [27],
since we used the provided bag-of-words features, whereas they use customized features.

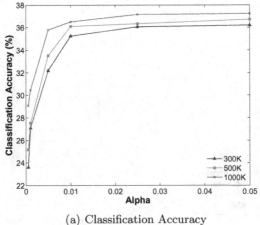

(a) Classification Accuracy

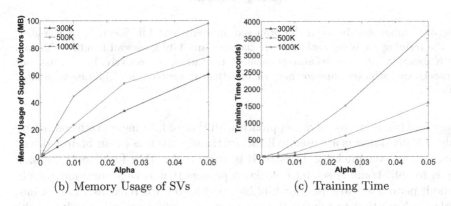

(b) Memory Usage of SVs   (c) Training Time

Fig. 2. Evaluation of support vector number on three sets of $300K$, $500K$ and $1000K$ training examples from the Flickr-tag data. Figure (a)-(c) show the classification accuracy, memory usage of support vectors and training time used under different α values, respectively.

used for $\geq 500K$ dataset. Our approach achieves similar accuracy to LibLinear on small datasets, and outperforms LibLinear's extension on large datasets.

Figure 3 demonstrates the following budget-driven features of our approach: First, the training process is user controllable. Users can terminate the training process anytime because of their time/memory limitation, and the partially trained model can still obtain reasonable accuracy. Second, the accuracy of our approach stabilizes after observing only $300K$ examples. Reading and training with more examples would not significantly improve performance. This confirms our assumption of redundancy of big data, and a fraction of instances can produce encouraging results for our online C-1SVM approach when training set is large enough.

Besides achieving promising classification accuracy on large datasets, our approach can converge fast and save vast memory. Figure 4 compares memory

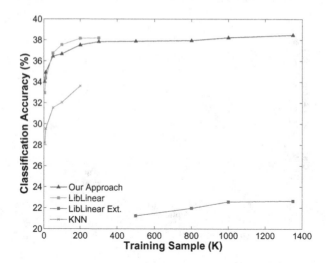

Fig. 3. Comparisons between the proposed approach and LibLinear, KNN classifier. As the training set is increasing, KNN classifier and LibLinear can handle $200K$ and $300K$ examples at most because of memory limitation, respectively. In contrast, our approach can achieves superior performance than the extension of LibLinear in large sets.

usage and training time of our approach with that of LibLinear extension, where only $\geq 500K$ datasets are used to illustrate the effectiveness of our budget-driven approach on extreme large datasets. It is worth noting that LibLinear extension needs to split training set into blocks, a process that is time-consuming itself. Default parameters are used for LibLinear extension, where dataset is split into 8 blocks. Note that to remove the impact of example observation order on the evaluation, results in Figure 3 and 4 are averaged on 5 runs.

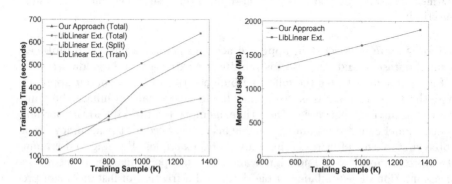

Fig. 4. Processing seconds memory usage on large subsets. Left and right side show comparisons on training time and memory usage, respectively. The processing time of LibLinear extension consists of data splitting and training.

5   Conclusions

A practical big data classification approach is proposed in this paper that is able to work under limited-budget environment. The proposed approach trains multiple C-1SVM models, one for each class. By employing online learning, the proposed approach is user controllable, capable of handling dynamic data, and can react quickly towards the newly observed examples. In terms of computational resources needed, our approach only requires keeping a fractional of examples in memory as support vectors, and the training can converge in as few as just one round of data-reading. In addition, only the most dominant support vectors are used in our approach, further saving memory. Experiments on a challenging social media dataset demonstrate that our approach possesses superior performance comparing to the state-of-the-art approaches, such as LibLinear, especially on large-scale data.

When a new training example arrives, the score function (Equation 1) compares it with all support vectors stored. This process is parallel friendly. In the future, we would like to explore the possibility of implementing our technique on GPUs to further reduce the training time needed.

References

1. Yahoo! large-scale flickr-tag image classification grand challenge. http://acmmm13.org/submissions/call-for-multimedia-grand-challenge-solutions/yahoo-large-scale-flickr-tag-image-classification-challenge/
2. Yahoo! webscope dataset ydata-flickr-ten-tag-images-v1_0. http://webscope.sandbox.yahoo.com/catalog.php?datatype=i
3. Barla, A., Odone, F., Verri, A.: Histogram intersection kernel for image classification. In: Proceedings of the 2003 International Conference on Image Processing, ICIP 2003, vol. 3, pp. III–513. IEEE (2003)
4. Blondel, M., Seki, K., Uehara, K.: Block coordinate descent algorithms for large-scale sparse multiclass classification. Machine Learning 93(1), 31–52 (2013)
5. Boullé, M.: A parameter-free classification method for large scale learning. The Journal of Machine Learning Research 10, 1367–1385 (2009)
6. Chang, K.W., Roth, D.: Selective block minimization for faster convergence of limited memory large-scale linear models. In: Proceedings of the 17th ACM SIGKDD International Conference on Knowledge Discovery and Data Mining, pp. 699–707. ACM (2011)
7. Cheng, L., Gong, M., Schuurmans, D., Caelli, T.: Real-time discriminative background subtraction. IEEE Transactions on Image Processing 20(5), 1401–1414 (2011)
8. Chu, C., Kim, S.K., Lin, Y.A., Yu, Y., Bradski, G., Ng, A.Y., Olukotun, K.: Map-reduce for machine learning on multicore. Advances in Neural Information Processing Systems 19, 281 (2007)
9. Crammer, K., Singer, Y.: On the algorithmic implementation of multiclass kernel-based vector machines. The Journal of Machine Learning Research 2, 265–292 (2002)
10. Dietterich, T.G., Bakiri, G.: Solving multiclass learning problems via error-correcting output codes. arXiv preprint cs/9501101 (1995)

11. Fan, R.E., Chang, K.W., Hsieh, C.J., Wang, X.R., Lin, C.J.: Liblinear: A library for large linear classification. The Journal of Machine Learning Research **9**, 1871–1874 (2008)

12. Gao, T., Koller, D.: Multiclass boosting with hinge loss based on output coding. In: Proceedings of the 28th International Conference on Machine Learning (ICML 2011), pp. 569–576 (2011)

13. Ghoting, A., Krishnamurthy, R., Pednault, E., Reinwald, B., Sindhwani, V., Tatikonda, S., Tian, Y., Vaithyanathan, S.: Systemml: Declarative machine learning on mapreduce. In: 2011 IEEE 27th International Conference on Data Engineering (ICDE), pp. 231–242. IEEE (2011)

14. Gong, M., Qian, Y., Cheng, L.: Integrated foreground segmentation and boundary matting for live videos. IEEE Trans., Image Processing (TIP) (2015)

15. Gong, M., Cheng, L.: Foreground segmentation of live videos using locally competing 1svms. In: 2011 IEEE Conference on Computer Vision and Pattern Recognition (CVPR), pp. 2105–2112. IEEE (2011)

16. Graf, H.P., Cosatto, E., Bottou, L., Dourdanovic, I., Vapnik, V.: Parallel support vector machines: The cascade svm. In: Advances in Neural Information Processing Systems, pp. 521–528 (2004)

17. Joachims, T.: Training linear svms in linear time. In: Proceedings of the 12th ACM SIGKDD International Conference on Knowledge Discovery and Data Mining, pp. 217–226. ACM (2006)

18. Langford, J., Li, L., Zhang, T.: Sparse online learning via truncated gradient. In: Advances in Neural Information Processing Systems, pp. 905–912 (2009)

19. Littlestone, N., Warmuth, M.K.: The weighted majority algorithm. Information and Computation **108**(2), 212–261 (1994)

20. Mantziou, E., Papadopoulos, S., Kompatsiaris, Y.: Scalable training with approximate incremental laplacian eigenmaps and pca. In: Proceedings of the 21st ACM International Conference on Multimedia, pp. 381–384. ACM (2013)

21. Nie, F., Huang, Y., Wang, X., Huang, H.: New primal svm solver with linear computational cost for big data classifications. In: Proceedings of the 31st International Conference on Machine Learning (ICML) (2014)

22. Qian, Y., Gong, M., Cheng, L.: Stocs: An efficient self-tuning multiclass classification approach. In: Advances in Artificial Intelligence - 28th Canadian Conference on Artificial Intelligence, Canadian AI 2015. Springer (2015)

23. Rai, P., Daumé III, H., Venkatasubramanian, S.: Streamed learning: One-pass svms. In: IJCAI, vol. 9, 1211–1216 (2009)

24. Rifkin, R., Klautau, A.: In defense of one-vs-all classification. The Journal of Machine Learning Research **5**, 101–141 (2004)

25. Shafer, J.C., Agrawal, R., Mehta, M.: Sprint: A scalable parallel classifier for data mining. In: Proceedings of the 22th International Conference on Very Large Data Bases, VLDB 1996, pp. 544–555. Morgan Kaufmann Publishers Inc., San Francisco (1996)

26. Shalev-Shwartz, S., Singer, Y., Srebro, N., Cotter, A.: Pegasos: Primal estimated sub-gradient solver for svm. Mathematical Programming **127**(1), 3–30 (2011)

27. Su, Y.C., Chiu, T.H., Wu, G.L., Yeh, C.Y., Wu, F., Hsu, W.: Flickr-tag prediction using multi-modal fusion and meta information. In: Proceedings of the 21st ACM International Conference on Multimedia, pp. 353–356. ACM (2013)

28. Tong, H.: Big data classification. Data Classification: Algorithms and Applications, p. 275 (2014)

29. Yu, H.F., Hsieh, C.J., Chang, K.W., Lin, C.J.: Large linear classification when data cannot fit in memory. ACM Transactions on Knowledge Discovery from Data (TKDD) **5**(4), 23 (2012)
30. Yuan, G.X., Ho, C.H., Lin, C.J.: Recent advances of large-scale linear classification. Proceedings of the IEEE **100**(9), 2584–2603 (2012)
31. Zhang, K., Lan, L., Wang, Z., Moerchen, F.: Scaling up kernel svm on limited resources: A low-rank linearization approach. In: International Conference on Artificial Intelligence and Statistics, pp. 1425–1434 (2012)

Inferring User Profiles in Online Social Networks Using a Partial Social Graph

Raïssa Yapan Dougnon[1], Philippe Fournier-Viger[1]($\boxtimes$), and Roger Nkambou[2]

[1] Dept. of Computer Science, Université de Moncton, Moncton, Canada
{eyd2562,philippe.fournier-viger}@umoncton.ca
[2] Dept. of Computer Science, Université du Quebec à Montréal, Montréal, Canada
nkambou.roger@uqam.ca

Abstract. Most algorithms for user profile inference in online social networks assume that the full social graph is available for training. This assumption is convenient in a research setting. However, in real-life, the full social graph is generally unavailable or may be very costly to obtain or update. Thus, several of these algorithms may be inapplicable or provide poor accuracy. Moreover, current approaches often do not exploit all the rich information that is available in social networks. In this paper, we address these challenges by proposing an algorithm named PGPI (Partial Graph Profile Inference) to accurately infer user profiles under the constraint of a partial social graph and without training. It is to our knowledge, the first algorithm that let the user control the trade-off between the amount of information accessed from the social graph and the accuracy of predictions. Moreover, it is also designed to use rich information about users such as group memberships, views and likes. An experimental evaluation with 11,247 Facebook user profiles shows that PGPI predicts user profiles more accurately and by accessing a smaller part of the social graph than four state-of-the-art algorithms. Moreover, an interesting result is that profile attributes such as status (student/professor) and gender can be predicted with more than 90% accuracy using PGPI.

Keywords: Social networks · Inference · User profiles · Partial graph

1 Introduction

In today's society, online social networks have become extremely popular. Various types of social networks are used such as friendship networks (e.g. Facebook and Twitter), professional networks (eg. LinkedIn and ResearchGate) and networks dedicated to specific interests (e.g. Flickr and IMDB). A concern that is often raised about social networks is how to protect users' personal information [7]. To address this concern, much work has been done to help users select how their personal information is shared and with whom, and to propose automatic anonymization techniques [7]. The result is that personal information is better protected but also that less information is now available to companies to provide targeted advertisements and services to their users. To address this issue,

© Springer International Publishing Switzerland 2015
D. Barbosa and E. Milios (Eds.): Canadian AI 2015, LNAI 9091, pp. 84–99, 2015.
DOI: 10.1007/978-3-319-18356-5_8

an important sub-field of social network mining is now interested in developing automatic techniques to infer user profiles using the disclosed public information.

Various methods have been proposed to solve this problem such as relational Naïve Bayes classifiers [11], label propagation [8,10], majority voting [4] and approaches based on linear regression [9], Latent-Dirichlet Allocation [2] and community detection [12]. It was shown that these methods can accurately predict hidden attributes of user profiles in many cases. However, these work suffers from two important limitations. First, the great majority of these approaches assumes the unrealistic assumption that the full social graph is available for training (e.g. label propagation). This is a convenient assumption used by researchers for testing algorithms using large public datasets. However, in real-life, the full social graph is often unavailable (e.g. on Facebook, LinkedIn and Google Plus) or may be very costly to obtain [2]. Moreover, even if the full social graph is available, it may be unpractical to keep it up to date. The result is that several of these methods are inapplicable or provide poor accuracy if the full social graph is not available (see the experiment section of this paper). A few approaches does not assume a full social graph such as majority-voting [4]. However, this latter is limited to only exploring immediate neighbors of a node, and thus do not let the user control the trade-off between the number of nodes accessed and prediction accuracy, which may lead to low accuracy. It is thus an important challenge to develop algorithms that let the user choose the best trade-off between the number of nodes accessed and prediction accuracy.

Second, another important limitation is that several algorithms do not consider the rich information that is generally available on social networks. On one hand, several algorithms consider links between users and user attributes but do not consider other information such as group memberships, "likes" and "views" that are available on social networks such as Facebook and Twitter [1,4,6,8,10–12]. On the other hand, several algorithms consider the similarity between user profiles and information such as "likes" but not the links between users and other rich information [2,9,13]. But exploiting more information may help to further increase accuracy.

In this paper, we address these challenges. Our contributions are as follows. First, we propose a new lazy algorithm named PGPI (Partial Graph Profile Inference) that can accurately infer user profiles under the constraint of a partial graph and without training. The algorithm is to our knowledge, the first algorithm that let the user select how many nodes of the social graph can be accessed to infer a user profile. This lets the user choose the best trade-off between accuracy versus number of nodes visited.

Second, the algorithm is also designed to be able to use not only information about friendship links and profiles but also group memberships, likes and views, when the information is available. Third, we report results from an extensive empirical evaluation against four state-of-the-art algorithms when applied to 11,247 Facebook user profiles. Results show that the proposed algorithm can provide a considerably higher accuracy while accessing a much smaller number of nodes from the social graph. Moreover, an interesting result is that profile attributes such as status (student/professor) and gender can be predicted with more than 90% accuracy using PGPI.

The rest of this paper is organized as follows. Section 2, 3, 4, 5 and 6 respectively presents the related work, the problem definition, the proposed algorithm, the experimental evaluation and the conclusion.

2   Related Work

Several work have been done to infer user profiles on online social networks. We briefly review recent work on this topic. Davis Jr et al. [4] inferred location of Twitter users based on relationships to other users with known locations. They used a dataset of about 50,000 users. Their approach consists of performing a majority vote over the locations of directly connected users. An important limitation is that this approach is designed to predict a single attribute based on a single attribute from other user profiles. Having the same goal, Jurgens [8] designed a variation of the popular *label propagation* approach to predict user locations on Twitter/Foursquare social networks. A major limitation is that it is an iterative algorithm that requires the full social graph since it propagates known labels to unlabeled nodes through links between nodes. Li et al. [10] developed an iterative algorithm to deduce LinkedIn user profiles based on relation types. The algorithm is similar to label propagation and requires a large training set to discover relation types.

He et al. [6] proposed an approach consisting of building a Bayesian network based on the full social graph to then predict user attribute values. The approach considers similarity between user profiles and links between users to perform predictions, and was applied to data collected from LiveJournal. Recently, Dong et al. [5] proposed a similar approach based on using graphical-models to predict the age and gender of users. Their study was performed with 1 billion phone and SMS data and 7 millions user profiles. Chaudhari [3] also proposed a graphical model-based approach to infer user profiles, designed to only perform predictions when the probability of making an accurate prediction is high. The approach has shown high accuracy on a dataset of more than 1M Twitter users and a datasets from the Pokac social network. A limitation of these approaches however, it that they all assumes that the full social graph is available for training. Furthermore, they only consider user attributes and links but do not consider additional information such as likes, views and group membership.

A community detection based approach for user profile inference was proposed by Mislove [12]. It was applied on more than 60K Facebook profiles with friendship links. It consists of applying a community detection algorithm and then to infer user profiles based on similarity to other members of the same community. Lindamood et al. [11] also inferred user profiles from Facebook data. They applied a modified Naïve Bayes classifier on 167K Facebook profiles with friendship links, and concluded that if links or attributes are erased, accuracy of the approach can greatly decrease. This raises the challenges of performing accurate predictions using few data. Recently, Blenn et al. [1] utilized bird flocking, association rule mining and statistical analysis to infer user profiles in a dataset of 3 millions Hyves.nl users. However, all these work assume that the full social graph is available for training and they only use profile information and social links to perform predictions.

Chaabane et al. [2] proposed an approach based on Latent Dirichlet Allocation (LDA) to infer Facebook user profiles. The approach extracts a probabilistic model from music interests and additional information provided from Wikipedia. A drawback is that it requires a large training dataset, which was very difficult and time-consuming to obtain according to the authors [2]. Kosinski et al. [9] also utilized information about user preferences to infer Facebook user profiles. The approach consists of applying Singular Value Decomposition to a large matrix of users/likes and then applying regression to perform prediction. A limitation of this work is that it does not utilize information about links between users and requires a very large training dataset.

Quercia et al. [13] developed an approach to predict the personality of Twitter users. The approach consists of training a decision tree using a large training sets of users tagged with their personality traits, and then use it to predict personality traits of other users. Although this approach was successful, it uses a very limited set of information to perform predictions: number of followers, number of persons followed and list membership count.

3 Problem Definition

As outlined above, most approaches assume a large training set or full social graph. The most common definition of the problem of inferring user profiles is the following [1,3,6,8,10–12].

Definition 1 (social graph). A *social graph* $\mathcal{G}$ is a quadruplet $\mathcal{G} = \{N, L, V, A\}$. N is the set of nodes in the social graph. $L \subseteq N \times N$ is a binary relation representing the link (edges) between nodes. Let be m attributes to describe users of the social network such that $V = \{V_1, V_2, ...V_m\}$ contains for each attribute i, the set of possible attribute values V_i. Finally, $A = \{A_1, A_2, ...A_m\}$ contains for each attribute i a relation assigning an attribute value to nodes, that is $A_i \subseteq N \times A_i$.

Example 1. Let be a social graph with three nodes $N = \{John, Alice, Mary\}$ and friendship links $L = \{(John, Mary), (Mary, John), (Mary, Alice), (Alice, Mary)\}$. Consider two attributes *gender* and *status*, respectively called attribute 1 and 2 to describe users. The set of possible attribute values for gender and status are respectively $V_1 = \{male, female\}$ and $V_2 = \{professor, student\}$. The relations assigning attributes values to nodes are $A_1 = \{(John, male), (Alice, female), (Mary, female)\}$ and $A_2 = \{(John, student), (Alice, student), (Mary, professor)\}$.

Definition 2 (Problem of inferring user profiles in a social graph). The problem of inferring the user profile of a node $n \in N$ in a social graph $\mathcal{G}$ is to correctly guess the attribute values of n using the other information provided in the social graph.

We also consider an extended problem definition where we consider additional information available in social networks such as Facebook (views, likes and group memberships).

Definition 3 (extended social graph). An *extended social graph* $\mathcal{E}$ is a tuple $\mathcal{E} = \{N, L, V, A, G, NG, P, PG, LP, VP\}$ where N, L, V, A are defined as previously. G is a set of groups that a user can be a member of. The relation $NG \subseteq N \times G$ indicates the membership of users to groups. P is a set of publications such as pictures, texts, videos that are posted in groups. PG is a relation $PG \subseteq P \times G$, which associates a publication to the group(s) where it was posted. LP is a relation $LP \subseteq N \times P$, which indicates the publication(s) liked by each user (e.g. the "likes" on Facebook). VP is a relation $VP \subseteq N \times P$, which indicates the publication(s) viewed by each user (e.g. the "views" on Facebook). An observation is that $LP \subseteq VP$.

Example 2. Let be two groups $G = \{book\_club, music\_lovers\}$ such that $NG = \{(John, book\_club), (Mary, book\_club), (Alice, music\_lovers)\}$. Let be two publications $P = \{picture1, picture2\}$ published in the groups $PG = \{(picture1, book\_club), (picture2, music\_lovers)\}$. The publications viewed by users are $VP = \{(John, picture1), (Mary, picture1), (Alice, picture2)\}$ while the publications liked by users are $LP = \{(John, picture1), (Alice, picture2)\}$.

Definition 4 (Problem of inferring user profiles in an extended social graph). The problem of inferring the user profile of a node $n \in N$ in an extended social graph $\mathcal{E}$ is to correctly guess the attribute values of n using the information in the social graph.

But the above definitions assume that the full social graph may be used to perform predictions. In this paper, we define the problem of inferring user profiles using a limited amount of information as follows.

Definition 5 (Problem of inferring user profiles using a partial (extended) social graph). Let $maxFacts \in \mathbf{N}^+$ be a parameter set by the user. The problem of inferring the user profile of a node $n \in N$ using a partial (extended) social graph $\mathcal{E}$ is to accurately predict the attribute values of n by accessing no more than $maxFacts$ facts from the social graph. A *fact* is a node, group or publication from N, G or P (excluding n).

4 The Proposed Algorithm

In this section, we present the proposed PGPI algorithm. We first describe a version PGPI-N that infer user profiles using only nodes and links. Then, we present an alternative version PGPI-G designed for predicting user profiles using only group and publication information (views and likes). Then, we explain how these two versions are combined in the full PGPI algorithm.

4.1   Inferring User Profiles Using Nodes and Links

Our proposed algorithm PGPI-N for inferring profiles using nodes and links is inspired by the *label propagation* family of algorithms, which was shown to provide high accuracy [8,10]. These algorithms suffer however from an important limitation. They are iterative algorithms that require the full social graph for training to propagate attribute values [8].

PGPI-N adapts the idea of label propagation for the case where at most $maxFacts$ facts from the social graph can be accessed to make a prediction. This is possible because PGPI-N is a lazy algorithm (it does not require training), unlike label propagation. To predict an attribute value of a node n, PGPI-N only explores the neighborhood of n, which is restricted by a parameter $maxDistance$.

The PGPI-N algorithm (Algorithm 1) takes as parameter a node n_i, an attribute k to be predicted, the $maxFacts$ and $maxDistance$ parameters and a social graph $\mathcal{G}$. It outputs a predicted value v for attribute k of node n_i. The algorithm first initializes a map M so that it contains a key-value pair $(v, 0)$ for each possible value v for attribute k. The algorithm then performs a breadth-first search. It first initializes a queue Q to store nodes and a set $seen$ to remember the already visited nodes, and the node n_i is pushed in the queue.

Then, while the queue is not empty and the number of accessed facts is less than $maxFacts$, the algorithm pops the first node n_j in the queue. Then, the formula $F_{i,j} = W_{i,j}/dist(n_i, n_j)$ is calculated. $W_{i,j}$ and $dist(n_i, n_j)$ are respectively used to weight the influence of node n_j by its similarity and distance to n_i. $W_{i,j}$ is defined as the number of attribute values common to n_i and n_j divided by the number of attributes m. $dist(x, y)$ is the number of edges in the shortest path between n_i and n_j. Then, $F_{i,j}$ is added to the entry in map M for the attribute value of n_j for attribute k. Then, if $dist(x, y) \leq maxDistance$, each node n_h linked to n_j that was not already visited is pushed in the queue and added to the set $seen$. Finally, when the while loop terminates, the attribute value v associated to the largest value in M is returned as the predicted value for n_i. Note that in our implementation, if the $maxFacts$ limit is reached, the algorithm do not perform a prediction to reduce the probability of making an inaccurate prediction.

4.2   Inferring User Profiles Using Groups and Publications

PGPI-G is a lazy algorithm designed to predict user attribute values using only group and publication information (views and likes). The algorithm is inspired by majority voting algorithms (e.g. [4]), which have been used for predicting user profiles based on links between users. PGPI-G adapts this idea for groups and publications and also to handle the constraint that at most $maxFacts$ facts from the social graph can be accessed to make a prediction.

The PGPI-G algorithm (Algorithm 2) takes as parameter a node n_i, an attribute k to be predicted, the $maxFacts$ and $maxDistance$ parameters and an extended social graph $\mathcal{E}$. It outputs a predicted value v for attribute k of

Algorithm 1. The PGPI-N algorithm

> **input** : n_i: a node, k: the attribute to be predicted, $maxFacts$ and
> $maxDistance$: the user-defined thresholds, $\mathcal{G}$: a social graph
> **output**: the predicted attribute value v

1   $M = \{(v, 0)|v \in V_k\}$;
2   Initialize a queue Q;
3   $Q.push(n_i)$;
4   $seen = \{n_i\}$;
5   **while** Q is not empty and $|accessedFacts| < maxFacts$ **do**
6     $n_j = Q.pop()$;
7     $F_{i,j} \leftarrow W_{i,j}/dist(n_i, n_j)$;
8   Add $F_{i,j}$ to the entry of value v in M such that $(n_j, v) \in A_k$;
9   **if** $dist(n_i, n_j) \leq maxDistance$ **then**
10   **foreach** $node$ $n_h \neq n_i$ such that $(n_h, n_j) \in L$ and $n_h \notin seen$ **do**
11         $Q.push(n_h)$;
12         $seen \leftarrow seen \cup \{n_h\}$;
13   **end**
14   **end**
15 **end**
16 **return** a value v such that $(v, z) \in M \wedge \nexists (v', z') \in M | z' > z$;

node n_i. The algorithm first initializes a map M so that it contains a key-value pair $(v, 0)$ for each possible value v for attribute k. Then, the algorithm iterates over each member $n_j \neq n_i$ of each group g where n_i is a member, while the number of accessed facts is less than $maxFacts$. For each such node n_j, the formula $Fg_{i,j}$ is calculated to estimate the similarity between n_i and n_j and the similarity between n_i and g. In this formula, $commonLikes(n_i, n_j)$ and $commonViews(n_i, n_j)$ respectively denotes the number of publications liked by both n_i and n_j, while $commonGroups(n_i, n_j)$ is the number of groups common to n_i and n_j. Finally, $commonPopularAttributes(n_i, g)$ is the number of attribute values of n_i that are the same as the most popular attribute values for members of g. Then, $Fg_{i,j}$ is added to the entry in map M for the attribute value of n_j for attribute k. Finally, the attribute value v associated to the largest value in M is returned as the predicted value for n_i. Note that in our implementation, if the $maxFacts$ limit is reached, the algorithm do not perform a prediction to reduce the probability of making an inaccurate prediction.

4.3 Inferring User Profiles Using Nodes, Links, Groups and Publications

The PGPI-N and PGPI-G algorithms have similar design. Both of them are lazy algorithms that update a map M containing key-value pairs and then return the attribute value v associated to the highest value in M as the prediction. Because of this similarity, the algorithms PGPI-N and PGPI-G can be easily combined. We name this combination PGPI. The pseudocode of PGPI is shown in Fig. 3,

---

Algorithm 2. The PGPI-G algorithm

---

input : n_i: a node, k: the attribute to be predicted, $maxFacts$: a user-defined
threshold, $\mathcal{E}$: an extended social graph
output: the predicted attribute value v

1   $M = \{(v,0)|v \in V_k\}$;
2 **foreach** group $g|(n_i,g) \in NG$ s.t. $|accessedFacts| < maxFacts$ **do**
3   **foreach** person $n_j \neq n_i \in g$ s.t. $|accessedFacts| < maxFacts$ **do**
4     $Fg_{i,j} \leftarrow W_{i,j} \times commonLikes(n_i,n_j) \times commonViews(n_i,n_j) \times$
    $(commonGroups(n_i,n_j)/|\{(n_i,x)|(n_i,x) \in NG\}|)\times$
5     $commonPopularAttributes(n_i,g)$;
6   Add $Fg_{i,j}$ to the entry of value v in M such that $(n_j,v) \in A_k$;
7   **end**
8 **end**
9 **return** a value v such that $(v,z) \in M \wedge \nexists (v',z') \in M|z' > z$;

---

and is obtained by inserting lines 2 to 15 of PGPI-N before line 8 of PGPI-G. Moreover, a new parameter named *ratioFacts* is added. It specifies how much facts of the *maxFacts* facts can be respectively used by PGPI-N and by PGPI-G to make a prediction. For example, *ratioFacts* = 0.3 means that PGPI-G may use up to 30% of the facts, and thus that PGPI-N may use the other 70%. In our experiment, the best value for this parameter was 0.5.

5   Experimental Evaluation

We performed several experiments to assess the accuracy of the proposed PGPI-N, PGPI-G and PGPI algorithms for predicting attribute values of nodes in a social network. Experiments were performed on a computer with a fourth generation 64 bit Core i5 processor running Windows 8.1 and 8 GB of RAM.

We compared the performance of the proposed algorithms with four state-of-the-art algorithms. The three first are Naïve Bayes classifiers [7]. Naïve Bayes (NB) infer user profiles strictly based on correlation between attribute values. Relational Naïve Bayes (RNB) consider the probability of having friends with specific attribute values. Collective Naïve Bayes (CNB) combines NB and RNB. To be able to compare NB, RNB and CNB with the proposed algorithms, we have adapted them to work with a partial graph. This is simply done by training them with *maxFacts* users chosen randomly instead of the full social graph. The last algorithm is label propagation (LP) [8]. Because LP requires the full social graph and does not consider the *maxFacts* parameter, its results are only used as a baseline. For algorithm specific parameters, the best values have been empirically found to provide the best results.

Experiments were carried on a real-life dataset containing 11,247 user profiles collected from the Facebook social network in 2005 [14]. Each user is described according to seven attributes: a student/faculty status flag, gender, major, second major/minor (if applicable), dorm/house, year, and high school.

Algorithm 3. The PGPI algorithm

---

input : n_i: a node, k: the attribute to be predicted, $maxFacts$: a user-defined threshold, $\mathcal{E}$: an extended social graph, $ratioFacts$: the ratio of facts to be used by PGPI-G

output: the predicted attribute value v

1 $M = \{(v,0)|v \in V_k\}$;

2 **foreach** group $g|(n_i,g) \in NG$ s.t. $|accessedFacts| < maxFacts \times ratioFacts$ **do**

3   **foreach** person $n_j \neq n_i \in g$ s.t. $|accessedFacts| < maxFacts$ **do**

4          $Fg_{i,j} \leftarrow W_{i,j} \times commonLikes(n_i,n_j) \times$
$commThepseudocodeofPGPIisshowninFig.3.onViews(n_i,n_j) \times$
$(commonGroups(n_i,n_j)/|\{(n_i,x)|(n_i,x) \in NG\}|) \times$

5          $commonPopularAttributes(n_i,g)$;

6   Add $Fg_{i,j}$ to the entry of value v in M such that $(n_j,v) \in A_k$;

7   **end**

8 **end**

9 Initialize a queue Q;

10 $Q.push(n_i)$;

11 $seen = \{n_i\}$;

12 **while** Q is not empty and $|accessedFacts| < maxFacts$ **do**

13      $n_j = Q.pop()$;

14      $F_{i,j} \leftarrow W_{i,j}/dist(n_i,n_j)$;

15   Add $F_{i,j}$ to the entry of value v in M such that $(n_j,v) \in A_k$;

16   **if** $dist(n_i,n_j) \leq maxDistance$ **then**

17   **foreach** node $n_h \neq n_i$ such that $(n_h,n_j) \in L$ and $n_h \notin seen$ **do**

18              $Q.push(n_h)$;

19              $seen \leftarrow seen \cup \{n_h\}$;

20   **end**

21   **end**

22 **end**

23 **return** a value v such that $(v,z) \in M \wedge \nexists(v',z') \in M|z' > z$;

These attributes respectively have 6, 2, 65, 66, 66, 15 and 1,420 possible values. The dataset is a social graph rather than an extended social graph, i.e it does not contain information about group memberships, views and likes. But this information is needed by PGPI-G and PGPI. To address this issue, we have generated synthetic data about group memberships, views and likes. Our data generator takes several parameters as input. To find realistic values for the parameters, we have observed more than 100 real groups on the Facebook network. Parameters are the following. We set the number of groups to 500 and each group to contain between 20 and 250 members, since we observed that most groups are small. Groups are randomly generated. But similar users are more likely to be member of a same group. Furthermore, we limit the number of groups per person to 25. We generated 10,000 publications that are used to represent content shared by users in groups such as pictures, text and videos. A publication may appear in three to five groups. A user has a 50% probability of viewing a publication and a 30% probability of liking a publication that he has viewed. Lastly, a user can like and view respectively no more than 50 and 100 publications, and a user is more likely to like a publication that users with a similar profile also like.

To compare algorithms, the prediction of an attribute value is said to be a *success* if the prediction is accurate, a *failure* if the prediction is inaccurate or otherwise a *no match* if the algorithm was unable to perform a prediction. Three performance measures are used. The *accuracy* is the number of successful predictions divided by the total number of prediction opportunities. The *coverage* is the number of success or failures divided by the number of prediction opportunities. The *product of accuracy and coverage* (PAC) is the product of the accuracy and coverage and is used to assess the trade-off obtained for these two measures. In the following, the accuracy and coverage are measured as the average for all seven attributes, except when indicated otherwise.

PAC w.r.t Number of Facts. We have run all algorithms while varying the *maxFacts* parameter to assess the influence of the number of accessed facts on the product of prediction accuracy and coverage. Fig. 1 shows the overall results. In this figure, the PAC is measured with respect to the number of accessed facts. It can be observed that PGPI provides the best results when 132 to 700 facts are accessed. For less than 132 facts, CNB provides the best results followed by PGPI. PGPI-N provides the second best results for 226 to 306 facts. Note that no results are provided for PGPI-N for more than 306 facts. The reason is that PGPI-N relies solely on links between nodes to perform predictions and the dataset does not contains enough links to let us further increase the number of facts accessed by this algorithm.

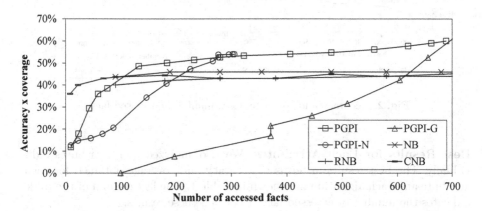

Fig. 1. Product of accuracy and coverage w.r.t. number of accessed facts

The algorithm providing the worst results is LP (not shown in figures). LP provides a maximum accuracy and coverage of respectively 43.2% and 100%. This is not good considering that LP uses the full social graph of more than

10,000 nodes. For the family of Naïve Bayes algorithms, CNB provides the best results until 90 facts are accessed. Then, NB is the best. RNB always provides the worst results among Naïve Bayes algorithms.

Accuracy and Coverage w.r.t Number of Facts. The left and right parts of Fig. 2 separately show the results for accuracy and coverage. In terms of accuracy, it can be observed that PGPI-N, PGPI-G and PGPI always provides much higher accuracy than NB, RNB, CNB and LP (from 9.2% to 36.8% higher accuracy). Thus, the reason why PGPI algorithms have a lower PAC than NB for 132 facts or less is because of coverage and not accuracy. In terms of coverage, the coverage is only shown in Fig. 2 for PGPI-N, PGPI-G and PGPI because the coverage of the other algorithms remains always the same (100% for NB, RNB and CNB, and 94.4% for LP). PGPI has a much higher coverage than PGPI-N and PGPI-G, even when it accesses much less facts. For example, PGPI has a coverage of 87% when it accesses 132 facts, while PGPI-N and PGPI-G have a similar coverage when using respectively 267 and 654 facts. The coverage of PGPI exceeds 95% when using about 400 facts.

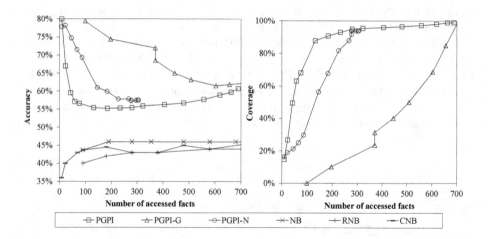

Fig. 2. Accuracy and coverage w.r.t. number of accessed facts

Best Results for Each Attributes. We also analyzed each attributes individually. The best results in terms of accuracy and coverage obtained for each attribute and each algorithm are shown in Table 1. The last column of the table indicates the number of accessed facts to obtain these results.

In terms of accuracy, the best accuracy was achieved by PGPI algorithms for all attributes. For *status,gender, major, minor, residence, year* and *school*, PGPI algorithms respectively obtained a maximum accuracy of 90.9%, 94.6%, 75.1%, 60.5%, 70.5% and 12.9%. The reason for the low accuracy for *school* is that there are 1,420 possible values and that users are rarely linked to other users with the same value.

The best results in terms of PAC are written in bold. PGPI provides the best PAC for all but one attribute. The second best algorithm is PGPI-N, which provides the second or third best PAC for five of seven attributes. Moreover, it is interesting to see that PGPI-N achieves this result using less than half the amount of facts used by PGPI. The third best algorithm is PGPI-G, which provides the second or third best PAC for four attributes.

Table 1. Best accuracy/coverage results for each attribute

algorithm	status	gender	major	minor	residence	year	school	$|facts|$
PGPI acc.	**90.8%**	**87.7%**	**31.5%**	**75.1%**	**59.59%**	**66.86%**	10.3%	684
PGPI cov.	**100%**	**98.9%**	**99%**	**99%**	**99%**	**99%**	99%	684
PGPI-N acc.	89.4%	90.9%	24.5%	73.9%	60.5%	66.77%	**12.9%**	272
PGPI-N cov.	94.4%	94.0%	94.4%	94.4%	94.4%	94.4%	**94.4%**	272
PGPI-G acc.	90.9%	94.6%	35.2%	70.8%	53.0%	70.5%	8.8%	689
PGPI-G cov.	96.6%	72%	72%	72%	72%	72%	72%	689
NB acc.	88.0%	51.1%	16.6%	74.4%	55.6%	26.0%	10.6%	189
NB cov.	100%	100%	100%	100%	100%	100%	100%	189
RNB acc.	80.2%	57.7%	15.0%	74.2%	55.0%	26.2%	9.8%	580
RNB cov.	100%	100%	100%	100%	100%	100%	100%	580
CNB acc.	88.0%	47.3%	9.0%	74.0%	54.8.%	26.0%	10.6%	189
CNB cov.	100%	100%	100%	100%	100%	100%	100%	189
LP acc.	83.0%	50.4%	16.7%	56.3%	49.1%	35.3%	10.1%	10K
LP cov.	94.4%	94.4%	94.4%	94.4%	94.4%	94.4%	94.4%	10K

Influence of _maxFacts_ and _ratioFacts_ Parameters. In the previous experiments, we have analyzed the relationship between the number of accessed facts and the PAC, accuracy and coverage. However, we did not discuss how to choose the _maxFacts_ parameter and how it influences the number of accessed facts for each algorithm.

For Naïve Bayes algorithms and LP, the _maxFacts_ parameter determines the number of nodes to be used for training, and thus directly determines the number of accessed facts.

For PGPI algorithms, we show the influence of _maxFacts_ on the number of accessed facts in Fig. 3. It can be observed that PGPI algorithms generally explore less nodes than _maxFacts_. For PGPI-N, the reason is that some social network users have less than _maxFacts_ friends whithin _maxDistance_ and thus, PGPI-N cannot use more than _maxFacts_ facts to perform a prediction. For PGPI-G, the reason is that some users are members of just a few groups. For PGPI, the number of accessed facts is greater than for PGPI-N but less than for PGPI-G. The reason is that PGPI combines both algorithms and uses the _ratioFacts_ parameter, as previously explained, to decide how much facts can be used by PGPI-N and PGPI-G. The value that we used in previous experiments for this parameter is 0.5, which means to use 50% of the facts for PGPI-G, and thus to use the other half for PGPI-N.

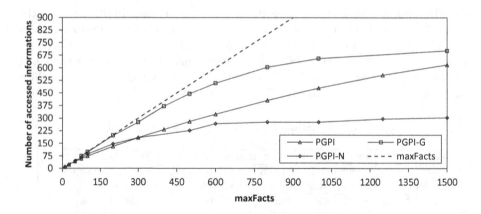

Fig. 3. Number of accessed facts w.r.t. *maxFacts*

A question that can be raised is how to set the *ratioFacts* parameter for the PGPI algorithms. In Fig. 4, we show the influence of the *ratioFacts* parameter on the product of accuracy and coverage. We have measured the PAC w.r.t the number of accessed facts when *ratiofacts* is respectively set to 0.3, 0.5 and 0.7. Note that setting *ratioFacts* to 0 and 1 would be respectively equivalent to using PGPI-N and PGPI-G. In Fig. 4, it can be observed that setting *ratioFacts* to 0.5 allows to obtain the highest PAC. Furthermore, in general a value of 0.5 also gives a high PAC w.r.t the number of accessed facts. A value of 0.3 sometimes gives a higher PAC than 0.5. However, the highest PAC value for 0.3 on overall is lower than that for 0.5. A value of 0.7 almost always gives a lower PAC.

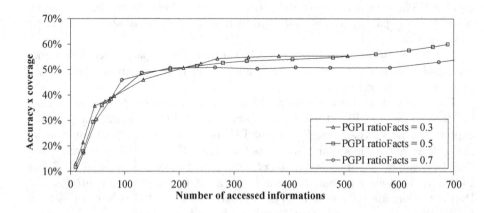

Fig. 4. Product of accuracy and coverage w.r.t. *ratioFacts*

Best Results When Using the Full Social Graph. We also performed an experiment using the full social graph ($maxFacts = \infty$). The best algorithm in this case is PGPI with an accuracy and coverage of 64.7% and 100%, while the second best algorithm (PGPI-G) achieves 62% and 98.4%.

Influence of the *maxDistance* **Parameter.** Recall that the PGPI-N and PGPI algorithms utilize a parameter called *maxDistance*. This parameter indicates that two nodes are in the same neighborhood if there are no more than *maxDistance* edges in the shortest path between them. In the previous experiments, we have empirically set this parameter to obtain the best results. We now discuss the influence of this parameter on the results of PGPI-N and PGPI.

We have investigated the influence of *maxDistance* on the product of accuracy and coverage, and also on the number of accessed facts. Results are shown in Fig. 5 for PGPI-N. It can be observed that as *maxDistance* is increased, the PAC also increases until it reaches its peak at around *maxDistance* = 6. After that, it remains unchanged. The main reason why it remains unchanged is that the influence of a user on another user is divided by their distance in the PGPI-N algorithm. Thus, farther nodes have a very small influence on each other. Another interesting observation is that as *maxDistance* increases, the number of accessed facts also increases. Thus, if our goal is to use less facts, then *maxDistance* should be set to a small value, whereas if our goal is to achieve the maximum PAC, *maxDistance* should be set to a large value. Lastly, we can also observe that the number of accessed facts does not increase after *maxDistance* = 7. The reason is that the Facebook dataset is quite sparse (i.e. users have few friendship links).

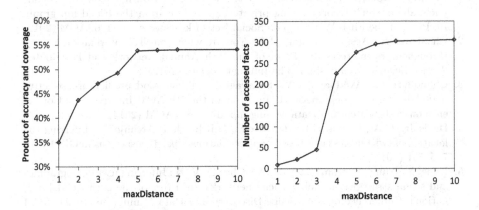

Fig. 5. Product of accuracy and coverage (left) and number of accessed facts (right) w.r.t. *maxDistance* for PGPI-N

Discussion. Overall, PGPI algorithms provide the best accuracy. A key reason is that they are lazy algorithms. They use local facts close to the node to be predicted rather than relying on a general training phase performed beforehand. A second important reason is that more rich information is used by PGPI-G and PGPI (groups, publications). A third reason is that PGPI-N calculates the similarity between users while label propagation does not.

It is interesting to note that although PGPI-G relis on synthetic data in our experiment, PGPI-N does not and performs very well.

6   Conclusion

We have proposed a lazy algorithm named PGPI that can accurately infer user profiles under the constraint of a partial social graph. It is to our knowledge, the first algorithm that let the user control the amount of information accessed from the social graph and the accuracy of predictions. Moreover, the algorithm is designed to use not only node links and profiles as basis to perform predictions but also group memberships, likes and views, if the information is available. An extensive experimental evaluation against four state-of-the-art algorithms shows that PGPI can provide a considerably higher accuracy while accessing a much smaller number of nodes from the social graph to perform predictions. Moreover, an interesting result is that profile attributes such as status (student/professor) and gender can be predicted with more than 90% accuracy using PGPI.

For future work, we plan to further improve the PGPI algorithms by proposing new optimizations and to perform experiments on additional datasets.

References

1. Blenn, N., Doerr, C., Shadravan, N., Van Mieghem, P.: How much do your friends know about you?: reconstructing private information from the friendship graph. In: Proc. of the Fifth Workshop on Social Network Systems, pp. 1–6. ACM (2012)
2. Chaabane, A., Acs, G., Kaafar, M.A.: You are what you like! information leakage through users interests. In: Proc. of the 19th Annual Network and Distributed System Security Symposium. The Internet Society (2012)
3. Chaudhari, G., Avadhanula, V., Sarawagi, S.: A few good predictions: selective node labeling in a social network. In: Proc. of the 7th ACM International Conference on Web Search and Data Mining, pp. 353–362. ACM (2014)
4. Davis Jr., C.A., Pappa, G.L., de Oliveira, D.R.R., de L Arcanjo, F.: Inferring the location of twitter messages based on user relationships. Transactions in GIS 15(6), 735–751 (2011)
5. Dong, Y., Yang, Y., Tang, J., Yang, Y., Chawla, V. N.: Inferring user demographics and social strategies in mobile social networks. In: Proc. of the 20th ACM International Conference on Knowledge Discovery and Data Mining, pp. 15–24. ACM (2014)
6. He, J., Chu, W.W., Liu, Z.V.: Inferring privacy information from social networks. In: Mehrotra, S., Zeng, D.D., Chen, H., Thuraisingham, B., Wang, F.-Y. (eds.) ISI 2006. LNCS, vol. 3975, pp. 154–165. Springer, Heidelberg (2006)
7. Heatherly, R., Kantarcioglu, M., Thuraisingham, B.: Preventing private information inference attacks on social networks. IEEE Transactions on Knowledge and Data Engineering. 25(8), 1849–1862 (2013)

8. Jurgens, D.: Thats what friends are for: Inferring location in online social media platforms based on social relationships. In: Proc. of the 7th International AAAI Conference on Weblogs and Social Media, pp 273–282. AAAI Press (2013)

9. Kosinski, M., Stillwell, D., Graepel, T.: Private traits and attributes are predictable from digital records of human behavior. National Academy of Sciences **110**(15), 5802–5805 (2013)

10. Li, R., Wang, C., Chang, K. C. C.: User profiling in an ego network: co-profiling attributes and relationships. In: Proc. of the 23rd International Conference on World Wide Web, pp. 819–830. ACM (2014)

11. Lindamood, J., Heatherly, R., Kantarcioglu, M., Thuraisingham, B.: Inferring private information using social network data. In: Proc. of the 18th International Conference on World Wide Web, pp. 1145–1146. ACM (2009)

12. Mislove, A., Viswanath, B., Gummadi, K. P., Druschel, P.: You are who you know: inferring user profiles in online social networks. In: Proc. of the 3rd ACM International Conference on Web Search and Data Mining, pp. 251–260. ACM (2010)

13. Quercia, D., Kosinski, M., Stillwell, D., Crowcroft, J.: Our Twitter profiles, our selves: Predicting personality with Twitter. In: Proc. of the 3rd IEEE International Conference on Social Computing, pp. 180–185. IEEE Press (2011)

14. Traud, A.L., Mucha, P.J., Porter, M.A.: Social structure of Facebook networks. Physica A: Statistical Mechanics and its Applications. **391**(16), 4165–4180 (2012)

An Improved Machine Learning Approach for Selecting a Polyhedral Model Transformation

Ray Ruvinskiy and Peter van Beek[✉]

Cheriton School of Computer Science, University of Waterloo,
Waterloo, ON N2L 3G1, Canada
vanbeek@cs.uwaterloo.ca

Abstract. Algorithms in fields like image manipulation, signal processing, and statistics frequently employ tight CPU-bound loops, whose performance is highly dependent on efficient utilization of the CPU and memory bus. The *polyhedral model* allows the automatic generation of loop nest transformations that are semantically equivalent to the original. The challenge, however, is to select the transformation that gives the highest performance on a given architecture. In this paper, we present an improved machine learning approach to select the best transformation. Our approach can be used as a stand-alone method that yields accuracy comparable to the best previous approach but offers a substantially faster selection process. As well, our approach can be combined with the best previous approach into a higher level selection process that is more accurate than either method alone. Compared to prior work, the key distinguishing characteristics to our approach are formulating the problem as a classification problem rather than a regression problem, using static structural features in addition to dynamic performance counter features, performing feature selection, and using ensemble methods to boost the performance of the classifier.

1 Introduction

Loops are a fundamental part of many scientific computing algorithms, with applications in image manipulation, sound and signal processing, and statistical simulations, among others. Loops in such algorithms are usually *tight*—they are computationally intensive and run for many iterations without blocking to perform high-latency operations such as disk or network I/O.

To achieve high performance, such loops must be structured so as to effectively utilize architectural features including instruction level parallelism, branch prediction, instruction and data prefetch capabilities, and multiple cores. Manually transforming the code to take advantage of architectural features results in code that is difficult to read and maintain. It also requires CPU architecture expertise that an application developer may not have. Previous work has focused on methods for automatically selecting transformations that preserve the semantics of the original code while optimizing the runtime of the loop nest on a given CPU architecture (see the description of related work in Section 3).

© Springer International Publishing Switzerland 2015
D. Barbosa and E. Milios (Eds.): Canadian AI 2015, LNAI 9091, pp. 100–113, 2015.
DOI: 10.1007/978-3-319-18356-5_9

We focus on the polyhedral model, a mathematical framework wherein loop nests can be represented with polyhedra and loop transformations are expressed as algebraic operations on polyhedra [14,15]. This enables the generation of a search space of loop transformations that preserve the semantics of the original loop. Park et al. [13] use supervised machine learning to select a transformation from the overall transformation space, using as features hardware performance counter values obtained while profiling the candidate program. In particular, they use regression models to predict the speed-up of an arbitrary transformation from the search space over the original program. Park et al. propose a *five-shot approach*, where binaries generated by applying each of the best five transformations suggested by the model are run, and the transformation that has the fastest runtime is chosen. Their five-shot approach, however, can result in an excessively long selection phase.

We present an improved machine learning approach that achieves performance comparable to the best results reported by Park et al. without the need for a five-shot selection process—it is sufficient to use the single transformation selected by our classifier—and the selection process is substantially faster. As well, our approach can be combined with Park et al.'s five-shot approach into a higher level *six*-shot selection process that is more accurate than either method alone. The key contributing factors to the improvements offered by our method include: formulating the problem as a classification problem rather than a regression problem, using static structural features in addition to dynamic performance counter features, performing feature selection, and using ensemble methods to boost the performance of the classifier.

2 Background

In this section, we review the necessary background in the polyhedral model. For more background on this topic, see, for example, [15].

The polyhedral model is a mathematical framework used to represent loops as *polyhedra* and facilitate loop transformations [4,14,15]. Algebraically, a polyhedron encompasses a set of points in a $\mathbb{Z}^n$ vector space satisfying the inequalities,

$$\mathcal{D} = \{x \mid x \in \mathbb{Z}^n, Ax + a \geq 0\},$$

where A is a matrix while x and a are column vectors. In the polyhedral model, A is a matrix of constants, x is a vector of iteration variables, a is a vector of constants, 0 is the zero vector, and the inequality operator ($\geq$) compares the vectors element-wise.

To be mapped to the polyhedral model, a block of statements must be restricted such that it is only enclosed in *if* statements and *for* loops. Pointer arithmetic is disallowed (but array accesses are allowed). Function calls must be inlined and loop bounds are restricted to affine functions of loop iterators and global variables. While these limitations appear onerous at first glance, such computational kernels play a large role in scientific and signal processing [2,6].

Loop bounds determine the *iteration domains* of the statements in consideration. An iteration domain represents the values taken on by the loop iterators for all iteration instances. The following example shows a simple nested loop and its associated iteration domain.

Example 1. Given the source code,

```
int A[n][m];
int B[m][n];
for (i = 0; i < n; i++)
  for (j = 0; j < m; j++)
    if (i < j)
      A[i][j] = n*m*B[j][i];
```

the iteration domain $\mathcal{D}_S$ of statement S in the polyhedral model becomes,

$$
\mathcal{D}_S = \left\{ \begin{pmatrix} i \\ j \end{pmatrix} \middle| \begin{pmatrix} i \\ j \end{pmatrix} \in \mathbb{Z}^2, \begin{pmatrix} 1 & 0 \\ -1 & 0 \\ 0 & 1 \\ 0 & -1 \\ -1 & 1 \end{pmatrix} \begin{pmatrix} i \\ j \end{pmatrix} + \begin{pmatrix} 0 \\ n-1 \\ 0 \\ m-1 \\ -1 \end{pmatrix} \geq 0 \right\}.
$$

The first two inequalities reflect the bounds imposed by the outer loop; the next two inequalities are the bounds imposed by the inner loop; and the last inequality reflects the logic of the conditional.

In the polyhedral model, a *timestamp* is associated with each statement instance. The relative order of the timestamps represents dependencies between the execution order of the instances. If two instances share a timestamp, no ordering between them need be enforced and they can be executed in parallel.

A multi-dimensional affine *schedule* specifies the execution order of each statement instance that respects the dependency information [14,15]. There are often many possible schedules, and each schedule, or polyhedral transformation, can be viewed as a program transformation. It is possible to algorithmically generate the set of schedules, where each schedule preserves the semantics of the original loop nest. Some schedules can be described as a sequence of well-known loop optimizations including: tiling [18], fusion [9], unrolling, prevectorization [10], and parallelization. However, there are also valid schedules that are not representable as a sequence of known loop optimizations.

Example 2. Consider the following semantically equivalent program transformation of the source code shown in Example 1,

```
for (c0 = 0; c0 <= n/32; c0++)
  for (c1 = 0; c1 <= m/32; c1++)
    for (i = 32*c0; i < min(n,32*(c0+1)); i++)
      for (j = 32*c1; j < min(m,32*(c1+1)); j++)
        if (i < j)
          A[i][j] = n*m*B[j][i];
```

The original source code shown in Example 1 could have poor cache behavior for larger values of n, possibly incurring a cache miss on every access of an element of matrix B. The transformed source code is an example of tiling. While more complicated, the transformed code can be algorithmically generated and could have significantly better cache performance.

Two schedules, or polyhedral transformations, can differ dramatically in performance on an architecture. However, selecting the appropriate polyhedral transformation is difficult even for machine architecture experts as the individual components can interact with each other and cannot be chosen independently.

3   Related Work

There are two main approaches to automatically selecting a high-performance polyhedral transformation: hand-crafted and machine learning approaches.

For hand-crafted approaches, Lim and Lam [11] propose an algorithm to select a transformation that maximizes parallelism and minimizes synchronization. However, their algorithm has super-exponential complexity in the worst case. Pouchet et al. [14,15] present search-based approaches to selecting a transformation: an exhaustive search and a genetic algorithm. Pouchet et al. [16] subsequently improve this approach to use a combination of analytic models, where they are available and effective, and empirical search for cases where analytic models do not account for important properties of the hardware that have a significant impact on performance. These search-based approaches all suffer from scalability concerns, as the size of the search space is large enough to make it impractical to explore the space iteratively and evaluate different transformations by compiling and running them.

For machine learning approaches, Park et al. [13] use regression models to predict the runtime of a polyhedral transformation of a source program and so select a transformation. The search space is limited to the transformations supported out-of-the-box by the Polyhedral Compiler Collection version 1.1 (PoCC) software: loop tiling, loop fusion, loop unrolling, loop prevectorization, and loop parallelization. Considering every combination of the transformations results in a search space of several hundred transformations. As in Cavazos et al. [3], Park et al. use hardware performance counters as features. Park et al. propose a technique called *five-shot* which requires running the program being optimized five times. For some programs their approach can be excessively time consuming, which may serve as a barrier to the uptake of the technique. Our approach builds on but improves Park et al. to either remove this deficiency or to improve the accuracy. In Section 5, we perform an extensive empirical comparison.

More recently, Park et al. [12] extend this work to include two additional loop optimizations—wavefront and SIMD vectorization—using a newer version of PoCC (version 1.2) and to propose a six-shot approach that consists of running the program six times, each time for a different regression classifier. Unfortunately, we were not successful at integrating version 1.2 into our experimental setup and no experimental results comparing the six-shot to the original five-shot proposal have been reported. Our experimental results reproduce and compare against the original five-shot approach using version 1.1 of PoCC. However, as there is evidence that the six-shot approach only gives incrementally better performance over the five-shot approach, and the six-shot approach retains the same inherent deficiency, we believe our comparison and conclusions still hold.

Additionally, there is a considerable body of work on using machine learning more generally in optimization in compilers including selecting a loop unroll factor [17], selecting compiler optimization settings [3], and selecting a loop tiling size [20]. In our approach we adapt some of the features previously proposed for selecting loop tile size [20].

4   An Improved Machine Learning Approach

Our proposal is an extension of the work by Park et al. [13]. The key distinguishing characteristics to our approach compared to Park et al.'s work are using static structural features in addition to dynamic performance counter features (Section 4.1), formulating the problem as a classification problem rather than a regression problem (Section 4.2), performing feature selection to exclude features that hold little or no predictive value for the class (Section 4.4), and using ensemble methods to boost the performance of the classifier (Section 4.5).

4.1   Initial Feature Set

As in previous work, we use hardware performance counters as features [3,13]. Over 90 program counters are available on the architecture used in our experimental setup. We selected a subset of those we deemed to be promising (see Table 1). Hardware performance counter values are collected by running a binary compiled from the unmodified program source on representative input. These values are then normalized by taking ratios and manually discretized. We also added static memory access features that are extracted from the program source structure and reflect the program's memory access patterns, as proposed by Yuki et al. [20].

4.2   Class Value

Rather than using regression models to predict speed-up, we express the problem as a binary classification problem by considering which of two transformations results in faster runtime for a given source program. More formally, given a source program P and two transformations, T_A and T_B, the transformation pair (T_A, T_B) is labeled 1 if P has a faster runtime having had T_A applied rather than having had T_B applied; otherwise, the pair is labeled 0. A complete training example then consists of the feature values for the benchmark, the transformation pair, and the class of the transformation pair. A decision tree classifier is learned from the data.

Table 1. Performance counter features c_i and memory access features m_i

feature	description
c_0	ratio of L1 cache misses to L1 cache accesses
c_1	ratio of L2 cache misses to L2 cache accesses
c_2	ratio of L3 cache misses to L3 cache accesses
c_3	ratio of total CPU cycles to retired instructions
c_4	ratio of cycles when no instructions were retired to total CPU cycles
c_5	ratio of L3 cache accesses to total CPU cycles
c_6	ratio of L3 cache misses to total CPU cycles
c_7	ratio of all resource-related stall cycles to total CPU cycles
c_8	ratio of load buffer stall cycles to total CPU cycles
c_9	ratio of stall cycles due to reservation station being full to total CPU cycles
c_{10}	ratio of stall cycles to reorder buffer being full
c_{11}	ratio of conditional branch instructions executed to retired instructions
c_{12}	ratio of mispredicted branches executed to retired instructions
m_0	number of prefetched memory reads
m_1	number of non-prefetched memory reads
m_2	number of loop-invariant memory reads
m_3	number of prefetched memory writes
m_4	number of non-prefetched memory writes
m_5	number of loop-invariant memory writes

Once the classifier is constructed, it is used to choose a polyhedral transformation as follows. For a previously unseen program P, the classifier labels all transformation pairs (T_A, T_B), for all transformations T_A and T_B in the transformation search space. Subsequently, for every transformation T_A, the number of times a tuple (T_A, T_B), where T_B is any transformation other than T_A, is predicted to have the label 1 is recorded as $score_{T_A}$. The transformation with the highest score is selected as the best transformation suggested by the classifier. In effect, a voting algorithm is used, with transformations being voted on within the context of every transformation pair combination. The transformation with the most votes is considered the winner. This is referred to as *pairwise preference ranking* or *round robin ranking* [5]. Other approaches to combining pairwise preferences exist (e.g., algorithms to calculate class probabilities [19]) but are not explored in the context of this paper.

4.3 Data Collection

Benchmarks from Polybench/C[1], a suite of computational kernels with loop nests, are used to train and test the classifier. The PAPI[2] library is used to collect performance counter values for each benchmark. Each benchmark is compiled using the gcc compiler version 4.7.2 at the highest standard optimization

---

[1] http://www.cs.ucla.edu/~pouchet/software/polybench
[2] http://icl.cs.utk.edu/papi/index.html

level, $-O3$. No other gcc flags are used. The benchmarks are then executed. The wall time runtime and selected hardware performance counter values are recorded. Each benchmark is run enough times for the sum of the runtimes of the executions to reach at least 10 seconds.

Subsequently, the Polyhedral Compiler Collection version 1.1 (PoCC)[3] is used to generate the transformation search space. PoCC has numerous options for various optimizations. We considered different values for the fusion, OpenMP, tiling, vectorization, and loop unrolling options. In addition, we included the identity transformation, where the original source code is not altered.

PoCC generates one source file per transformation per benchmark. The source files are compiled, generating a binary. The binary is executed and the wall time runtime is recorded. A training example is then generated for every pair of transformations in each benchmark. The format of a training example is $c_0, \ldots, c_{12}$, $m_0, \ldots, m_5$, $identity_A$, $tiling_A$, $openmp_A$, $vectorization_A$, $unroll_A$, $identity_B$, $tiling_B$, $openmp_B$, $vectorization_B$, $unroll_B$, $class$. The features $c_0, \ldots, c_{12}$ are the discretized performance counter values for the benchmark, and $m_0, \ldots, m_5$ are the memory feature values for the benchmark. The features $identity_A, \ldots, unroll_A$, are the feature values for the first transformation in the pair, and the features $identity_B, \ldots, unroll_B$, are the feature values for the second transformation in the pair. *Identity* is a binary feature, and its value is 1 if the transformation is an identity transformation and 0 otherwise. *Tiling* is an enumerated feature corresponding to the tiling setting used in the transformation. The tiling setting consists of the tiling factors for the top three nested loops. Possible tiling factors are 1 (no tiling) or 32. This gives 8 values for the tiling setting in total. *Openmp* and *vectorization* are binary features reflecting their presence or absence in the transformation. The *unroll* value of unroll is the loop unrolling factor used (8, 4, or 0 for no unrolling). The feature *class* is the binary classification. If the runtime of the first transformation in the pair is less than the runtime of the second transformation, *class* is 1; otherwise, it is 0. The transformations T_A, T_B and T_B, T_A are treated as distinct pairs, and a separate training example is generated for each. The two pairs will belong to opposite classes.

4.4   Feature Selection

Feature selection (see, e.g., [7] and references therein) is used to select features that have predictive value. A classifier is generated using every combination of two features and evaluated for every benchmark. The classifiers are then sorted in order of the number of benchmarks where the classifier suggested a transformation with a better runtime than the expected runtime of a randomly selected transformation for the benchmark (see lines 2–8 of Algorithm 1). The random selection process is repeated 100 times. In contrast to Park et al. [12], who found that on average feature selection only degraded performance, feature selection is an integral component of our overall approach.

---

[3] http://www.cs.ucla.edu/~pouchet/software/pocc

Algorithm 1. Feature selection and classifier evaluation.

---

1 **foreach** $b \in$ *Benchmarks* **do**

2   **foreach** $f_1, f_2 \in$ *Features* **do**

3   **foreach** $b' \in (Benchmarks - \{b\})$ **do**

4   Learn a classifier using f_1, f_2 and data from *Benchmarks* $- \{b, b'\}$;

5   Test the classifier using data from b';

6   **end foreach**

7   Tally number of benchmarks for which the classifier using the feature combination f_1, f_2 was judged effective;

8   **end foreach**

9   Sort feature combinations in descending order by number of benchmarks for which they were judged effective;

10   Take top 21 feature combinations and construct a decision tree from each combination using data from *Benchmarks* $- \{b\}$;

11   Construct an ensemble classifier that classifies the data using the 21 trees and predicts the class by a majority vote of all 21 trees;

12   Use the ensemble classifier to predict best transformation for b;

13   Report the accuracy of the predicted transformation for b;

14 **end foreach**

The selected features and the number of times a feature occurred in a feature pair that was selected for the classifier ensemble are the following.

feature	c_0	c_1	c_3	c_4	c_5	c_6	c_7	c_{12}	m_1	m_3	m_4	m_5
number	8	4	3	1	6	2	1	3	4	3	2	1

The two features that appear the most often are the number of L1 cache misses normalized with respect to the number of L1 cache accesses and the number of L3 cache accesses normalized with respect to total CPU cycles. On the other hand, the number of L3 cache accesses normalized with respect to L3 cache misses does not appear at all. It is not immediately obvious why L1 cache misses are significant while L3 cache misses are not and why overall L3 cache accesses are significant while overall L1 cache accesses are not. This may be related to architectural peculiarities. It is also interesting to note that prefetched memory writes are significant, while prefetched memory reads are not. Again, this may be related to the architecture on which the experiments were run. Such subtleties regarding feature significance are not intuitive, and they become apparent only as a result of the feature selection process.

4.5   Classifier Ensembles

The performance of a classifier can often be boosted by generating multiple classifiers (or *base-learners* from smaller feature sets) and combining their results [1, pp. 419–421]. Such combinations of classifiers are known as *ensembles*. The simplest way to combine the outputs of an ensemble of classifiers is *voting*, and the simplest and most widely-used voting technique is *simple voting*, where all classifiers are equally weighted [1, pp. 424-425].

A simple voting approach is used to construct a classifier ensemble for predicting the label of a transformation pair (T_A, T_B). The top n classifiers as determined in the feature selection stage vote to predict the class. (In addition to ensembles of decision trees containing pairs of features, we also considered ensembles of decision stumps but these did not lead to improved performance.) Each classifier's vote is weighted equally. A value of 21 is used for n. The value is determined empirically, and it is found that a value of 21 performs better than a lower value, while values between 21 and 40 all perform relatively equally well (see lines 9–11 of Algorithm 1).

5   Experimental Evaluation

We present an empirical evaluation of our proposal against a baseline and the current state-of-the-art approach. Our experimental results were obtained on an Intel Core i7: Nehalem microarchitecture, Lynnfield Performance Desktop, model 870, 4 cores, and clock rate of 2.93 GHz. We show that on a benchmark suite of computational kernels with loop nests, our method is competitive for accuracy (Section 5.1), offers a substantially faster selection process (Section 5.2), and can be combined with the best previous approach into a higher level selection process that is more accurate than either method alone (Section 5.3).

The results presented for linear regression (LR) and support vector machine regression (SVM), for both the one-shot and five-shot approaches, are our best efforts to reproduce the results reported in Park et al. [13]. One-shot results evaluate the transformation predicted to be the best by the given method. Five-shot results take the best runtime of the top five results as predicted by the given method; i.e., the binary corresponding to each of the top five results is run and the binary that has the best runtime is used.

5.1   Evaluation of Selection Accuracy

A nested leave-one-benchmark-out approach [8, pp. 245-247] was used to evaluate our overall approach (see Algorithm 1). For each benchmark, the data sets of all other benchmarks are used as the training data, while the data set of the benchmark in question is used as the test data. C4.5[4] is used to generate decision trees from the training data.

Following Park et al., as a measure of accuracy we use percentage of optimal: the ratio of the runtime of the benchmark binary obtained by applying the optimal transformation in the search space to the runtime of a benchmark binary obtained by applying a transformation selected by the approach being evaluated. The ratio is expressed as a percentage, with a value of 100% indicating that the transformation selected is the optimal transformation in the search space.

---

[4] http://www.rulequest.com/Personal

Table 2. Percentage of optimal on benchmark kernels for the identity transformation; and the transformations chosen by the linear regression (LR), support vector machine (SVM), and decision tree vote (DTV) methods

benchmark	identity	LR 1-shot	LR 5-shot	SVM 1-shot	SVM 5-shot	DTV
2mm	15.9	76.9	100.0	96.7	100.0	77.8
3mm	19.0	99.4	99.4	75.1	89.0	100.0
adi	73.3	46.5	54.1	40.8	48.8	100.0
bicg	73.6	68.1	68.1	33.7	100.0	100.0
cholesky	68.3	99.6	99.6	95.6	99.2	95.6
correlation	5.1	54.8	60.4	60.4	99.4	98.8
covariance	4.8	58.9	73.0	38.1	57.5	100.0
doitgen	46.6	32.0	67.7	66.6	100.0	31.7
durbin	70.5	98.7	99.4	98.6	98.7	97.9
dynprog	59.2	83.8	84.5	77.6	81.3	77.6
fdtd-2d	50.6	98.7	98.7	17.5	62.8	55.0
fdtd-apml	70.0	97.9	97.9	100.0	100.0	99.3
floyd-warshall	71.3	62.9	74.9	100.0	100.0	100.0
gemm	11.4	98.2	100.0	69.1	83.5	96.9
gemver	28.8	65.7	100.0	96.7	98.6	65.7
gesummv	73.7	66.8	67.0	100.0	100.0	83.7
gramschmidt	9.3	44.3	44.3	13.3	43.7	43.5
jacobi-1d-imper	77.2	93.8	94.4	94.7	99.1	99.1
jacobi-2d-imper	49.2	80.3	80.6	98.8	98.9	68.2
lu	65.7	78.5	79.0	41.5	41.5	100.0
ludcmp	69.4	83.6	99.6	95.9	99.6	86.0
mvt	31.9	98.0	98.0	90.8	99.6	53.3
reg_detect	70.4	86.7	99.7	98.2	98.2	98.2
seidel-2d	42.4	49.6	50.2	61.4	61.4	100.0
symm	70.0	98.4	98.5	97.3	98.3	98.5
syr2k	35.4	51.3	52.1	51.5	84.6	90.0
syrk	19.2	41.4	44.8	27.8	90.5	89.5
trisolv	80.8	57.3	57.7	62.6	100.0	100.0
trmm	25.2	57.4	58.5	35.9	100.0	90.6
average	47.9	73.4	79.4	70.2	87.4	86.1

The optimal transformation is found by applying every transformation in the search space to the benchmark source code, running the resulting binaries, recording the runtimes, and selecting the transformation corresponding to the binary with the lowest runtime. As a baseline, we use the runtime of the stock benchmark with no modifications to the source code, referred to as identity.

Table 2 contrasts the accuracy, expressed as percentage-of-optimal, of the identity transformation, the one-shot and five-shot approaches for linear and SVM regression, and our decision tree voting approach. Table 3 shows the speed-up over the identity transformation for every benchmark for the five-shot SVM

Table 3. Speed-up on benchmark kernels over the identity transformation using the SVM five-shot (SVM) and decision tree vote (DTV) methods

benchmark	SVM	DTV	benchmark	SVM	DTV
2mm	6.3	4.9	gesummv	1.4	1.4
3mm	4.7	5.3	gramschmidt	4.7	4.7
adi	0.7	1.4	jacobi-1d-imper	1.3	1.3
bicg	1.4	1.4	jacobi-2d-imper	2.0	1.4
cholesky	1.5	1.4	lu	0.6	1.5
correlation	19.4	19.3	ludcmp	1.4	1.2
covariance	11.9	20.7	mvt	3.1	1.7
doitgen	2.2	0.7	reg_detect	1.4	1.4
durbin	1.4	1.4	seidel-2d	1.5	2.4
dynprog	1.4	1.3	symm	1.4	1.4
fdtd-2d	1.2	1.1	syr2k	2.4	2.2
fdtd-apml	1.4	1.4	syrk	4.7	5.2
floyd-warshall	1.4	1.4	trisolv	1.2	1.2
gemm	7.3	8.5	trmm	4.0	3.6
gemver	3.4	2.3	median	1.5	1.4
			average	3.3	3.5

regression and the one-shot decision tree voting. The results of our reproduction of Park et al.'s work are similar to the results reported in their paper; thus, we have some confidence in their correctness. Our one-shot decision tree vote and the five-shot SVM approach perform about equally well.

5.2 Evaluation of Selection Speed

Table 4 contrasts the runtime (seconds) needed to select a transformation using the support vector machine and decision tree vote methods. For both methods, the untransformed code is first run to gather the feature values. These feature values are then fed into a classifier to predict the best transformations. Gathering the feature values is the most time consuming phase as enabling the program counter features slows the execution, in proportion to the number of program counters enabled. Park et al. [12] use 56 program counter features[5]. We effectively used feature selection to narrow this down to 8 program counters and 4 structural features. The speedups range from 3.7 to 15.8, with an average of 7.2. The maximum wall clock speed up was for the 3mm benchmark. Park et al. took 1154.3 seconds (approximately 18 minutes) whereas ours took 182.8 seconds (approximately 3 minutes), for a speedup ratio of six times faster.

---

[5] Park et al. [13] do not provide details on which program counters were used in their experiments, so we are relying instead on the expanded version of their paper [12] where they use 56 program counters, 47 of which are available on our architecture.

Table 4. Runtime (seconds) on benchmark kernels needed to select a transformation using the SVM five-shot (SVM) and decision tree vote (DTV) methods. Also shown is the ratio of the two runtimes. The average ratio is 7.2.

benchmark	SVM	DTV	ratio	benchmark	SVM	DTV	ratio
2mm	857.8	131.6	6.5	gesummv	20.4	3.3	6.2
3mm	1154.3	182.8	6.3	gramschmidt	487.1	75.9	6.4
adi	79.0	9.8	8.1	jacobi-1d-imper	20.7	4.0	5.2
bicg	17.6	2.5	7.0	jacobi-2d-imper	15.2	1.8	8.4
cholesky	15.4	1.9	9.1	lu	96.5	6.1	15.8
correlation	264.2	44.4	6.0	ludcmp	51.3	13.7	3.7
covariance	310.3	46.1	6.7	mvt	18.4	2.1	8.8
doitgen	68.4	10.3	6.6	reg_detect	17.8	1.8	9.9
durbin	50.3	8.2	6.1	seidel-2d	30.8	4.1	7.5
dynprog	72.8	10.6	6.9	symm	686.4	91.5	7.5
fdtd-2d	90.8	13.0	7.0	syr2k	222.3	33.4	6.7
fdtd-apml	74.7	9.1	8.2	syrk	163.1	25.0	6.5
floyd-warshall	81.3	9.8	8.3	trisolv	31.9	7.5	4.3
gemm	361.7	69.8	5.2	trmm	164.0	27.6	5.9
gemver	19.8	2.3	8.6				

5.3 A Six-Shot Selection Process

The per-benchmark breakdown in Table 2 reveals that while, on average, our one-shot DTV method is competitive with Park et al.'s five-shot SVM method, the different approaches do better on different benchmarks. For examples, DTV obtains 100% on lu when SVM obtains only 41.5%, and SVM obtains 100% on doitgen when DTV obtains only 31.7%.

That the two methods have different strengths suggests that combining them would be advantageous. We define a *six*-shot selection process which consists of: (i) run the five best transformations predicted by Park et al.'s SVM approach and record the best transformation, the transformed binary that has the fastest runtime, (ii) run the transformed binary predicted by our one-shot DTV approach, and (iii) report the fastest of the two transformations from steps (i) and (ii). The result is a six-shot selection process that is more accurate than either method alone: The accuracy is boosted from 87.4% and 86.1% for the SVM and DTV methods, respectively, to within 95.0% of optimal, on average. Note that adding step (ii) has a proportionally negligible effect on the overall runtime.

6 Conclusion

We presented an improved machine learning approach within the polyhedral framework to select the transformation of a loop nest that gives the highest performance on a given architecture. On a benchmark suite of computational kernels our DTV method achieves accuracy results competitive with Park et al.'s [13] five-shot SVM regression approach, the best previous approach, while speeding up the selection process by a factor of 7.2 on average. As well, when the DTV approach is combined with Park et al.'s five-shot approach into a higher level six-shot selection process the new six-shot selection process is more accurate than either method alone. On the benchmark suite, the accuracy of the combined six-shot process, as measured by percentage from optimal, was boosted to 95.0% as compared to 86.1% and 87.4% when our method and Park et al.'s method, respectively, are used as stand-alone methods.

References

1. Alpaydin, E.: Introduction to Machine Learning, 2nd edn. The MIT Press (2010)
2. Benabderrahmane, M.-W., Pouchet, L.-N., Cohen, A., Bastoul, C.: The polyhedral model is more widely applicable than you think. In: Gupta, R. (ed.) CC 2010. LNCS, vol. 6011, pp. 283–303. Springer, Heidelberg (2010)
3. Cavazos, J., Fursin, G., Agakov, F., Bonilla, E., O'Boyle, M.F.P., Temam, O.: Rapidly selecting good compiler optimizations using performance counters. In: Proceedings of CGO 2007, pp. 185–197 (2007)
4. Feautrier, P.: Automatic parallelization in the polytope model. In: Perrin, G.-R., Darte, A. (eds.) The Data Parallel Programming Model. LNCS, vol. 1132, pp. 79–103. Springer, Heidelberg (1996)
5. Fürnkranz, J., Hüllermeier, E.: Pairwise preference learning and ranking. In: Lavrač, N., Gamberger, D., Todorovski, L., Blockeel, H. (eds.) ECML 2003. LNCS (LNAI), vol. 2837, pp. 145–156. Springer, Heidelberg (2003)
6. Girbal, S., Vasilache, N., Bastoul, C., Cohen, A., Parello, D., Sigler, M., Temam, O.: Semi-automatic composition of loop transformations for deep parallelism and memory hierarchies. Intl J. of Parallel Programming **34**, 2006 (2006)
7. Guyon, I., Elisseeff, A.: An introduction to variable and feature selection. J. Mach. Learn. Res. **3**, 1157–1182 (2003)
8. Hastie, T., Tibshirani, R., Friedman, J.: The Elements of Statistical Learning: Data mining, Inference and Prediction, 2nd edn. Springer (2009)
9. Kennedy, K., McKinley, K.S.: Maximizing loop parallelism and improving data locality via loop fusion and distribution. In: Banerjee, U., Gelernter, D., Nicolau, A., Padua, D.A. (eds.) LCPC 1993. LNCS, vol. 768, pp. 301–320. Springer, Heidelberg (1994)
10. Larsen, S., Amarasinghe, S.: Exploiting superword level parallelism with multimedia instruction sets. In: Proceedings of PLDI 2000, pp. 145–156 (2000)
11. Lim, A.W., Lam, M.S.: Maximizing parallelism and minimizing synchronization with affine transforms. In: Proceedings of POPL 1997, pp. 201–214 (1997)
12. Park, E., Cavazos, J., Pouchet, L.N., Bastoul, C., Cohen, A., Sadayappan, P.: Predictive modeling in a polyhedral optimization space. Intl J. of Parallel Programming **41**, 704–750 (2013)
13. Park, E., Pouche, L.N., Cavazos, J., Cohen, A., Sadayappan, P.: Predictive modeling in a polyhedral optimization space. In: Proc. of CGO 2011, pp. 119–129 (2011)
14. Pouchet, L.N., Bastoul, C., Cohen, A., Cavazos, J.: Iterative optimization in the polyhedral model: Part II, multi-dimensional time. In: Proceedings of PLDI 2008, pp. 90–100 (2008)

15. Pouchet, L.N., Bastoul, C., Cohen, A., Vasilache, N.: Iterative optimization in the polyhedral model: Part I, one-dimensional time. In: Proceedings of CGO 2007, pp. 144–156 (2007)
16. Pouchet, L.N., Bondhugula, U., Bastoul, C., Cohen, A., Ramanujam, J., Sadayappan, P., Vasilache, N.: Loop transformations: convexity, pruning and optimization. SIGPLAN Not. **46**, 549–562 (2011)
17. Stephenson, M., Amarasinghe, S.: Predicting unroll factors using supervised classification. In: Proceedings of CGO 2005, pp. 123–134 (2005)
18. Wolf, M.E., Lam, M.S.: A data locality optimizing algorithm. In: Proceedings of PLDI 1991, pp. 30–44 (1991)
19. Wu, T.F., Lin, C.J., Weng, R.C.: Probability estimates for multi-class classification by pairwise coupling. J. Mach. Learn. Res. **5**, 975–1005 (2004)
20. Yuki, T., Renganarayanan, L., Rajopadhye, S., Anderson, C., Eichenberger, A.E., O'Brien, K.: Automatic creation of tile size selection models. In: Proceedings of CGO 2010, pp. 190–199 (2010)

Dynamic Budget-Constrained Pricing in the Cloud

Eric Friedman, Miklós Z. Rácz[⊠], and Scott Shenker

International Computer Science Institute and UC Berkeley, Berkeley, USA
{ejf,shenker}@icsi.berkeley.edu, racz@stat.berkeley.edu

Abstract. We introduce a new model of user-based dynamic pricing in which decisions occur in real time and are strongly influenced by the budget constraints of users. This model captures the fundamental operation of many electronic markets that are used for allocating resources. In particular, we focus on those used in data centers and cloud computing where pricing is often an internal mechanism used to efficiently allocate virtual machines. We study the allocative properties and dynamic stability of this pricing model under a standard framework of cloud computing systems which leads to highly degenerate systems of prices. We show that as the size of the system grows the user-based budget-constrained dynamic pricing mechanism converges to the standard Walrasian prices. However, for finite systems, the prices can be non-degenerate and the allocations unfair, with large groups of users receiving allocations significantly below their fair share. In addition, we show that improper choice of price update parameters can lead to significant instabilities in prices, which could be problematic in real cloud computing systems, by inducing system instabilities and allowing manipulations by users. We construct scaling rules for parameters that reduce these instabilities.

1 Introduction

Price-based mechanisms provide simple, powerful, and robust tools to allocate resources in complex systems. They are easy to design, as they adaptively set prices for each resource and then allow users to purchase their optimal bundle of resources at those prices. Furthermore, unlike traditional algorithmic methods, they adapt easily to changes or additions in the architecture of the underlying system. For these reasons they are widely used for internal pricing to optimize resource allocation in cloud computing and data centers.

While the static/equilibrium theory of price mechanisms is well established, their dynamics are not as well understood. However, in modern electronic markets and computer systems the real-time behavior is crucial. The most well understood dynamics of price mechanisms are studies of the Tatonnement [14], a fictitious price adjustment mechanism where users reveal their true preferences to a sequence of hypothetical prices. While this and other previously studied dynamic models of pricing (e.g., [4,6]) can be informative, they do not capture the key issues that arise in many computational settings. They both overlook

© Springer International Publishing Switzerland 2015
D. Barbosa and E. Milios (Eds.): Canadian AI 2015, LNAI 9091, pp. 114–121, 2015.
DOI: 10.1007/978-3-319-18356-5_10

important dynamical details as well as ignoring budget constraints—an important aspect of real systems using user-based markets.

In this paper we construct a dynamic model of real time pricing that is driven by users' budget constraints. We show that in the limit of a large number of users (relative to the number of different types of resources) budget-constrained pricing (BCP) leads to the standard (static) price equilibrium. However, with a finite number of users there are discrepancies and instabilities which we study both analytically and via simulations.

For concreteness, we focus on a specific resource allocation problem arising in data centers and cloud computing: allocating virtual machines (VMs) to users, where the goal is to simultaneously maintain high efficiency and max-min fairness [7]. Unlike some well-known systems like Amazon's Elastic Cloud Compute Center (EC3) where VMs are rented out to maximize profit, and allocative fairness is secondary, these clouds are typically either used by a single large organization, where max-min fairness is based on internal divisions in the company [2], or shared by many organizations over the long term and max-min fairness ensures that each receives its correct share of the resources [7,8].

In such systems (e.g., [9]) there are typically tens to hundreds of users and thousands to millions of VMs. The key constraint is that the VM be compatible with the user's request. For example, many users have tasks that can only run under a specific operating system (Linux/Windows), a VM with specific hardware, such as a GPU or other properties, such as a public IP address.

These user-based markets use scrip [3,6,11,12] (money that has no value outside the system) which is replenished at a regular rate. Given this dynamic supply of scrip, budgeting is of primary importance and budget constraints can dominate user behavior, i.e., if a user spends all its money it cannot run any more jobs until a replenishment arrives.

We study the behavior of these user-based budget-constrained markets and show that they are surprisingly effective—they can attain both high efficiency and max-min fairness. This is true in the large market limit, but can also be attained for finite markets subject to some key design principles that we uncover.

In particular, our analysis shows that these user-based budget-constrained markets provide a solution to a problem that arises in the analysis of these systems using a classical market. This problem arises because in equilibrium the classical Walrasian prices are highly degenerate for this economy. In particular, the equilibrium prices in such an economy are all the same which can lead to problems in allocations, i.e., when prices of different VMs are the same but the optimal allocation requires a specific allocation, it is unclear how users would implement this allocation when they are unaware of its details.

We show that this problem is largely resolved by the budget-constrained behavior which leads to slight differences in prices providing the incentives to move away from over-demanded VMs. These price differences can be either transitory in the case of tie breaking or permanent in the case of more popular machines.

Our model is also of interest to the study of pricing and markets in economic theory as it provides a new alternative to the Tatonnement. In particular, our "quasi-static" approximation provides a tractable model of price adjustments that may be more widely applicable.

2   Model

Our model consists of two parts: (1) a standard model of internal cloud computing centers (CCCs) where VMs are allocated to users, coupled with (2) a user-based budget-constrained dynamic pricing mechanism, which governs the prices of VMs, and consequently also affects the allocation of VMs to users. The main contribution of our paper is studying the effects of this user-based budget-constrained dynamic pricing mechanism.

Let us first introduce the standard model of CCCs that we consider. The set of VMs, M, is divided into classes according to the partition C, and let $m = |M|$. The set of users is denoted by N, with $n = |N|$. Each user i has a set of allowable VM classes, $C_i \subseteq C$, on which she can run her tasks. In order to avoid trivialities, assume that every class is demanded by at least one user.

Let X represent an allocation of VMs where $X_i(t)$ is the set of VMs allocated to user i at time t, and let $x_i(t) = |X_i(t)|$. For analytic tractability[1], we assume that all tasks take the same amount of time on average and are distributed i.i.d. according to an exponential distribution with rate α. We assume that users receive equal utility for each completed job, and are indifferent between different VMs within their allowable types. Consequently, $x_i(t)$ is the instantaneous utility of user i at time t, since we assume that users are only allocated their allowable classes and we are focusing on the average value over time of this. In particular, we will be interested in the equal utility allocation, $x_i(t) = m/n$, and more generally the max-min fair allocations [7,8]. Note that one can easily check the feasibility of the max-min fair allocations (or any other specified allocation) by solving a maximum flow problem on a bipartite graph.

In order to dynamically allocate the VMs we consider a simple dynamic pricing mechanism that captures the essence of many pricing mechanisms used in CCCs. Unlike traditional (non-price-based) allocation mechanisms for CCCs [7], user-based pricing mechanisms are simple decentralized mechanisms that can easily adapt to changes in the system, as the optimization is facilitated directly by the users. However, in addition to the added responsibility this imposes on users, it may also lead to instabilities, as we will see later.

In our model we assume that each VM class c has a current price $p_c(t)$ at time t. Users receive allotments, s units, of scrip at small time intervals (the lengths of which we assume for tractability are i.i.d. exponentially distributed with rate γ), and their current budget at time t is given by $b_i(t)$.

---

[1] We admit that this assumption is at odds with standard run time distributions of tasks in CCCs, which tend to be heavy tailed [1], but believe that by greatly facilitating the analysis and providing insights into problems that would not be tractable otherwise, it provides a useful approximation.

When a VM completes a task it becomes available and users can request to run their next task on that VM. In order to focus on budgetary issues rather than strategic purchasing decisions, we make the simplifying assumption that a user will always request a VM if the class c of this VM is allowable for them, i.e., $c \in C_i$, and if they have sufficient budget, i.e., $b_i(t) \geq p_c(t)$. In particular, this assumes that users are not constrained by capacities—they want as many machines as possible; they are constrained only by their budgets. Let $U(t)$ be the set of users requesting a VM from class c that became available at time t, and let $u(t) = |U(t)|$. If $u(t) > 0$ then the VM is allocated uniformly at random to one of the users in $U(t)$, and the chosen user's budget is decreased by $p_c(t)$ (i.e., they pay for the machine immediately); otherwise the VM becomes inactive for an exponentially distributed time with rate β. In both cases the price is updated as described below. When an inactive VM becomes available again it is readvertised to the users, some of which may now have enough budget to purchase the VM.

The price is updated according to a simple multiplicative rule:

$$p_c(t) \to \exp(\epsilon(u(t) - 1))p_c(t), \tag{1}$$

where $\epsilon > 0$. This rule is chosen for its tractability and its value in constructing polynomial time algorithms [4], but does not affect our main results as long as the price decreases for $u(t) = 0$ and increases for $u(t) > 1$. One could also use an additive rule, as is common in the economics literature [13]: $p_c(t) \to p_c(t) + \epsilon(u(t) - 1)$. For such rules, if a class of VMs is consistently not demanded then its price will fall, while if it is overdemanded, then its price will rise.

3 Quasi-Static Approximations, Asymptotic Fairness and Walrasian Equilibrium

The first key property of price mechanisms is their efficiency; however in the general setting of CCCs efficiency is easy to attain—just allocate any available VM to any user who can use it. So the main reason CCCs use pricing mechanisms is equity (fairness). In our setting the standard (static) Walrasian equilibrium [14] attains the equal share solution if it is feasible, or more generally, it attains the max-min fair solution. We now argue that in the large n limit the equilibrium from user-based budget-constrained dynamic pricing also does so.

Recall that the Walrasian equilibrium [14] is defined as follows: Let m_c be the number of VMs of class c available. User i has $b_i = 1$ units of scrip which she uses to purchase as many VMs as she can afford at the current prices p. Let $X_i(p)$ be the set of VMs that user i purchases at price p, and let $\Xi_i(p)$ be the set of all optimal purchase sets. A price allocation (p, X) is an equilibrium if $X_i(p) \in \Xi_i(p)$, i.e., all users are purchasing an optimal bundle, and $\bigcup_i X_i(p) = M$ with $\bigcap_i X_i(p) = \emptyset$, i.e., all VMs are allocated, with each VM allocated to a single user. One can easily compute in polynomial time the Walrasian equilibrium allocation and prices by solving a max-flow problem.

Although the utilities in the Walrasian equilibrium allocation are unique [4], note that the equilibrium prices are typically highly degenerate.

Theorem 1. *If VM classes c and c' are both demanded by at least two users and there exists a user who purchases VMs from both class c and c' in a Walrasian equilibrium, then $p_c = p_{c'}$.*

Proof: The fact that c and c' are demanded by at least two users guarantees that the equilibrium prices for these classes are positive: $p_c, p_{c'} > 0$. Then if, say, $p_c > p_{c'} > 0$, then it is suboptimal for user i to request goods from class c; she should simply make all her purchases from the lower priced class c'.

Thus, in equilibrium the prices provide little information. This raises the question of how the users 'know' which classes of VMs to purchase. For example, consider an example with $k + 1$ classes of VMs and $k + 1$ groups of users, where users of type 0 only like class 0 VMs, while for $1 \leq j \leq k$ users of type j like VMs from classes 0 and j. We refer to this as the "finicky" example, as users in group 0 only want the overdemanded VMs. In equilibrium users of group $j > 0$ typically purchase VMs from classes 0 and j. However, since the equilibrium prices are the same for both classes, they have no way of "knowing" how to divide their demands among the two classes of VMs. For example, if there are only 2 classes, i.e., $k = 1$, and there are 100 machines in each class and 10 users in each group then the type 1 users should each purchase exactly 10 VMs in class 1. However, if there were 150 machines in class 0 and 50 in class 1 then the prices would be unchanged (and still equal) yet somehow the type 1 users would be expected to purchase 5 VMs in each class. In more complex situations this issue is exacerbated, and one would not expect users to purchase the correct bundles relying solely on price information.

This price degeneracy also arises in the pricing model in the large n limit under suitable parameter choices, which we discuss now. A crucial quantity for the smooth running of user-based budget-constrained dynamic pricing is the rate at which machines appear and are available to be bought by the users. The total rate is $\alpha(m - R(t)) + \beta R(t)$, where $R(t)$ is the reserve size at time t, which comes from jobs finishing and idle machines being readvertised. Since this rate is $\Omega(m)$, in order to avoid drastic price fluctuations, it is natural to scale the price adjustment constant as $\epsilon = \tilde{\epsilon}/m$. This leads to the price changing a constant amount over a constant amount of time, provided that $\beta R(t) = O(m)$ and $\tilde{\epsilon} = \Theta(1)$. If $\beta R(t) = \omega(m)$, then the prices change (in fact, decrease) rapidly, adjusting to the budget shortage of the users. Thus choosing $\beta = \Theta(m)$ leads to the reserve being of constant size, i.e., this leads to *efficiency*.

The main observation is that the stochastic process describing our dynamic pricing model has *two time-scales*. On the *"fast time-scale"* t, the machine allocations fluctuate, but the prices are essentially unchanged; while on the *"slow time-scale"* $\tau = t/\epsilon$, the prices also change by a constant amount. There has been significant work on multi-time-scale stochastic processes [15]. The main results are that, under suitable conditions, (i) for the fast-time dynamics, the quantity that changes over the slow time-scale can be considered constant, while (ii) for the slow-time dynamics the quantity that changes over the fast time-scale can be "averaged out", i.e., we can assume it is in its stationary distribution.

The state-of-the-art for multi-time-scale stochastic processes (see [15] and the references therein) can deal with processes on a countable state space, and also

diffusions. However, processes which combine these have received little attention to date (see [10]), and their multi-time-scale behavior is as of yet not well
understood. In our dynamic pricing model prices change often by small amounts,
while allocations comprise a jump process, and so it falls into the latter category
with no known results on systems of this type. A complete rigorous analysis
of dynamic pricing is outside of the scope of the present article, but based on
analogies with known simpler systems, we make conjectures about its long-term
behavior. These are further supported by simulation results in Section 4.

In order to analyze the large system size limit, we consider the following
limiting process using replica economies [5]. Given some CCC with m, n, C and
C_i's, we define the d-replicant of the economy by taking d copies of every user,
with the same preferences, and d copies of every VM, of the same class. This
provides a large economy which has essentially the same Walrasian equilibrium,
with same prices and allocations for each VM class and user.

For fast time-scales, we can consider the prices as fixed and compute the
average demand for each good under this assumption. From this we can estimate
the rate and direction of the price changes at the current set of prices, which
allows us to construct a vector field of price adjustments to understand the
slow-time dynamics of the prices.

Conjecture 1. Consider the dynamic price mechanism under the quasi-static
assumptions (i.e., $\tilde{\epsilon} = o(1)$), $\beta = \Theta(d)$, and a d-replicated economy. Assume that
there is a quasi-static equilibrium set of prices such that for two VM classes, c
and c', the prices satisfy $p_{c'} - p_c = \Omega(d^{-1/2})$. Then, any user who purchases
$O(1)$ VMs from class c, purchases at most $o(d^{-1/2})$ VMs from class c' with high
probability (i.e., with probability tending to 1 as $d \to \infty$; henceforth w.h.p.).

Heuristic Argument. The key idea is that once the user has sufficient budget
to buy a VM from class c, then with high probability she will get the chance
to do so before her budget increases sufficiently to be able to purchase a VM
from class c'. Let $\Delta := p_{c'} - p_c$ and assume that user i has a budget between
p_c and $p_c + \Delta/2$. Then the expected time until user i purchases a VM of class
c is $O(1/(d\alpha))$; this uses the fact that there is a constant probability of a user
being able to purchase a VM that they request, since typically there are only a
constant number of users with sufficient budget to purchase a VM (if this were
not the case, the price mechanism would raise the prices). However, assuming
$\Delta = \Omega(d^{-1/2})$, the expected time until the budget of user i could reach $p_{c'}$ is
$\Omega(1/(d^{1/2}\gamma s))$. Thus for d sufficiently large, user i can rarely afford VMs from
the more expensive class c' and the ratio of purchases of VMs in class c to those
in class c' is $o(d^{-1/2})$.

Conjecture 2. Under the quasi-static assumptions, $\beta = \Theta(d)$, and in a d-replicated
economy, in the limit as $d \to \infty$, there exists a unique equilibrium of the dynamic
price mechanism which is the same as the Walrasian equilibrium.

Heuristic Argument. Conjecture 1 shows that in the equilibrium the VMs and
users will partition based on (almost) equal prices, i.e., each cluster of VMs has

prices that are within $o(d^{-1/2})$ of each other w.h.p. and has a related subset of users for whom those VMs are the least expensive of their allowable VMs so they only buy goods from that subset of VM classes. Theorem 1 shows that the Walrasian equilibrium also must partition in this manner. The choice of $\beta = \Theta(d)$ guarantees efficiency: w.h.p. only $O(1)$ VMs are in the reserve. From this one can see that the two equilibria must be identical, and moreover the Walrasian equilibrium is known to exist in this setting.

Conjecture 3. Consider the dynamic price mechanism under the quasi-static assumptions detailed above (i.e., $\tilde{\epsilon} = o(1)$), and $\beta = \Theta(d)$. Let $p^{(d)}(t)$ denote the prices at time t for the d-replicated economy. For any fixed t, in the limit as $d \to \infty$, starting from any initial condition $p > 0$, the prices $p^{(d)}(t)$ will converge in probability to the Walrasian equilibrium.

Heuristic Argument. To simplify the notation assume that in the Walrasian equilibrium all the prices p^* are the same and that for every class c there exists another class c' such that some user purchases VMs from both classes. Now consider a set of prices $p > 0$ where the largest price p_c is unique and greater than p_c^*. Following the arguments in Conjecture 1, one can see that w.h.p. for sufficiently large d the only users who purchase VMs in class c are those for whom these are their only allowable machines. By assumption there exist other users who would also be purchasing these VMs in the Walrasian equilibrium. Thus, the demand for these VMs is lower under p then under $p*$ in the equilibrium which implies that the price adjustment process would cause this price to fall. Similarly, if there was a unique lowest price $p_{c'} < p_{c'}^*$ then a similar argument would show that that price would rise under the price adjustment process. Thus, assuming these unique highest and lowest prices, we see that the highest price (greater than the equilibrium price) would fall and the lowest price (smaller than the equilibrium price) would rise over time, leading to convergence at the equilibrium prices using a contraction mapping in the L_∞ norm.

4 Simulation Results

We performed detailed simulations on the effects of user-based budget-constrained dynamic pricing. Due to space constraints the details and figures will appear in the full version of the paper; here we present our main conclusions.

For one, large β is necessary for efficiency, but there is a tradeoff between fairness and efficiency. The choice of $\beta = \kappa m$ for an appropriately chosen small constant κ (e.g., $\kappa = 0.1$) combines both efficiency and fairness. There is also a tradeoff between efficiency and price stability, which depends on the price update parameter ϵ. Small ϵ leads to stable prices, but this decreases efficiency, and a further drawback of it is that prices are slow to adapt to sudden changes in the environment. Finally, as the system size grows, prices and allocations converge to their respective values in equilibrium, provided the parameters are chosen appropriately. However, finite size effects are strong, resulting in price differences for finite systems, which can be unfair.

Acknowledgments. The authors would like to thank Ali Ghodsi, Elchanan Mossel and Ion Stoica for helpful discussions. This research has been supported by grants NSF-1216073, NSF-1161813, and by DOD ONR grant N000141110140.

References

1. Ananthanarayanan, G., Ghodsi, A., Shenker, S., Stoica, I.: Disk-locality in data-center computing considered irrelevant. In: Proceedings of the 13th USENIX Workshop on Hot Topics in Operating Systems, pp. 1–5 (2011)
2. Bhattacharya, A.A., Culler, D., Friedman, E., Ghodsi, A., Shenker, S., Stoica, I.: Hierarchical scheduling for diverse datacenter workloads. In: Proceedings of ACM Symposium on Cloud Computing, SoCC 2013 (2013)
3. Chun, B., Culler, D., Roscoe, T., Bavier, A., Peterson, L., Wawrzoniak, M., Bowman, M.: Planetlab: an overlay testbed for broad-coverage services. ACM SIG-COMM Computer Communication Review **33**(3), 3–12 (2003)
4. Cole, R., Fleischer, L.: Fast-Converging tatonnement algorithms for one-time and ongoing market problems. In: Proceedings of the 40th Annual ACM Symposium on Theory of Computing, pp. 315–324. ACM (2008)
5. Debreu, G., Scarf, H.: A limit theorem on the core of an economy. International Economic Review **4**(3), 235–246 (1963)
6. Friedman, E.J., Halpern, J.Y., Kash, I.: Efficiency and nash equilibria in a scrip system for P2P networks. In: Feigenbaum, J., Chuang, J., Pennock, D. (eds.) ACM Conference on Electronic Commerce (EC), pp. 140–149 (2006)
7. Ghodsi, A., Zaharia, M., Shenker, S., Stoica, I.: Choosy: max-min fair sharing for datacenter jobs with constraints. In: Proceedings of the 8th ACM European Conference on Computer Systems, pp. 365–378. ACM (2013)
8. Hahne, E.L., Gallager, R.G.: Round robin scheduling for fair flow control in data communication networks. NASA STI/Recon Tech. Rep. N 86, 30047 (1986)
9. Hindman, B., Konwinski, A., Zaharia, M., Ghodsi, A., Joseph, A.D., Katz, R.H., Shenker, S., Stoica, I.: Mesos: a platform for fine-grained resource sharing in the data center. In: Proceedings of NSDI (2011)
10. Il'in, A.M., Khasminskii, R.Z., Yin, G.: Singularly Perturbed Switching Diffusions: Rapid Switchings and Fast Diffusions. Journal of Optimization Theory and Applications **102**(3), 555–591 (1999)
11. Sweeney, J., Sweeney, R.J.: Monetary theory and the great Capitol Hill Baby Sitting Co-op crisis: comment. J. of Money, Credit and Banking **9**(1), 86–89 (1977)
12. Timberlake, R.H.: Private production of scrip-money in the isolated community. Journal of Money, Credit and Banking **19**(4), 437–447 (1987)
13. Uzawa, H.: Walras' Tatonnement in the Theory of Exchange. The Review of Economic Studies **27**(3), 182–194 (1960)
14. Walras, L.: Elements of Pure Economics Or The Theory of Social Wealth (1954)
15. Yin, G., Zhang, Q.: Continuous-Time Markov Chains and Applications: A Two-Time-Scale Approach, vol. 37. Springer (2012)

Visual Predictions of Traffic Conditions

Jason Rhinelander[1], Mathew Kallada[2], and Pawan Lingras[1]([✉])

[1] Saint Mary's University, Halifax, Canada
jason.rhinelander@smu.ca, pawan@cs.smu.ca
[2] Dalhousie University, Halifax, Canada
kallada@cs.dal.ca

Abstract. The proliferation of internet connected cameras means that drivers can easily access camera images to find out the current traffic conditions as part of their daily commute planning. This paper describes our novel system that will enhance the commute planning process by presenting the current traffic camera image and forecast the expected traffic conditions. The forecast traffic conditions are presented as database reference images. Image processing is applied to each camera image to extract quantitative attributes related to weather, lighting conditions, and traffic activity. The resulting data set of images is then clustered to identify different categories of traffic conditions based on the extracted features. Our prediction system uses a time-series of image attributes to forecast the traffic condition category in the near future (up to 9 hours). The commuter can then not only look at the current camera image, but also view a traffic condition forecast image.

Keywords: K-means clustering · Random forest classifier · Traffic forecasting · Image processing

1 Introduction

The deployment of network connected cameras has increased in recent years due to relatively inexpensive technology and mass manufacture, and as a result, the amount of real-time image data involving traffic conditions has been growing significantly. A picture can communicate implicit information that cannot be easily captured by natural written or auditory language [10]. Related work in the area of traffic prediction can be accomplished in various ways; by means of Global Positioning System (GPS) information and other ubiquitous technologies [3,7,8], fusion of other forms of data such as weather radar and traffic maps services [5,15], or image processing of a region of interest [14]. We present a novel traffic condition analysis and forecasting system that solely makes use of traffic camera imagery.

Image based descriptions are efficient for interpreting geographic conditions at a specific point in time. A number of image processing techniques, such as texture analysis [13] and activity analysis [4], make it possible to describe various weather and traffic features of each unique traffic image.

© Springer International Publishing Switzerland 2015
D. Barbosa and E. Milios (Eds.): Canadian AI 2015, LNAI 9091, pp. 122–129, 2015.
DOI: 10.1007/978-3-319-18356-5_11

Clustering is applied to the weather and traffic features and a large data set of images is categorized into a meaningful small number of categories such as, "clear sky with heavy traffic" or "cloudy with moderate traffic". However, instead of showing the textual natural language description for the category (or even worse, showing semantically obtuse values of the extracted image features), the categories are represented visually by displaying the most representative (the medoid) image from the category.

A classifier is trained using a recent history of the image features is used to predict the traffic conditions in the near future. A commuter accessing this forecasting system will be able to see the current camera image as well as a forecast traffic image (the respective cluster medoid). Our novel approach is demonstrated and validated based on an image data set from Nova Scotia Webcams, focusing on traffic condition forecasting on the Mackay Bridge in Halifax, Nova Scotia.

2   Methodology

We make use of a two stage approach for the prediction of traffic conditions using visual information. In the first stage, image processing is applied to each image so that useful identifying features can be extracted. Clustering is then applied to each image based on the extracted features for that image. The corresponding medoid image for each cluster is then used as a visual guide for any image falling into that particular cluster. It is important to note that semantic tags describing traffic conditions are used in our paper for illustration only. A user of our system would visually observe the medoid image as the representation of the forecast.

The second stage starts by using the medoid of each cluster to assign a class label for each of the possible traffic conditions that exist throughout the database of images. A Random Forest (RF) [2] classifier is trained using the features of the current image and four previous images as an input vector. The training output class label is the medoid x-minutes in advance of the current image. Once trained, the classifier will predict the future medoid class label (which can be displayed to the user) based on current and past traffic conditions.

2.1   Study Data

Study data was provided by Nova Scotia Webcams consisting of 71 cameras in total, with the size of the MacKay Bridge data set being 4.3GB. The present study focuses on a single camera stream that monitors the traffic conditions on the MacKay bridge that links Halifax to Dartmouth. The camera collects a snapshot every 10 minutes with a resolution of 960×600 pixels and was collected over a 21 month period from May 29th, 2012 to January 18th, 2014. It is one of the heaviest traffic corridors in the region and cause of congestion during peak travel hours. We restricted our image analysis to the hours between 7 am and 7 pm which corresponds to peak activity.

2.2   Image Processing and Feature Extraction

Varying levels of image processing are used in the first stage of our traffic condition prediction algorithm. Two Regions Of Interest (ROIs), are defined for each image. The first ROI is the traffic region encompassing the bridge area, and the second ROI is in the sky region of the image for weather analysis. Pixel-level operations such as average red, green and blue pixel intensity operations are computed for the weather ROI, and grey scale conversion is performed on both ROIs.

Texture based feature extraction is performed on both grey scale ROIs by the calculation of a gray-level co-occurrence matrix (GLCM) [13]. The features *contrast, dissimilarity, homogeneity, energy,* and *correlation* for the weather and traffic ROIs are extracted from the respective GLCM.

The contrast and dissimilarity features tend to be strongly correlated with high spatial frequencies in the texture of the ROI. Homogeneity is a feature that has smaller values for larger grey level differences across the ROI. As its name implies, the feature is sensitive to regular patterns. Higher values for the energy feature result from uniform or periodic grey levels across the ROI. Chaotic or random grey levels result in lower energy values. Finally, a high correlation feature value implies that there is a strong linear relationship between pixel values.

The GLCM is an effective method for the extraction of texture features from an image [1], and we use these features to distinguish between different textures in the traffic and weather ROIs. Detection of traffic activity within a sequence of images can be achieved by image subtraction. When one image is subtracted from a previously captured image, differences are extracted by means of a threshold on the resulting image. A Sobel filter is then applied to the image difference where "blobs" are formed and counted. The procedure can be repeated with the multiple previous images. The mean "blob" count and "blob" standard deviation are both useful statistics and determine the activity level in the current image.

Table 1 describes a list of attributes that were extracted from each image, their semantics, and their range of values (if applicable). Figure 1 shows one of the camera view points during separate times of the year along with the range of values that the corresponding extracted features had for each image.

2.3   Clustering and Classification

Clustering allows for the unsupervised grouping of objects together. A number of cluster validity measures including Davies-Bouldin (DB) index [6] have been proposed to evaluate clustering schemes. The objective of the k-means clustering algorithm [9,11] is to find a locally optimal solution. The k-means algorithm creates k clusters from a total of n objects. Since k-means is a greedy approach, we run ten trials of k-means and pick the trial where the clusters have the smallest cluster inertia [12].

As mentioned previously, instead of providing centroid values of semantically obtuse image attributes, we visually analyzed the medoid images in each cluster.

Table 1. List of feature attributes extracted from traffic camera images

Attribute	Semantics	Value range
weather_b	The average blue channel intensity level for the sky region.	1.693-254.69
weather_homogeneity	The homogeneity feature from the weather GLCM.	13243-14021.5
blobs_found	The average number of blobs found in the image.	0-122.24
weather_g	The average green channel intensity level for the weather ROI.	1.69-255.0
weather_energy	The energy feature from the weather GLCM.	9845.13-14021.0
weather_dissimilarity	The dissimilarity feature from the weather GLCM.	1-1558
energy	The energy feature from the traffic GLCM.	1879.4-89879.5
computation_time	The time taken in processing the current image.	2.41-85.2
weather_contrast	The contrast feature from the weather GLCM.	1-1558
hour	The hour component (24 hour day cycle) of the time when the image was acquired.	7-19
homogeneity	The homogeneity feature from the traffic GLCM.	568848.5-574079.5
area	The average total area of all detected traffic blobs.	0-26550.4
weather_r	The average green channel intensity level for the weather ROI.	1.69-254.8
time	The time that the image was captured (yyyymmdd, y:year, m:month, d:day)	20120101-20121231
area_std	The standard deviation in the total area of all detected traffic blobs across the 10 previous images.	0-28500.5
dissimilarity	The dissimilarity feature from the traffic GLCM.	1-10463
correlation	The correlation feature from the traffic GLCM.	0.9999982581-1.0
day	The day of the month that the image was taken.	1-(31,30,28)
contrast	The contrast feature from the traffic GLCM.	1-10463
blob_std	The standard deviation in the total number of blobs detected across the 10 previous images.	0-122.243

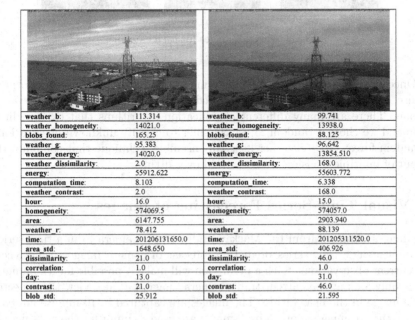

weather_b:	113.314	weather_b:	99.741
weather_homogeneity:	14021.0	weather_homogeneity:	13938.0
blobs_found:	165.25	blobs_found:	88.125
weather_g:	95.383	weather_g:	96.642
weather_energy:	14020.0	weather_energy:	13854.510
weather_dissimilarity:	2.0	weather_dissimilarity:	168.0
energy:	55912.622	energy:	55603.772
computation_time:	8.103	computation_time:	6.338
weather_contrast:	2.0	weather_contrast:	168.0
hour:	16.0	hour:	15.0
homogeneity:	574069.5	homogeneity:	574057.0
area:	6147.755	area:	2903.940
weather_r:	78.412	weather_r:	88.139
time:	201206131650.0	time:	201205311520.0
area_std:	1648.650	area_std:	406.926
dissimilarity:	21.0	dissimilarity:	46.0
correlation:	1.0	correlation:	1.0
day:	13.0	day:	31.0
contrast:	21.0	contrast:	46.0
blob_std:	25.912	blob_std:	21.595

Fig. 1. Two images taken from the same camera on different days in 2012. The difference in corresponding image attributes is especially evident in the weather features.

A medoid image is the image whose image attribute values are the closest to the centroid values of the cluster. Visual inspection of the medoid images of the cluster aided us in creating the descriptive labels for use in this paper. Figure 2 shows the medoid images for each of the categories. It can be seen that each medoid image offers a unique visual reference for the weather and traffic conditions on the bridge.

Fig. 2. Medoid images for each category

Once attribute extraction and clustering of the data set has been performed, we use a classification algorithm to predict to which cluster an observation belongs. There are many different classification algorithms that exist in literature and we chose a random forest classifier to classify unknown observations. A random forest (RF) classifier [2] is an ensemble technique that combines the output of multiple decision trees to avoid over fitting.

3   Visual Time-Series Forecasting of Images

The images obtained from the Mackay bridge traffic camera form a time-series. We implement a time-series prediction forecast of camera images using recent and current observations. Such a prediction will make it possible for a commuter to not only look at the current traffic camera image, but also view the expected traffic image in the near future. The images are captured in 10-minute intervals and the current image + previous 4 images are used as input for the classifier. At time t, the classifier input is composed of $input_t = \{x_t, x_{t-1}, x_{t-2}, x_{t-3}, x_{t-4}\}$ resulting in a total of 100 image features.

Multiple random forest classifiers (using the scikit-learn [12] library) are trained with the following input and label pairs: $(input_t, y_t), (input_t, y_{t+1}), \dots,$ $(input_t, y_{t+54})$. The first classifier predicts the current condition and each classifier in the sequence forecasts the cluster label for 10 minutes into the future. The final classifier, $(input_t, y_{t+54})$ forecasts the traffic conditions 540 minutes (9 hours) ahead of the current time t.

Figure 3 shows three examples of captured images and corresponding forecast images. Forecasting is illustrated for 40, 30, and 20 minutes ahead of the current captured image. Qualitatively, one can see that the forecast images are reasonably similar to the actual images.

Fig. 3. Current conditions versus forecast medoid images

Table 2 shows the accuracy of predictions. A 10-fold cross-validation was used to gather statistics about the accuracy of the classification system. The results confirm our earlier observation that the classifiers that predict image categories in the near future (0-20 minutes) are relatively more accurate than those that predict distant future (120-540 minutes).

Table 2. Accuracy of Visual Forecasting of Traffic Conditions (10-fold cross-validation)

Classifier Tested	Classification Accuracy
Classifier +0 minutes	98.98%
Classifier +10 minutes	97.88%
Classifier +20 minutes	97.55%
Classifier +30 minutes	97.34%
Classifier +70 minutes	97.03%
Classifier +90 minutes	96.97%
Classifier +120 minutes	96.55%
Classifier +240 minutes	96.17%
Classifier +360 minutes	96.06%
Classifier +480 minutes	95.60%
Classifier +540 minutes	95.43%

4 Summary and Conclusions

This paper describes the design, development, and implementation of a novel visual forecasting method for traffic planning. Traffic camera images are first analyzed using image processing techniques to extract weather, lighting conditions, and traffic activity features.

The features for each image are then clustered into semantically meaningful categories that reflect the driving conditions. Instead of representing each category as values of image attributes, each category is represented visually using the medoid image of each cluster.

Finally, we use a time-series of images to predict possible traffic conditions in the near future. The results can be presented as semantically meaningful labels, as well as medoid images of traffic conditions over the next nine hour period.

Since there is a strong time correlation between the current conditions and forecast conditions, the proposed system is most accurate for shorter forecasting time horizons, with a slight decrease in accuracy as the forecast time becomes larger. A sequence of forecast images will help a commuter have a visual, as well semantic understanding of current and future traffic for better commute planning.

References

1. Baraldi, A., Parmiggiani, F.: An investigation of the textural characteristics associated with gray level cooccurrence matrix statistical parameters. IEEE Transactions on Geoscience and Remote Sensing **33**(2), 293–304 (1995)
2. Breiman, L.: Random forests. Machine Learning **45**(1), 5–32 (2001)
3. Castro, P.S., Zhang, D., Li, S.: Urban traffic modelling and prediction Using large scale taxi GPS traces. In: Kay, J., Lukowicz, P., Tokuda, H., Olivier, P., Krüger, A. (eds.) Pervasive 2012. LNCS, vol. 7319, pp. 57–72. Springer, Heidelberg (2012)

4. Chen, T.H., Lin, Y.F., Chen, T.Y.: Intelligent vehicle counting method based on blob analysis in traffic surveillance. In: Second International Conference on Innovative Computing, Information and Control, ICICIC 2007, pp. 238–238, September 2007

5. Chong, C.S., Zoebir, B., Tan, A.Y.S., Tjhi, W.C., Zhang, T., Lee, K.K., Li, R.M., Tung, W.L., Lee, F.B.S.: Collaborative analytics for predicting expressway-traffic congestion. In: Proceedings of the 14th Annual International Conference on Electronic Commerce, ICEC 2012, pp. 35–38. ACM, New York (2012)

6. Davies, D.L., Bouldin, D.W.: A cluster separation measure. IEEE Transactions on Pattern Analysis and Machine Intelligence (2), 224–227 (1979)

7. Giannotti, F., Nanni, M., Pedreschi, D., Pinelli, F.: Trajectory pattern analysis for urban traffic. In: Proceedings of the Second International Workshop on Computational Transportation Science, IWCTS 2009, pp. 43–47. ACM, New York (2009)

8. Gidófalvi, G., Borgelt, C., Kaul, M., Pedersen, T.B.: Frequent route based continuous moving object location- and density prediction on road networks. In: Proceedings of the 19th ACM SIGSPATIAL International Conference on Advances in Geographic Information Systems, GIS 2011, pp. 381–384. ACM, New York (2011)

9. Hartigan, J.A., Wong, M.A.: Algorithm as 136: A k-means clustering algorithm. Journal of the Royal Statistical Society. Series C (Applied Statistics) 28(1), 100–108 (1979)

10. Ma, X., Boyd-Graber, J., Nikolova, S., Cook, P.R.: Speaking through pictures: Images vs. icons. In: Proceedings of the 11th International ACM SIGACCESS Conference on Computers and Accessibility, pp. 163–170. ACM, New York (2009)

11. MacQueen, J.: Some methods for classification and analysis of multivariate observations. In: Proceedings of the Fifth Berkeley Symposium on Mathematical Statistics and Probability. Statistics, vol. 1, pp. 281–297. University of California Press (1967)

12. Pedregosa, F., Varoquaux, G., Gramfort, A., Michel, V., Thirion, B., Grisel, O., Blondel, M., Prettenhofer, P., Weiss, R., Dubourg, V., Vanderplas, J., Passos, A., Cournapeau, D., Brucher, M., Perrot, M., Duchesnay, E.: Scikit-learn: Machine Learning in Python. Journal of Machine Learning Research 12, 2825–2830 (2011)

13. Rampun, A., Strange, H., Zwiggelaar, R.: Texture segmentation using different orientations of glcm features. In: Proceedings of the 6th International Conference on Computer Vision / Computer Graphics Collaboration Techniques and Applications, MIRAGE 2013, pp. 17:1–17:8. ACM, New York (2013)

14. Shuai, Z., Oh, S., Yang, M.H.: Traffic modeling and prediction using camera sensor networks. In: Proceedings of the Fourth ACM/IEEE International Conference on Distributed Smart Cameras, ICDSC 2010, pp. 49–56. ACM, New York (2010)

15. Tostes, A.I.J., de L. P. Duarte-Figueiredo, F., Assunção, R., Salles, J., Loureiro, A.A.F.: From data to knowledge: City-wide traffic flows analysis and prediction using bing maps. In: Proceedings of the 2nd ACM SIGKDD International Workshop on Urban Computing, UrbComp 2013, pp. 12:1–12:8. ACM, New York (2013)

Privacy-Aware Wrappers

Yasser Jafer[1(✉)], Stan Matwin[1,2,3], and Marina Sokolova[1,2,4]

[1] School of Electrical Engineering and Computer Science,
University of Ottawa, Ottawa, Canada
{yjafer,sokolova}@uottawa.ca, stan@cs.dal.ca
[2] Institute for Big Data Analytics, Dalhousie University, Halifax, Canada
[3] Institute of Computer Science, Polish Academy of Sciences, Warsaw, Poland
[4] Faculty of Medicine, University of Ottawa, Ottawa, Canada

Abstract. We introduce a Privacy-aware Wrapper (*PW*) system which incorporates privacy into the functionality of wrappers. It ensures that privacy gain is achieved through the selected features without significantly impacting the performance of the models if compared with the performance of the original dataset.

Keywords: Privacy · Wrappers · Data publishing · Data mining · Classification · Feature selection

1 Introduction

In general, privacy preserving data publishing does not make assumptions about how the data will be used. This leads to a general purpose anonymization and consequently overprotection thus impacting the usefulness of the released data. As such, different privacy-preserving methods have been proposed in the past which sustain the data utility for certain data mining tasks ([1-3]).

We assume a scenario that consists of a Data Holder (DH) who holds the original data on the on hand and, a Data Recipient (DR) who wants the data in order to apply certain data mining task on the other hand. In our work, we assume that DR wants to classify the dataset. DH publishes a customized dataset which takes the intended analysis task of DR into consideration. Since the dataset is going to be published, in practice, it will also be available to the attacker. However, since the dataset is tailored for a given analysis task, while guaranteeing privacy according to a given privacy model, if the attacker uses the same dataset to do say clustering analysis instead, the results would be misleading and of less or no value. DH and DR could be a hospital and a research center respectively.

The goal of automatic feature selection is to obtain the most optimal feature set that provides for algorithm's best performance. In our work, we incorporate privacy considerations into the very functionality of wrappers. Wrappers assess subsets of attributes based on their usefulness for a given predictor. Our aim is to minimize the release of potentially privacy breaching attributes.

© Springer International Publishing Switzerland 2015
D. Barbosa and E. Milios (Eds.): Canadian AI 2015, LNAI 9091, pp. 130–138, 2015.
DOI: 10.1007/978-3-319-18356-5_12

Our experimental results show that, in some cases we are able to eliminate all of the potentially privacy breaching attributes and therefore the associated risk. In other cases, we are able to further reduce the release of potentially privacy breaching attributes. These objectives are attained while maintaining a good utility comparable to the utility of the original dataset.

2   Related Works

The notion of privacy is attached to a sanitization technique based on a given attack model. The attack model may depend on the attacker's ability to uniquely identify an individual in the released dataset (i.e. *identity disclosure* [4]). It could also depend on the ability of attacker in determining/predicting the sensitive information of the individuals (i.e. *attribute disclosure*[5]). The models addressing these attacks compare privacy disclosure in terms of prior and posterior probability before and after releasing the dataset. A more rigorous notion of privacy is called ε-differential privacy [6] which guarantees that adding or removing a single record does not substantially influence the outcome of any analysis. ε-differential privacy compares privacy disclosure with or without existence of a record in the dataset. One popular method that addresses identity disclosure is k-anonymity. A table is considered k-anonymous if the quasi-identifier values of each tuple are indistinguishable from k-1 other tuples. Quasi-identifiers (QIs) are attributes that when "combined", could be linked to external datasets and potentially identify individuals. It was shown in [7] that automatic feature selection could be utilized towards privacy protection in k-anonymity. The idea is that, the more attributes included in the QI set, the more distortion is needed in order to achieve k-anonymity because the records in a group must agree on more attributes [8]. When we reduce the number of released QI attributes it becomes more difficult to uniquely identify an individual (because according to the very definition of QIs, all attributes in the QI set combined are needed in order to single out an individual). Furthermore, less distortion is required in to achieve the same privacy level. This will eventually lead to preserving more utility [7]. Our work is inspired by this observation.

In [9], automatic feature selection was applied prior to anonymization in order to increase the utility of differentially private data publishing. Privacy-aware filter-based feature selection was addressed in [10]. That work considered incorporating privacy into filters. The authors introduced a trade-off measure λ in order to control the amount of privacy and efficacy (i.e. classification accuracy). Although, in general, filters are faster than wrappers, they have lower performance. Such better performance reported by wrappers is mainly due to the interaction with the learning model. We show in our work that this advantage of wrappers could be well utilized for privacy preserving purposes.

3   Privacy-Aware Wrappers

Let's assume that $Acc(Opt\text{-}accuracy)$ is the accuracy corresponding to the optimal feature set. That is, the classifier that is built using only selected attributes by wrapper feature selection. Let's also assume that $Acc(Org)$ represents the accuracy of building classifiers with the complete feature set (without feature selection).

Acc(Opt-accuracy) – *Acc(Org)* represents a range of acceptable accuracy of a privacy-aware feature set that is obtained via privacy-aware wrappers.

Dataset **D** consists of a set of *n* training instances where each instance *T* is an element of the set $F_1 \times F_2 \times \ldots \times F_m$, and F_i is the domain of the *i*th feature. The original feature vector consists of *X* attributes $\{X_1, X_2, \ldots, X_m\}$ in addition to the class attribute *C*. The accuracy of the original attribute set is denoted as *Acc(D[X, C])*. Assume that (**QI'**) refers to a set of QI attributes that are eliminated/replaced. *MAX(QI')* refers to a set of maximum number of QI attributes that get eliminated from the final selected feature set. Our objective is therefore to: *find the maximum number of QI attributes which could be removed (via elimination and/or correlation) such that, the model built with the resulting dataset (with fewer projected features) does not degrade the classification accuracy significantly.* More formally,

find *MAX* (**QI'**) | (*Acc(D[(X-QI'), C])* ≥ *Acc(D[X, C])* ‖ *Diff* (*Acc(D[X, C])*, *Acc(D[(X-QI'), C])*)) ! = statistically significant)

Our proposed Privacy-aware Wrapper (*PW*) system is depicted in Figure 1. The inputs to the system include the dataset **D**, **e**, **FS**, **PA**, **c**, and **d**. These inputs are discussed in more details as we explain the functionality of the system. **PA** refers to the list of potentially privacy breaching attributes. In our model **PA**= *QI*. This is a special case where all QI attributes are given equal privacy weight and therefore it does not matter which attribute in the QI set gets eliminated. It is possible, however, to associate different privacy weights with the attributes in **PA**. As such, priority of removal/replacement is given to attributes with higher privacy weight. The evaluation criteria **e** refers to the classification accuracy of a chosen classifier. **FS** in our case refers to the specific type of wrapper and the search technique employed. The initial evaluation of the dataset is performed in the *Evaluation* block. The output of the *Evaluation* block includes two sets of attributes, namely **WS** and **R**. **R** refers to a ranked list of attributes in the original feature space. This rank is obtained by individually considering the ability of features in predicting the target class. **WS** refers to the list of selected attributes due to applying wrapper-based feature selection to the dataset **D**. In general, **WS** is a subset of the original attribute set, i.e. **WS** ⊂ **R**. At this point, two possible outcomes may occur. If **WS** ∩ **PA** = 0, this is an indication that, none of the attributes in the **PA** set were selected by the wrapper and the algorithm would terminate since there are no additional potentially privacy breaching attributes to be eliminated.

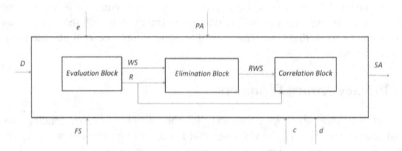

Fig. 1. Privacy-aware Wrapper (*PW*) System

In most scenarios this is not the case. That is, wrappers eliminate some of the *PA* attributes, but some others will still be present in the *WS* feature set which need to be handled in the upcoming blocks.

Next block in the system is the *Elimination* block and its goal is to remove more QI attributes as long as such removal does not degrade the performance of the resulting models significantly. The input to this block includes *WS* and *R*. Using the rank of individual attributes in the original set, a ranked list of *WS* is obtained, i.e., *RWS*. Then using the *PA* list, the QI attributes in the *RWS* set are identified. In the following step, starting from the bottom of the *RWS* list, the QI attributes are removed and the classification model is built with the remaining attributes. If the new accuracy is higher, equal, or (not significantly lower) than the original accuracy, then it is safe to remove this particular QI attribute. The *RWS* list is updated accordingly until all of the QI attributes are visited.

Following the above step, if *RWS* ∩ *PA* = 0, it indicates that all of the QI attributes were eliminated in the *Elimination* block. The algorithm terminates and, *SA* which is equal to *RWS*, is released. In other words, the *Correlation* block is executed only if after executing the *Elimination* block *RWS* ∩ *PA* is still ≠ 0.

Let us consider the *Correlation* block in more details. Similar to the *Elimination* block, our goal is to reduce the number of QI attributes that could not be eliminated previously. One immediate observation is that removing of any of these QI attributes at this step would result in a significant reduction in the performance. After all, this is the main reason these attributes could not be removed in the previous step. However, it is possible that these QI attributes are highly correlated with other attributes in the original feature space, i.e. *X*.

The correlation measure *c* could be any correlation criteria such as Mutual Information (*MI*), Symmetric Uncertainty (*SU*), and so on, and is considered the choice of the user. In this work, we choose *c* to be the Symmetric Uncertainty (*SU*) metric. *SU* is a modified information gain measure in order to estimate the degree of association between discrete features [11]. It measures the correlation between two features and more specifically the amount of information a given attribute provides about another attribute.

$$SU = 2.0 \times \left[\frac{H(X) + H(Y) - H(X,Y)}{H(X) + H(Y)} \right] \qquad (1)$$

In formula (1), $H(X)$ and $H(Y)$ refer to the entropy of attributes X and Y respectively, and $H(X,Y)$ refers to the joint entropy of X and Y. Since *SU* is symmetric it can be used to measure feature-feature correlations in which there is no notion of one attribute being the "class" attribute [12]. The discretization factor, *d* is used and is set to either true or false. When it is set to true the type of discretization technique needs to be specified. It is required for the *SU* measure that the attributes be categorical. We set *d* to true and discretize the numeric features using the technique of [13].

The input to the *Correlation* block includes *R* and *RWS*. Similar to the pervious block, we start with the bottom of the *RWS* list searching for the QI attributes. Whenever such attribute is found, we finds its correlation with all of the attributes that have higher rank in the *R* list and are neither in the *RWS* set nor in the *PA* set. That is, *NR* = *R* − *PA* − *RWS*. We then find the attribute(s) in the *NR* set that have higher rank compared with the rank of an encountered QI attribute in the *RWS* set, and construct the *HNR* set. Following this step, we find the attribute in *HNR* that has maximum correlation with our QI attribute. In the following step, we replace our QI attribute with that attribute and build the classifier with the updated *RWS*. The feature vector in this case would be *D*[*RWS*-{*candidate QI attribute to be removed*}+ {replacement attribute from the HNR}]. If the new accuracy is higher, equal, or (not significantly lower) than the original accuracy, then it is OK to finalize the removal of the given QI attribute and the replacement is considered acceptable. If not, we repeat the process with the next attribute in the *HNR* set that has the highest correlation with the QI attribute. If the replacement was successful we stop the search, break out of the loop, and continue the procedure with the next QI attribute in the *RWS* set. If after scanning all of the attributes in the *HNR* set we could not replace the QI attribute, such QI attribute will remain in the final subset.

4   Experiments

We used a number of real datasets from the UCI repository [14]. In most of these datasets, a subset of features was selected as the QI attributes set [15]. For two datasets, i.e. *CMC* and *Wisconsin breast cancer*, all of the attributes except the target class were selected as the QI set.

We used four classifiers, namely, C4.5, N.B., KNN, and SVM. They belong to different categories of models. C4.5 is considered a logical model whereas N.B. is considered a probabilistic model. KNN and SVM on the other hand, are a geometric models[16].We applied *t*-test to compare our results. All the models were built using 10-fold cross validation. Our results are shown in Table 1. The accuracy values in the baseline column refer to the results of the classifiers built using the complete original attribute set, i.e., prior to applying any feature selection on the dataset. WFS refers to the classification accuracy of a dataset that consists only of attributes selected by wrapper-based feature selection. *PW* refers to the accuracy of the dataset consists of attributes selected by the privacy-aware wrapper system. We show the number of QI attributes in the baseline dataset, in the dataset resulting from applying WFS, and in the dataset resulting from applying *PW*. The *p*-value corresponding to the *t*-test is shown in the final column as well. The *p*-value corresponds to the *PW* algorithm results compared to the baseline.

Let's consider the results of the Pima dataset, case of C4.5. Initially there are 3 QI attributes in this dataset. The baseline accuracy is 73.82%. This is the accuracy of building a C4.5 classifier from the original dataset without any modification. If we apply WFS, one of the three QI gets eliminated. WFS results in a dataset with optimal accuracy (75.787) which is higher than the baseline accuracy. By applying *PW* we are able to further reduce the number of QI attributes without significant reduction in the accuracy compared with the baseline accuracy. With *PW* we achieve lower accuracy compared with WFS which is expected. Our algorithm reduces the QI attributes in a step-by-step approach until any further reduction of the QI attributes results in an accuracy that is significantly lower than the baseline accuracy. In our case we notice that, even if all QI attributes are removed, the resulting accuracy is acceptable. In such case, using *PW* we can remove all QI attributes. The privacy implication is that the resulting dataset can be released without any anonymization or modification (case of identity disclosure). We see the same behavior in case of N.B., KNN, and SVM. In fact such behavior is seen for other datasets/classifiers combinations in Table 1.

Table 1. Comparison results of the performance and privacy obtained from the original dataset, WFS, and *PW*. ⊕/⊖ corresponds to statistically significant increase/decreases.

Dataset (No. tuples)	Alg.	Baseline		WFS		PW		
		No. QI	Acc%	No. QI	Acc %	No. QI	Acc %	*p*-value
Pima (768)	C4.5	3	73.82	2	75.787	1	74.356	0.6306
						0	**73.174**	0.7133
	N.B.	3	76.302	2	77.734	1	77.218	0.2964
						0	**75.915**	0.6970
	KNN	3	72.656	2	73.567	1	70.704	0.0808
						0	**71.732**	0.6445
	SVM	3	77.343	3	77.343	2	76.823	0.3732
						1	75.651	0.0613
						0	**75.912**	0.1197
German credit (1000)	C4.5	6	70.700	2	73.100	1	73.500	0.1636
						0	**73.000**	0.2382
	N.B.	6	75.400	3	76.200	2	76.000	0.5462
						1	76.000	0.5203
						0	**75.900**	0.5366
	KNN	6	73.300	1	75.100	0	**72.900**	0.7828
	SVM	6	75.100	2	75.800	1	74.800	0.4344
						0	**74.600**	0.6141
Liver patients (583)	C4.5	2	68.781	0	71.012	N/A	N/A	N/A
	N.B.	2	55.743	0	71.871	N/A	N/A	N/A
	KNN	2	64.665	2	71.526	1	71.355	0.5864
						0	**67.924**	0.1702
	SVM	2	71.355	0	71.355	N/A	N/A	N/A
Heart stat logs (270)	C4.5	2	76.673	0	85.154	N/A	N/A	N/A
	N.B.	2	83.702	1	86.292	0	**85.555**	0.0957
	KNN	2	78.888	0	84.444	N/A	N/A	N/A
	SVM	2	84.074	0	84.074	N/A	N/A	N/A
CMC (1473)	C4.5	9	52.140	3	55.533	2	**50.369**	0.1703
						1	47.591⊖	0.0046
	N.B.	9	50.781	3	55.397	2	50.706	0.9648
						1	**47.861**	0.1690
	KNN	9	45.281	3	52.682	2	51934⊕	0.0001
						1	**47.998**	0.1273
	SVM	9	48.201	9	48.201	8	45.892⊖	0.0443
						7	45.689⊖	0.0443
						6	45.572⊖	0.0345
Wisc. breast cancer (683)	C4.5	9	93.411	2	95.168	1	**91.790**	0.1776
	N.B.	9	97.365	6	97.804	5	97.952⊕	0.0367
						4	97.364	0.9966
						3	**96.632**	0.2126
						2	93.557⊖	0.0005
						1	91.790⊖	0.0005
	KNN	9	95.315	7	96.339	6	96.339⊕	0.0438
						5	95.900	0.2214
						4	95.607	0.6372
						3	95.461	0.7826
						2	**94.289**	0.1536
						1	92.532⊖	0.0007
	SVM	9	96.193	6	97.071	5	96.485	0.5095
						4	**95.603**	0.3965
						3	94.722⊖	0.0229
						2	94.727⊖	0.0497
						1	92.647⊖	0.0029
Adult (45222)	C4.5	6	85.573	3	85.779	2	85.785	0.1043
						1	85.745	0.2637
						0	**85.551**⊖	0.8894
	N.B.	6	82.709	6	82.727	5	82.579⊖	0.0148
						4	81.327⊖	9.4e-09
						3	80.324⊖	3.6e-13
						2	80.061⊖	1.5e-12
	KNN	6	81.354	1	85.206	0	**82.719** ⊕	4.0e-05

The German credit dataset shows similar behavior as of the Pima dataset. It is possible, that all of the QI attribute get removed by WFS anyways. See the results of the liver patients dataset. In three cases of C4.5, N.B, and SVM, WFS removes both of the QI attributes identified for this dataset.

Let us consider the CMC dataset. With the exception of SVM, for other three classifiers WFS excludes six attributes. *PW* eliminates more QI attributes while maintaining acceptable accuracy. In the case of C4.5 removal of one extra QI attribute is possible. However, removing of additional attribute reduces the accuracy significantly. This is shown by $\ominus$ next to the accuracy of 47.5%. Therefore, the best acceptable result is 50.3%. For the N.B. classifier, the best achieved result is 47.9% where two additional attributes get excluded. As for the KNN classifier, exclusion of two more extra attributes does not have a negative impact on accuracy significantly. As for SVM, it is not possible to remove any QI attribute without significantly degrading the classification accuracy. We only show three attempts of removal. Wisconsin breast cancer dataset shows similar results.

Therefore, there are no more QI attributes to be removed by the *PW* method. This is shown by N/A in the corresponding cells. The only exception is KNN.

For the Adult dataset, in the case of C4.5, by applying *PW* it is possible to remove/replace three extra QI attributes and achieve accuracy of 85.5% which does not significantly defer from the baseline accuracy. The same is applied for the case of KNN classifier. One exception is the N.B. classifier. In this case, all attributes are selected by the wrapper feature selection. Therefore WFS does not eliminate any of the QI attributes. Any removal of the attributes in the QI set significantly degrades the classification accuracy. The correlation block cannot be implemented since all attribute are selected in the first place, i.e., R = *RWS, PA* $\cap$ *RWS = PA*, and *NR* is empty.

5   Conclusion and Future Work

In this work we incorporated privacy into the very process of wrapper-based feature selection. Our results showed that, compared with basic wrappers, our proposed *PW* system, was either able to eliminate all of the QI attributes or was able to exclude more QI attributes compared with a non privacy-aware wrapper. These objectives are attained while maintaining a good utility that does not differ significantly from the utility of the original dataset. In the *PW* system, the complexity of the blocks following the *Evaluation* block (which implements a standard non privacy-aware wrapper) is considered minimal. In the future, we aim to empirically study the added complexity of the system.

It was mentioned in Section 2 that, reducing QI enhances the models that support identity disclosure. In the future, we will study the impact of reducing the number of QI attributes, hence changing the size of equivalent class (number of tuples that share the same QI values) in the context of attribute disclosure. This includes applying *PW* prior to anonymization techniques such as *l*-diversity [17] and *t*-closeness[5]. In all of the above models the assumption is that the attacker knows the QI of the victim. An existing challenge, however, is determining the QI attributes which is/remains an open issue in PPDP.

We also aim to design a wrapper whose evaluation function will be $E(S) = w1 \times Perf(S) + w2 \times Priv(S)$, where $Priv(S)$ is some innate measure of privacy of the attribute set S (without making any assumption or requiring any additional info as do QIs). For instance, $Priv(S)$ may be the probability of error in inferring the value of the sensitive attribute(s) of individuals in the dataset given S (and given background knowledge) and is referred to as empirical privacy[18]. *Perf*, on the other hand, refers to utility. Both *Perf* and *Priv* could be subject to constraints, e.g. requiring that accuracy does not differ significantly from $Acc(Org)$ and the privacy does not drop below a threshold α. Another future task is to consider larger datasets.

References

1. LeFevre, K., DeWitt, D., Ramakrishnan, R.: Workload-aware anonymization. In: Proceedings of the 12th ACM SIGKDD International Conference on Knowledge Discovery and Data Mining, pp. 277–286. ACM, Philadelphia (2006)
2. Fung, B.C., Wang, K., Yu, P.S.: Top-Down specialization for information and privacy preservation. In: Proceedings of the 21st International Conference on Data Engineering, pp. 205–216. IEEE Computer Society (2005)
3. Wang, K., Yu, P.S., Chakraborty, S.: Bottom-up generalization: a data mining solution to privacy protection. In: Proceedings of the Fourth IEEE International Conference on Data Mining, Bringhton, UK, pp. 249–256 (2004)
4. Sweeney, L.: Achieving k-anonymity privacy protection using generalization and suppression. Int. J. Uncertain. Fuzziness. Knowl.-Based Syst. **10**(5), 571–588 (2002)
5. Li, N., Li, T., Venkatasubramanian, S.: t-Closeness: privacy beyond k-anonymity and l-diversity. In: Proceedings of the Twenty Third International Conference on Data Engineering, Istanbul, Turkey, pp. 106–115 (2007)
6. Dwork, C.: Differential privacy. In: Bugliesi, M., Preneel, B., Sassone, V., Wegener, I. (eds.) ICALP 2006. LNCS, vol. 4052, pp. 1–12. Springer, Heidelberg (2006)
7. Jafer, Y., Matwin, S., Sokolova, M.: Task oriented privacy preserving data publishing using feature selection. In: Sokolova, M., van Beek, P. (eds.) Canadian AI 2014. LNCS, vol. 8436, pp. 143–154. Springer, Heidelberg (2014)
8. Fung, B.C., et al.: Privacy-preserving data publishing: A survey of recent developments. ACM Comput. Surv. **42**(4), 1–53 (2010)
9. Jafer, Y., Matwin, S., Sokolova, M.: Using Feature Selection to Improve the Utility of Differentially Private Data Publishing. Procedia Computer Science **37**, 511–516 (2014)
10. Jafer, Y., Matwin, S., Sokolova, M.: Privacy-aware filter-based feature selection. In: First IEEE International Workshop on Big Data Security and Privacy (BDSP 2014), Washington DC, USA (2014)
11. Press, W.H.: Numerical recipes in C: The art of scientific computing, xxii, 735 p. Cambridge University Press, Cambridge (1988)
12. Hall, M.A.: Correlation-based feature selection for machine learning. The University of Waikato (1999)
13. Fayyad, U.M., Irani, K.: Multi-interval discretization of continuous-valued attributes for classification learning. In: Proceedings of the 13th International Joint Conference on Artificial Intelligence, Chambery, France, pp. 1022–1027 (1993)
14. UCI repository. http://archive.ics.uci.edu/ml/ (cited 2013)

15. Keng-Pei, L., Ming-Syan, C.: On the Design and Analysis of the Privacy-Preserving SVM Classifier. IEEE Transactions on Knowledge and Data Engineering **23**(11), 1704–1717 (2011)
16. Flach, P.: Machine Learning The Art and Science of Algorithms that Make Sense of Data, p. 1 online resource (409 p.) digital, PDF file(s). Cambridge University Press, Cambridge (2012)
17. Machanavajjhala, A., et al.: L-diversity: privacy beyond k-anonymity. In: Proceedings of the 22nd International Conference on Data Engineering, Atlanta, Georgia, US, p. 24 (2006)
18. Cormode, G., et al.: Empirical privacy and empirical utility of anonymized data. In: 2013 IEEE 29th International Conference on Data Engineering Workshops (ICDEW). IEEE (2013)

NLP, Text and Social Media Mining

Analyzing Productivity Shifts in Meetings

Gabriel Murray[(✉)]

University of the Fraser Valley, Abbotsford, BC, Canada
gabriel.murray@ufv.ca
http://www.ufv.ca/cis/gabriel-murray/

Abstract. Group productivity can vary between and within meetings, and here we consider the case of productivity shifting within meetings. We divide a meeting into intervals and measure the productivity of each interval using the number of summary-worthy sentences contained therein. We evaluate the relationship between productivity and a variety of linguistic and structural features, using correlation and regression analysis. We then attempt to identify the point at which productivity shifts in meetings, using Bayesian changepoint analysis.

1 Introduction

A motivating intuition of this work is that extractive summarization can operate as a proxy for assessing productivity in meetings. If a meeting is highly productive, there should correspondingly be a high number of extracted sentences, relating to phenomena such as decisions, action items, and generally active, on-task discussion. But even a productive meeting may not be consistently productive throughout. For example, a meeting may have many extracted sentences from the first half of the meeting and then very few summary-worthy sentences in the second half, perhaps because participants were becoming tired or simply because the conversation continued long after all the decision items were addressed. Throughout this work, we assume that the number of extracted sentences (or more properly, extracted *dialogue act* units) reflects the group productivity level, both for the meeting as a whole and for intervals of a meeting.

Give that measurement of productivity, it is clear that productivity is not consistent within meetings. Figure 1 shows that extracted dialogue acts are more frequent at the beginnings of meetings and less frequent at the ends of meetings. This suggests that many meetings begin productively but become less productive as they go on.

In this paper, we analyze how productivity can shift *within* meetings. We divide meetings into intervals and perform correlation and regression analyses to determine which linguistic and structural features are closely associated with rising and falling productivity. We are also interested in why a meeting suddenly becomes less productive or suddenly becomes more productive. We use Bayesian changepoint analysis to find the spot in the meeting where productivity shifts, and then aim to learn the traits that characterize the less-productive portion of the meeting from the more-productive portion, using a logistic regression model.

© Springer International Publishing Switzerland 2015
D. Barbosa and E. Milios (Eds.): Canadian AI 2015, LNAI 9091, pp. 141–154, 2015.
DOI: 10.1007/978-3-319-18356-5_13

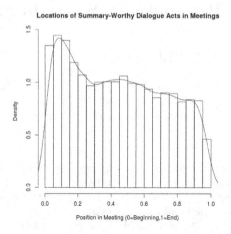

Fig. 1. Histogram + KDE Showing Meeting Position for Extracted Sentences

The central contributions of this work are as follows:

1. This is the first in-depth investigation of how productivity shifts within meetings.
2. We include detailed analysis of linguistic and structural features relating to productivity.
3. We introduce a novel application of Bayesian changepoint analysis for detecting changes in productivity.

The structure of the paper is as follows. Section 2 describes related work on extractive summarization and changepoint analysis. Section 3 introduces the features used in these experiments, with a correlation and regression analysis. Section 4 describes the application of Bayesian changepoint analysis to our particular problem, while Section 5 describes the logistic regression model. The experimental setup for the changepoint experiments is described in Section 6 while the main results are presented in Section 7. Finally, we discuss future work in Section 8 and offer our main conclusions in Section 9.

2   Related Work

The most closely related work to ours is on meeting summarization, an area that has seen increased attention in the past ten years, particularly as automatic speech recognition (ASR) technology has improved. These range from *extractive* (cut-and-paste) approaches [1–3] where the goal is to classify dialogue acts as important or not important, to *abstractive* systems [4–7] that include natural language generation (NLG) components intended to describe the meeting from a high-level perspective. Carenini et al. [8] provide a survey of techniques for summarizing conversational data. This work also relates to the task of identifying action items in meetings [9] and detecting decision points [10]. Renals et al. [11] provide a survey of various work that has been done analyzing meeting interactions.

Other research [12,13] has looked at productivity *across* meetings, i.e., determining how productive and unproductive meetings differ in terms of linguistic and structural features. However, that work does not examine how productivity shifts *within* meetings. For example, even meetings that are mostly productive will have periods where participants get off task and are less productive.

The classic example of applying Bayesian changepoint analysis involves a much-studied dataset of historical coal-mining accidents in Britain [14,15]. The changepoint analysis reveals that such accidents had a marked decrease after the introduction of new safety regulations in the late 1880's.

3   Correlation and Regression Analysis

To analyze productivity shifts within meetings, we divide each meeting into one-minute intervals. We then count the number of extracted dialogue acts (a feature *numSum*) for each interval. We also extract a variety of linguistic, structural and speaker-related features for each interval. We group them into feature categories, beginning with **term-weight (tf.idf)** features:

- *tfidfSum* The sum of $tf.idf$ term scores in the meeting portion.
- *tfidfAve* The average of $tf.idf$ term scores in the meeting portion.
- *conCoh* The conversation cohesion in the meeting portion, as measured by calculating the cosine similarity between all adjacent pairs of dialogue acts, and averaging. Each dialogue act is represented as a vector of $tf.idf$ scores.

Next are the features relating to meeting and dialogue act **length**:

- *DALength* The average length of dialogue acts in the meeting portion.
- *countDA* The number of dialogue acts in the meeting portion.
- *wTypes* The number of unique word types in the meeting portion (as opposed to word tokens).

There are several **entropy** features. If s is a string of words, and N is the number of words types in s, M is the number of word tokens in s, and x_i is a word type in s, then the word entropy *went* of s is:

$$went(s) = \frac{\sum_{i=1}^{N} p(x_i) \cdot -\log(p(x_i))}{(\frac{1}{N} \cdot -\log(\frac{1}{N})) \cdot M}$$

where $p(x_i)$ is the probability of the word based on its normalized frequency in the string. Note that word entropy essentially captures information about type-token ratios. For example, if each word token in the string was a unique type then the word entropy score would be 1. Given that definition of entropy, the derived **entropy** features are:

- *docEnt* The word entropy of the entire meeting portion.
- *speakEnt* This is the speaker entropy, essentially using speaker ID's instead of words. The speaker entropy would be 1 if every dialogue act were uttered by a unique speaker. It would be close to 0 if one speaker were very dominant.
- *domSpeak* Another measure of speaker dominance, this is calculated as the percentage of total meeting portion DA's uttered by the most dominant speaker.

We have one feature relating to **disfluencies**:

- *fPauses* The number of filled pauses in the meeting portion, as a percentage of the total word tokens. A filled pause is a word such as *um*, *uh*, *erm* or *mm* − *hmm*.

Finally, we use two features relating to **subjectivity / sentiment**. These features rely on a sentiment lexicon provided by the SO-Cal sentiment tool [16].

- *posWords* The number of positive words in the meeting portion.
- *negWords* The number of negative words in the meeting portion.

Figure 2 depicts a correlogram for the features, illustrating the correlations between all features. In the portion below the diagonal, the slope of the lines indicates a positive or negative correlation, and the shading indicates the strength of the correlation. The portion above the diagonal shows the confidence ellipses and smoothed lines. Of particular interest is the first row and first column *numSum*, corresponding to the number of extracted dialogue acts in each interval, our dependent variable in the regression analysis below. We can see that most of the features have a positive correlation with *numSum*, with *tfidfSum* having the strongest positive correlation. Only document entropy (*docEnt*) and speaker entropy (*speakEnt*) have negative correlations with the number of summary-worthy dialogue acts in each interval. Interestingly, the total number of dialogue acts in an interval does not strongly correlate with the number of summary-worthy dialogue acts in the interval.

We construct a Poisson regression model using *numSum* as the dependent variable and the other features as predictors. We can evaluate the fitted model using the *deviance* measure. The deviance is -2 times the log likelihood:

$$Deviance(\theta) = -2 \, log[\, p(y|\theta)\,]$$

A lower deviance indicates a better-fitting model. Adding a random noise predictor should decrease the deviance by about 1, on average, and so adding an informative predictor should decrease the deviance by more than 1. And adding k informative predictors should decrease the deviance by more than k. The deviance of our fitted model is 18049, compared with a baseline intercept-only deviance of 19046. Since we added 12 predictors, we should expect the deviance to decrease by at least 12 over the baseline, and in fact it decreased by about 1000.

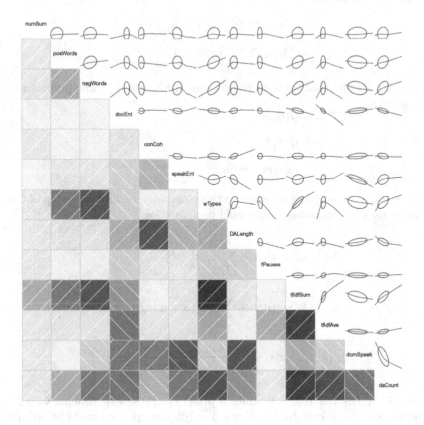

Fig. 2. Feature Correlogram

4   Bayesian Changepoint Analysis for Meeting Productivity

For changepoint estimation, our time-series data consists of the number of extracted (summary-worthy) dialogue acts in each one-minute interval of a meeting. So for a given meeting, our number of datapoints n will be equal to the length of the meeting in minutes. We hypothesize that in many meetings there will be a shift where the number of extracted sentences per minute markedly increases or decreases. Our first task is to locate the changepoint, if there is one, for each meeting.

In more general terms, we have count data $x_1, \cdots, x_n$ and a possible changepoint at some point k in the series. If it is determined that $k = n$, then there is no changepoint. If there *is* a changepoint k, we would then have two Poisson data-generating processes:

$$x_i \mid \lambda \sim \mathcal{P}(\lambda) \quad i = 1, \cdots, k$$
$$x_i \mid \phi \sim \mathcal{P}(\phi) \quad i = k+1, \cdots, n \tag{1}$$

The parameters that we want to estimate are λ, ϕ and k. The priors for each of these are:

$$\lambda \sim \mathcal{G}(\alpha, \beta)$$
$$\phi \sim \mathcal{G}(\gamma, \delta) \tag{2}$$
$$k \sim \text{discrete uniform on } [1, 2, \cdots, n]$$

We use Gibbs sampling to estimate these parameters, and each step of Gibbs sampling uses the following conditional probabilities:

$$\lambda \mid \phi, k \sim \mathcal{G}\left(\alpha + \sum_{i=1}^{k} x_i, \beta + k\right)$$

$$\phi \mid \lambda, k \sim \mathcal{G}\left(\gamma + \sum_{i=k+1}^{n} x_i, \delta + n - k\right) \tag{3}$$

$$p(k \mid x, \lambda, \phi) = \frac{L(x, \lambda, \phi \mid k)\, p(k)}{\sum_{l=1}^{n} L(x, \lambda, \phi \mid k_l)\, p(k_l)}$$

where $p(k)$ in the last line is a uniform prior. For these experiments we set the prior parameters directly ($\alpha = \beta = \gamma = \delta = 10$), though they could be estimated as well. In any case, we found that the final results were not particularly sensitive to the choice of hyperparameter. We run the Gibbs sampler for 50 replications of 10,000 iterations each, with a burn-in of 1000. We estimate k using the posterior mean over the 50 replications. Figure 3 shows an example of the Gibbs sampling updates for k, λ and ϕ for one particular meeting.

5   Classification of Productive/Unproductive Meeting Portions

Once the changepoint has been determined, we want to learn the differences between the two portions of the meeting: the portion before the changepoint, and the portion after. To do so, we differentiate between the following two cases:

– **Case 1 – Increasing Productivity:** These are meetings where the productivity increases after the changepoint. In such meetings, we want to learn the differences between the less-productive beginning of the meeting and more-productive end.

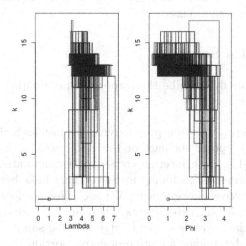

Fig. 3. Gibbs Sampling Updates

- **Case 2 – Decreasing Productivity:** These are meetings where the productivity decreases after the changepoint. In such meetings, we want to learn the differences between the more-productive beginning of the meeting and the less-productive end.

In both cases, we want to learn the linguistic and structural features that characterize the less-productive from the more-productive portion. But we treat these cases above as two separate machine-learning tasks because, for example, an unproductive beginning of a meeting may have much different characteristics than an unproductive end of a meeting. In fact, we will later see that this is the case.

In addition to the features described in Section 3, we also use:

- *speakEntF100* The speaker entropy for the first 100 dialogue acts of the meeting portion.
- *speakEntL100* The speaker entropy for the last 100 dialogue acts of the meeting portion.
- *shortDAs* The number of dialogue acts in the meeting portion shorter than 6 words.
- *longDAs* The number of dialogue acts in the meeting portion longer than 15 words.

For this binary classification task, we use a standard logistic regression model where our predictions $\theta^T X$ are constrained by the sigmoid function to fall between 0 and 1: $g = \frac{1}{1+e^{-\theta^T X}}$ We employ a leave-one-out procedure to maximize our training data, testing on each meeting individually after training on the rest. For both Case 1 and Case 2 meetings, we consider the more-productive portion to be the "positive" class.

6   Experimental Setup

In this section we briefly describe the corpora and evaluation methods used in these experiments.

Corpora. In analyzing meeting productivity, we use both the AMI [17] and ICSI [18] meeting corpora. As part of the AMI project on studying multi-modal interaction [11], both meeting corpora were annotated with extractive and abstractive summaries, including many-to-many links between abstractive sentences and extractive dialogue acts. We use these gold-standard summary annotations in the following experiments. There are 197 meetings in total. Since we are performing classification at the level of meeting sub-portions, each meeting contributes two datapoints: a more-productive portion and a less-productive portion. This gives us a total of 394 training examples, requiring the leave-one-out procedure.

Evaluation. For the logistic regression experiments, we evaluate our predictions on test data using precision/recall/F-score. We also evaluate the fitted models primarily in terms of the *deviance*, described earlier in Section 3. We also present the θ parameters of the fitted logistic regression models. For the logistic regression model, the θ parameters can be interpreted in terms of the *log odds*. For a given parameter value θ_n, a one-unit increase in the relevant predictor is associated with a change of θ_n in the log odds.

7   Results

We first present key results of the Bayesian changepoint analysis, followed by results of the classification task.

7.1   Bayesian Changepoint Results

Of the 197 meetings used in these experiments, only one meeting was determined to have a changepoint very close (within two minutes) to the end of the meeting, and three meetings had changepoints very close to the beginning of the meeting. We excluded those meetings from the remainder of the experiments and focused on the 193 meetings that featured a clear shift in productivity. Of those, 75 meetings fall into Case 1 (increasing productivity after the changepoint) and 118 fall into Case 2 (decreasing productivity after the changepoint).

Figure 4 shows an example of a changepoint in a Case 1 meeting, where the x-axis is divided into 1-minute intervals and the y-axis shows the cumulative count of extractive dialogue acts seen up to that interval of the meeting. In this particular meeting, the first 15 minutes featured only about 12 summary-worthy

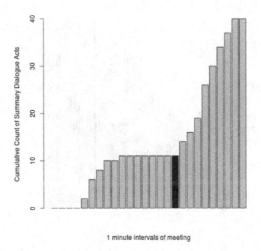

Fig. 4. Example of Meeting Changepoint (Increasing Productivity)

dialogue acts and there was a period of 8 minutes where nothing summary-worthy was said. The meeting then became much more productive after the changepoint, with about 30 extracted dialogue acts in the final 9 minutes.

In contrast, Figure 5 shows a Case 2 meeting changepoint where there is a dramatic decrease in productivity near the end of the meeting. There are no summary-worthy dialogue acts for nearly 20 minutes before the meeting is finally wrapped up.

Once the changepoint for each meeting has been determined, we classify meetings into Case 1 or Case 2 based on whether the meeting portion before the changepoint has a higher or lower number of extracted dialogue acts per minute than the portion after the changepoint.

For meetings that fall into Case 1 (increasing productivity), the changepoint tends to fall relatively early in the meeting. Specifically, the less productive early part of the meeting has about 270 dialogue acts on average, while the more productive later portion of the meeting has more than 700 dialogue acts on average. In contrast, for meetings that fall into Case 2 (decreasing productivity), the changepoint is closer to the middle of the meeting on average. Specifically, the more productive early portion of the meeting has about 588 dialogue acts on average, while the less productive later portion has 557 on average.

7.2   Classification Results

Table 1 shows the precision, recall and F-scores for Cases 1 and 2. The most interesting result here is that classification is easier for Case 1 than Case 2. In other words, when a meeting has a changepoint after which it becomes more

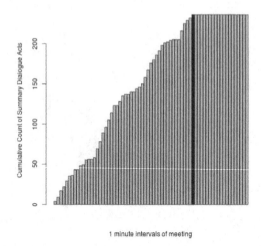

Fig. 5. Example of Meeting Changepoint (Decreasing Productivity)

productive, it is easier to discriminate the two sub-portions of the meeting in terms of linguistic and structural features, relative to the alternative case where the changepoint leads to the meeting becoming less productive.

Table 1. Prediction Results

Productivity Case	Precision	Recall	F-Score
Case 1: Increasing	0.73	0.77	0.75
Case 2: Decreasing	0.60	0.59	0.60

Table 2 shows the deviance scores for individual predictor models (each trained using a single feature) as well as the combined predictor model, compared with the null (intercept-only) baselines. For Case 1, we see that the best single predictor is the word entropy of the meeting portion, and that the number of word types in the meeting portion and the tf.idf average are also very useful. The combined model for Case 1 shows a marked decrease in deviance from the null baseline, from 206.6 to 135.4, much more than would be expected by adding random noise predictors.

For Case 2, none of the individual predictor models are particularly effective on their own, though conversation cohesion and number of short dialogue acts are the most useful features. Despite the individual predictor models not being particularly effective, the combined model for Case 2 shows a substantial decrease in deviance over the null baseline, from 325.9 to 287.4.

Table 2. Deviance Using Single and Combined Predictors

Feature	Case 1	Case 2
null (intercept)	206.6	325.9
tfidfSum	171.2	325.9
tfidfAve	161.4	326.0
conCoh	205.4	**315.6**
DALength	204.5	325.2
shortDAs	207.7	**315.2**
longDAs	203.3	326.1
countDA	167.9	326.8
wTypes	162.4	325.5
docEnt	**156.9**	327.1
domSpeak	180.6	326.9
speakEnt	199.1	327.2
speakEntF100	207.9	326.8
speakEntL100	207.9	326.4
fPauses	207.4	327.0
posWords	193.8	327.1
negWords	207.1	326.8
COMBINED-FEAS	**135.4**	**287.4**

Our final presented results are the parameter (coefficient) estimates of the logistic regression model, shown in Table 3 for both Cases 1 and 2. These are the parameters of the individual predictor models, each trained using a single feature. We are more interested here in the sign of the parameter than the magnitude of the parameter (though we note that all features were normalized to fall within the 0-1 range). It is also interesting to note cases where the sign of the parameter is flipped between Case 1 and Case 2.

As mentioned above, the most useful feature for discriminating the more-productive portion from the less-productive portion in Case 1 was the document entropy, and the θ coefficient for word entropy is negative, meaning that an increase in entropy is associated with a decrease in the log-odds of productivity. This is flipped in Case 2, where θ is positive for the entropy feature. Conversation cohesion was the most useful feature for classification within Case 2 meetings, and its sign is positive for Case 2, meaning that the more-productive portions tend to have higher conversational cohesion. This is flipped in Case 1, where θ is negative for the conversation cohesion feature.

8   Future Work

For future work, we plan to adopt methods for detecting multiple change-points [19]. Some meetings may, for example, have an unproductive portion at the beginning and another unproductive portion at the end, with a productive

Table 3. Single Predictor Parameter Estimates

Feature	θ (Case 1)	θ (Case 2)
tfidfSum	3.74	0.43
tfidfAve	5.53	0.64
conCoh	-7.76	21.28
DALength	1.92	1.20
shortDAs	-0.93	-6.48
longDAs	5.75	2.26
countDA	3.48	0.19
wTypes	4.53	0.59
docEnt	-24.16	0.59
domSpeak	-8.29	-0.66
speakEnt	-10.26	-0.20
speakEntF100	0.02	-1.65
speakEntL100	-0.41	-1.88
fPauses	5.79	2.75
posWords	63.569	-2.14
negWords	10.873	6.99

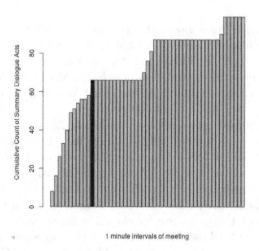

Fig. 6. Potentially Multiple Changepoints

section in the middle. For example, the meeting shown in Figure 6 seems to have at least two changepoints, if not more, as evidenced by the multiple plateaus in the cumulative plot. Our current model cannot handle multiple shifts in productivity. On the other hand, this type of plot was a rarity among the meetings; the assumption that a meeting would have zero or one changepoints was correct in the majority of cases, based upon inspection of these visualizations.

Ongoing work would also be aided by the creation of gold-standard annotations for productivity in meetings. In this work we have simply exploited existing resources [13] that act as a proxy for measuring productivity. Some of these are reliable indicators of productivity or metrics of whether participants are "on-task": for example, meeting dialogue acts linked to the *decision* or *action item* portions of the abstractive summary are by definition on-task as pertains to the goals of the group. However, it would be useful to have annotations indicating the degree to which each dialogue act is productive or contributes to the stated purpose of the meeting.

9   Conclusion

In this work we have analyzed how productivity shifts within meetings, using extractive summarization as a proxy task. We first carried out correlation and regression analyses to investigate which linguistic and structural features relate to productivity. We then developed a novel application of Bayesian change-point analysis to determine where in the meeting productivity shifts, subsequently learning the linguistic and structural features that discriminate the more-productive portion of the meeting from the less-productive portion.

References

1. Galley, M.: A skip-chain conditional random field for ranking meeting utterances by importance. In: Proc. of EMNLP 2006, 364–372, Sydney, Australia (2006)
2. Xie, S., Favre, B., Hakkani-Tür, D., Liu, Y.: Leveraging sentence weights in a concept-based optimization framework for extractive meeting summarization. In: Proc. of Interspeech 2009, Brighton, England (2009)
3. Murray, G., Carenini, G., Ng, R.: The impact of asr on abstractive vs. extractive meeting summaries. In: Proc. of Interspeech 2010, Tokyo, Japan, pp. 1688–1691 (2010)
4. Murray, G., Carenini, G., Ng, R.: Generating and validating abstracts of meeting conversations: a user study. In: Proc. of INLG 2010, Dublin, Ireland, pp. 105–113 (2010)
5. Mehdad, Y., Carenini, G., Tompa, F., Ng, R.: Abstractive meeting summarization with entailment and fusion. In: Proc. of ENLG 2013, Sofia, Bulgaria, pp. 136–146 (2013)
6. Liu, F., Liu, Y.: Towards abstractive speech summarization: Exploring unsupervised and supervised approaches for spoken utterance compression. IEEE Transactions on Audio, Speech and Language Processing **21**(7), 1469–1480 (2013)
7. Wang, L., Cardie, C.: Domain-independent abstract generation for focused meeting summarization. In: Proc. of ACL 2013, Sofia, Bulgaria, pp. 1395–1405 (2013)
8. Carenini, G., Murray, G., Ng, R.: Methods for Mining and Summarizing Text Conversations, 1st edn. Morgan Claypool, San Rafael (2011)

9. Purver, M., Dowding, J., Niekrasz, J., Ehlen, P., Noorbaloochi, S.: Detecting and summarizing action items in multi-party dialogue. In: Proc. of the 9th SIGdial Workshop on Discourse and Dialogue, Antwerp, Belgium (2007)
10. Hsueh, P.-Y., Moore, J.D.: Automatic Decision Detection in Meeting Speech. In: Popescu-Belis, A., Renals, S., Bourlard, H. (eds.) MLMI 2007. LNCS, vol. 4892, pp. 168–179. Springer, Heidelberg (2008)
11. Renals, S., Bourlard, H., Carletta, J., Popescu-Belis, A.: Multimodal Signal Processing: Human Interactions in Meetings, 1st edn. Cambridge University Press, New York (2012)
12. Murray, G.: Learning How Productive and Unproductive Meetings Differ. In: Sokolova, M., van Beek, P. (eds.) Canadian AI. LNCS, vol. 8436, pp. 191–202. Springer, Heidelberg (2014)
13. Murray, G.: Resources for analyzing productivity in group interactions. In: Proc. of LREC 2014 Workshop on Multi-Modal Corpora, Reykjavik, Iceland, pp. 39–42 (2014)
14. Jarrett, R.: A Note on the Intervals Between Coal-Mining Disasters. Biometrika 66(1), 191–193 (1979)
15. Gill, J.: Bayesian Methods: A Social and Behavioral Sciences Approach, 2nd edn. Chapman & Hall, London (2008)
16. Taboada, M., Brooke, J., Tofiloski, M., Voll, K., Stede, M.: Lexicon-based methods for sentiment analysis. Computational Linguistics 37(2), 267–307 (2011)
17. Carletta, J.E., et al.: The AMI Meeting Corpus: A Pre-announcement. In: Renals, S., Bengio, S. (eds.) MLMI 2005. LNCS, vol. 3869, pp. 28–39. Springer, Heidelberg (2006)
18. Janin, A., Baron, D., Edwards, J., Ellis, D., Gelbart, D., Morgan, N., Peskin, B., Pfau, T., Shriberg, E., Stolcke, A., Wooters, C.: The ICSI meeting corpus. In: Proc. of IEEE ICASSP 2003, Hong Kong, China, pp. 364–367 (2003)
19. Fearnhead, P.: Exact and efficient bayesian inference for multiple changepoint problems. Statistics and Computing 16(2), 203–213 (2006)

Temporal Feature Selection for Noisy Speech Recognition

Ludovic Trottier[✉], Brahim Chaib-draa, and Philippe Giguère

Department of Computer Science and Software Engineering,
Université Laval, Quebec (QC) G1V 0A6, Canada
`ludovic.trottier.1@ulaval.ca`,
`{chaib,philippe.giguere}@ift.ulaval.ca`
`http://www.damas.ift.ulaval.ca`

Abstract. Automatic speech recognition systems rely on feature extraction techniques to improve their performance. Static features obtained from each frame are usually enhanced with dynamical components using derivative operations (delta features). However, the susceptibility to noise of the derivative impacts on the accuracy of the recognition in noisy environments. We propose an alternative to the delta features by selecting coefficients from adjacent frames based on frequency. We noticed that consecutive samples were highly correlated at low frequency and more representative dynamics could be incorporated by looking farther away in time. The strategy we developed to perform this frequency-based selection was evaluated on the Aurora 2 continuous-digits and connected-digits tasks using MFCC, PLPCC and LPCC standard features. The results of our experimentations show that our strategy achieved an average relative improvement of 32.10% in accuracy, with most gains in very noisy environments where the traditional delta features have low recognition rates.

Keywords: Automatic speech recognition · Delta features · Feature extraction · Noise robustness

1 Introduction

Automatic speech recognition (ASR) is the transcription of spoken utterances into text. A system that performs ASR tasks takes an audio signal as input and classifies it into a series of words. In order to improve the performance of the system, feature extraction approaches are applied on the signal to provide reliable features. The three most frequently used features in ASR are the Mel frequency cepstral coefficients (MFCC), the perceptual linear predictive cepstral coefficients (PLPCC) and the linear predictive cepstral coefficients (LPCC) (see [18] for a review). These filter bank analysis extraction methods use various transformations, such as the Fourier transform, to convert a signal into a series of static vectors called *feature frames*. The coefficients in a feature frame are usually ordered from low-frequency to high-frequency and this observation will play a central role in our approach.

© Springer International Publishing Switzerland 2015
D. Barbosa and E. Milios (Eds.): Canadian AI 2015, LNAI 9091, pp. 155–166, 2015.
DOI: 10.1007/978-3-319-18356-5_14

Classical feature extraction methods enhance each feature frame with dynamical components by concatenating the first- and second-order derivatives. These *delta* features were proposed as a way to improve the spectral dynamics of static features [3]. Delta features improve the accuracy of the hidden Markov model (HMM) [23] by reducing the impacts of the state conditional independence [4]. However, it is known from signal processing theories that the derivative of a noisy signal amplifies the noise and reduces the quality of the extracted information [14]. This can be especially harmful in real world situations where noise affects the recognition, such as when driving a car [13].

We have proposed, in a preliminary approach, that the delta features could be replaced with a mere concatenation of adjacent (in time) coefficients based on frequency [19]. This approach will be referred to as Temporal Feature Selection (TFS). The suggestion that dynamical features should be dependent on frequency was motivated by the importance of modeling inter-frame dependencies for speech utterances. Signal processing theories suggest that information in a signal varies according to its frequency [14]. For example, implosive consonant will result in fast, high-frequency features, while vowels will produce slow-changing, lower-frequency features. It thus appears that dynamical features may be enhanced by measuring the variation of the signal's information with frequency.

In this paper, we extend our TFS method with a learning framework. Our framework uses the variance of the difference of adjacent feature frames as a way to identify the positions where more reliable dynamical information resides. We show experimentally that our dynamical features improve the accuracy over the classical delta features on the Aurora 2 [15] continuous-digits and connected-digits tasks.

The rest of the paper is organized as follows. Section 2 describes related approaches, section 3 contains background information about feature extraction, section 4 presents the TFS method, section 5 details the experimentations and section 6 concludes this work.

2   Related Work

To overcome the delta features' problem of susceptibility to noise, recent alternatives have been investigated. The delta-spectral cepstral coefficients (DSCC) have been proposed in replacement of the delta features to add robustness to additive noise [10]. Also, the discrete cosine transform (DCT) has been used in a distributed fashion (DDCT) prior to the calculation of the delta features [8]. Finally, a weighted sum combining the static and delta features have been proposed in replacement of the usual concatenation [20]. The main drawback of all these methods is that derivative operations are still part of their processing pipeline thus making the features prone to be corrupted by noise.

Additionally, splicing followed by decorrelation and dimensionality reduction has been used to enhance the inputs of deep neural networks (DNNs) [16]. Splicing consists in concatenating all feature frames (with delta features) in a *context window* of size c around each frame [1]. Moreover, it was showed that deeper layers allow more discriminative and invariant features to be learned [22]. While we acknowledge that deep learning is a promising avenue for feature extraction in ASR, we argue that better feature engineering methods could facilitate the DNN learning process.

In the context of linear feature transformations, some have looked at dimensionality reduction approaches such as Principal Component Analysis (PCA) [9], Linear Discriminant Analysis (LDA) [2], Heteroscedastic LDA (HLDA) [11] amd Heteroscedastic DA (HDA) [17]. These approaches are essential tools for speech feature extraction, but we argue that they may be avoided by using a better model for gathering the speech dynamics. Linear transformations of speech features have also been applied for decorrelation, such as Maximum Likelihood Linear Transform (MLLT) [7], Global Semi-tied Covariance (GSC) [6], and for speaker adaptation, such as feature-space Maximum Likelihood Linear Regression (fMLLR) [12] and Constrained MLLR (CMLLR) [5]. However, these feature selection techniques do not address the problem of modeling speech dynamics.

3   Background

In this section, we review in brief the steps for performing MFCC feature extraction on speech signals. The overview of the method is presented in Fig. 1. For additional details on this technique and other related approaches (such as PLPCC and LPCC), see [18].

Pre-emphasis: A speech waveform entering the pipeline is first filtered with a first order high pass filter. The goal of this transformation is to remove the low-frequency parts of the speech, as they tend to have similar and redundant adjacent values.

Windowing: The resulting signal is them divided into 20-40 milliseconds *frames*. A length of 25 ms is typical in speech processing. Assuming the signal is sampled with a frequency of 8 kHz, the frame length corresponds to $0.025 * 8000 = 200$ samples. Usually, the frames are overlapping by 15 ms (120 samples at 8 kHz), which means that a frame is extracted at every 10 ms (80 samples at 8 kHz) in the signal.

Periodogram Estimate of Power Spectrum: To perform the periodogram estimate of the power spectrum, the discrete Fourier transform (DFT) is applied on each frame to transform the waveform into its frequency domain. DFT assumes that each signal is periodic, which means that the beginning and the end of each frame should be connected. For a randomly selected frame, this hypothesis will not be respected and will lead to abrupt transitions. The discontinuities at the edges will be reflected in the spectrum by the presence of spectral leakage. To get a better resolution, the Hamming windowing function is applied to connect the edges in a smoother way. The length of the DFT is typically 512, but only the first 257 coefficients are kept since the other 255 are redundant due to the nature of the Fourier transform. Finally, the squared absolute value of the DFT is applied which gives the periodogram estimate of power spectrum.

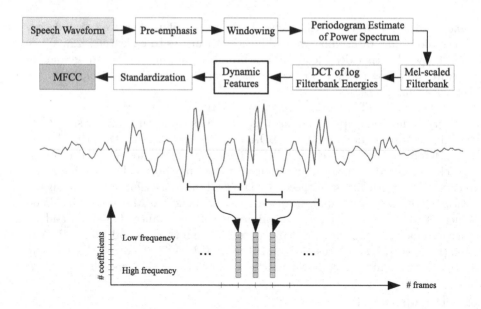

Fig. 1. MFCC extraction

Mel-scaled Filterbank: Each power spectral estimate is filtered using triangular Mel-spaced filterbanks (see [21] for more details). The filterbanks are described as 26 vectors of size 257 (assuming $K = 512$). Each vector contains mostly zeros excepted at certain regions of the spectrum and thus act as bandpass filters. Mapping the frequency to the Mel scale allows the features to match more closely the non linear perception of pitch of the human auditory system. To compute the filterbank energies, we can simply multiply the periodogram estimates by each filterbank and sum the values. The 26 numbers give an indication of the amount of energy in each filterbank.

DCT of log Filterbank Energies: Then, a type 2 discrete cosine transform (DCT-II) is performed on the log filterbank energies to decorrelate the values. The length of the DCT is usually 14 and the first coefficient is discarded. The resulting 13 coefficient correspond to the static features.

Dynamic Features: As explain in section 1, the impacts of the state conditional independence of the HMM can be reduced by gathering dynamical information. In most cases, the delta features are appended to the static features by computing discrete derivatives (more details in section 4.2).

Standardization: Finally, we subtract each coefficient with its sample mean and divide by its sample standard deviation. These statistics can be calculated once for all utterances or individually for each utterance.

As shown in Fig. 3, the signal is transformed into a series of vectors, one for each frame extracted during windowing. Each utterance has a different number of frames, depending on its length and its sampling frequency. We now present our approach that is an alternative method to the computation of delta features during the *Dynamic Features* step.

4   Temporal Feature Selection

4.1   Definition

Let $\Phi^{(n)} = \left(\phi_{:,1}^{(n)} \ldots \phi_{:,T_n}^{(n)} \right)$, $n = 1 \ldots N$, be a $D \times T_n$ matrix of D-dimensional static features. N is the total amount of utterances and T_n denotes the number of frames extracted from utterance n. For example, $\Phi^{(n)}$ could represent MFCC, as presented in section 3. We denote the column vector $\phi_{:,t}^{(n)}$ as the feature frame at position t. The classical method of computing the delta features uses the following equations:

$$\Delta\phi_{:,t}^{(n)} = \frac{\sum_{k=1}^{K} k \left(\phi_{:,t+k}^{(n)} - \phi_{:,t-k}^{(n)} \right)}{2\sum_{k=1}^{K} k^2}, \tag{1}$$

$$\Delta\Delta\phi_{:,t}^{(n)} = \frac{\sum_{k=1}^{K} k \left(\Delta\phi_{:,t+k}^{(n)} - \Delta\phi_{:,t-k}^{(n)} \right)}{2\sum_{k=1}^{K} k^2}, \tag{2}$$

where $K = 2$ is a typical value for the summation. Although relevant dynamical information can be extracted with Eq. 1 and 2, the use of subtractions makes Δ and $\Delta\Delta$ features susceptible to noise.

The TFS features are, in contrast, coefficients taken from adjacent feature frames based on the frame position offsets $\mathbf{z} = [z_1, \ldots, z_D]$. We define them as:

$$\tau\phi_{i,t}^{(n)} = \left(\phi_{i,t+z_i}^{(n)}, \phi_{i,t-z_i}^{(n)} \right), \tag{3}$$

where z_i is a strictly positive integer. The parametrization of $\mathbf{z}$ is essential to extract robust dynamics. By imposing frequency dependency, τ will be constituted of coefficients ϕ that are dissimilar, but not too much. The intuition is that too similar values do not increase the amount of information the feature frames carry, but increase its dimensionality, and this makes the speech recognition task harder. On the other hand, if the coefficients are too far apart, then their temporal correlation is meaningless.

4.2   Learning the TFS Features

We now present the proposed framework to learn the frequency dependent offsets $\mathbf{z}$. The method first computes the sample variance of the difference of neighboring feature frames. In other words, for each position t and utterance n, the difference between the feature frame $\phi_{:,t}^{(n)}$ and its corresponding j^{th} neighbor $\phi_{:,t+j}^{(n)}$ is computed. The variance of these differences is then calculated for $j \in \{1 \ldots M\}$,

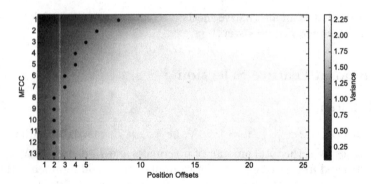

Fig. 2. Variance of the difference between a frame and its neighbors for MFCC features on the Aurora 2 [15] training dataset (best seen in colors). The coefficients are ordered from low frequency (1) to high frequency (13) for visual convenience (the proposed method does not require a specific ordering). The color refers to the variance of the difference Σ^M, where M was limited to 25 to reduce the computational burden.

where $M = \min\{T_1 \ldots T_N\} - 1$. We define the matrix containing those values as:

$$
\Sigma^M = \begin{bmatrix} \mathrm{Var}\left(\phi_{1,t}^{(n)} - \phi_{1,t+1}^{(n)}\right) & \cdots & \mathrm{Var}\left(\phi_{1,t}^{(n)} - \phi_{1,t+M}^{(n)}\right) \\ \vdots & & \vdots \\ \mathrm{Var}\left(\phi_{D,t}^{(n)} - \phi_{D,t+1}^{(n)}\right) & \cdots & \mathrm{Var}\left(\phi_{D,t}^{(n)} - \phi_{D,t+M}^{(n)}\right) \end{bmatrix}, \tag{4}
$$

where the variances are taken over all positions t and utterances n. The variance is then be computed as follows:

$$
\Sigma_{i,j}^M = \frac{1}{N_j^+} \sum_{n=1}^{N} \sum_{t=1}^{T_n - j} \left(\phi_{i,t}^{(n)} - \phi_{i,t+j}^{(n)} - \mu_{i,j}\right)^2, \tag{5}
$$

where $\mu_{i,j}$ corresponds to the mean of the difference:

$$
\mu_{i,j} = \frac{1}{N_j^+} \sum_{n=1}^{N} \sum_{t=1}^{T_n - j} \left(\phi_{i,t}^{(n)} - \phi_{i,t+j}^{(n)}\right), \tag{6}
$$

and N_j^+ is the total number of frames:

$$
N_j^+ = \sum_{n=1}^{N} T_n - j. \tag{7}
$$

The purpose of computing Σ^M is to find the frame position offsets $\mathbf{z}$. Using the parameter V_{thresh} as a variance threshold, $\mathbf{z}$ is computed using the following equation:

$$
z_i = \arg\min_j \left|\Sigma_{i,j}^M - V_{thresh}\right|, \tag{8}
$$

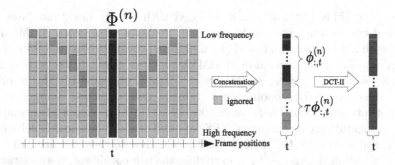

Fig. 3. Pipeline of processing for TFS features. After computing the frame position offsets **z** using Eq. 8, the features are concatenated and decorrelated with DCT-II.

where V_{thresh} is a hyper-parameter to choose. The frame position offsets **z** represented in Fig. 2 by the black dots are based on Eq. 8 for $V_{thresh} = 1$ and $M = 25$. In this particular example, $\mathbf{z} = [8, 6, 5, 4, 4, 3, 3, 2, 2, 2, 2, 2, 2]$. For $i = 1$, this implies that $\tau\phi_{1,t} = (\phi_{1,t+8}, \phi_{1,t-8})$.

What can be seen from this figure is that **z** depends on frequency. High frequency components have small offsets whereas low frequency components have large offsets. As explained in section 1, more reliable dynamical informations can be extracted from neighboring feature frames when frequency is taken into account. The relevant dynamical information of high frequency coefficients can only be extracted from nearly adjacent frames ($z_{13} = 2$). On the other hand, adjacent low frequency coefficients share most of their information and more time is needed to gather the relevant dynamics ($z_1 = 8$). Therefore, by using the variance of neighboring feature frames, **z** now incorporate the wanted characteristic of frequency dependency.

As in standard feature extraction, the features from each frame, $\phi_{:,t}^{(n)}$ and $\tau\phi_{:,t}^{(n)}$ are concatenated into a single column vector as shown in Fig. 3. Each resulting vector is then decorrelated with a type 2 DCT in order to accommodate it to the independence hypothesis of the Gaussian Mixture Model Hidden Markov Model (GMM-HMM) that was chosen as the inference method [4].

5   Experimental Results

5.1   Experimental Setup

The database that we used for our experiments is Aurora 2 [15] which contains a vocabulary of 11 spoken digits (*zero* to *nine* with *oh*). The digits are connected, thus they can be spoken in any order and in any amount (up to 7) with possible pauses between them. The training set contains 8,440 utterances, both test set A and B have 28,028 and test C has 14,014 utterances. The utterances are noisy and the signal-to-noise ratio (SNR) varies from -5 dB, 0 dB, ..., 20 dB, Inf dB (clean). Different kinds of noise are present such as train, airport, car, restaurant, etc. On average, an utterance lasts approximately 2 seconds.

Using the HTK [21] framework provided with the Aurora 2 database, we performed two experiments. In the first one, 18 states whole-word HMMs were trained with a 3 components GMM (with diagonal covariance) as the state emission density. There was a total of 11 HMMs (one per class). In the second one, the whole-word HMMs were replaced with 5 states phoneme HMMs. In other words, using the CMU pronouncing dictionary, each digit was mapped to its ARPAbet interpretation. There was a total of 19 HMMs (one per phoneme). In our experimentations, we compared TFS (-T) to delta (-D) and double delta (-A) dynamic features on MFCC, PLPCC and LPCC. For all these features, 13 coefficients, including the energy (-E), excluding the 0th coefficient, were extracted to be used as observations. For all experiments, a variance threshold $V_{thresh} = 1$ was used. The performance of each method was averaged over all test sets for each noise level separately.

5.2   Experimental Results

The performances in word accuracy of our method are reported in Table 1 and 2 for whole-word and phoneme HMM respectively. In each table, the 7 noise levels from the Aurora 2 database are ordered from clean signals (SNR Inf) to highly noisy signals (SNR -5). The average over all noise levels is reported on the right. The last column consists of the relative improvement of the method over the reference model.

Based on these results, our approach achieved an averaged relative improvement of 20.79% for whole-word HMM and 32.10% for phoneme HMM. Also, it can be observed that TFS features increased the accuracy on all noisy tasks, but did not improve the results for clean signals with whole-word HMM. Nonetheless, these results support our initial intuition that using a pure derivative approach leads to inferior performances.

The variation of the word accuracy of TFS, with respect to V_{thresh}, is shown in Fig. 4 for whole-word HMMs. The performance of the method is reported for the 7 noise levels of the database. The crosses indicate the best result the method achieved for each noise level. This figure demonstrates the behavior of the performance of our approach with respect to the parametrization of $\mathbf{z}$.

5.3   Discussion

The results of table 1 show that the proposed TFS method does not outperform the delta features for clean utterances. This limitation is consistent with the intuition given in section 1 that dynamical components extracted using derivatives on clean utterances are not affected by amplified noise. However, this appears to be the case only for whole-word HMMs. As suggested by the results of table 2, the TFS method improves the word accuracy even in the absence of noise when using phoneme HMMs. These non intuitive results may be related to the shorter duration of phonemes in comparison to whole-words. By selecting very distanced coefficients, TFS also incorporates information about adjacent phonemes and appears to simulate triphone modeling, where a HMM is

Table 1. Word accuracy (%) of TFS (-E-T) features on the Aurora 2 database using whole-word HMMs. The results are averaged according to the noise level. The reference models are suffixed with -E-D-A.

SNR (dB) / Features	Inf	20	15	10	5	0	-5	Avg.	R.I. (%)
(a) MFCC Based									
MFCC-E-D-A	98.54	97.14	96.02	93.27	84.86	57.47	23.35	78.66	-
MFCC-E-T $z = [8,6,5,4,4,3,3,2,2,2,2,2,2]$	97.64	97.46	96.68	94.39	88.03	71.31	38.93	**83.49**	**22.63**
(b) PLPCC Based									
PLPCC-E-D-A	98.65	97.56	96.48	93.85	85.93	59.83	25.36	79.66	-
PLPCC-E-T $z = [8,6,5,4,4,3,3,3,3,3,2,2,2]$	97.40	97.36	96.60	94.52	88.43	71.98	40.23	**83.79**	**20.30**
(c) LPCC Based									
LPCC-E-D-A	98.30	96.82	95.59	92.28	81.87	54.52	22.96	77.48	-
LPCC-E-T $z = [8,6,5,5,4,4,3,3,2,2,2,2,2]$	96.74	96.90	95.90	93.30	85.91	67.86	36.39	**81.86**	**19.45**

Table 2. Word accuracy (%) of TFS (-E-T) features on the Aurora 2 database using phoneme HMMs. The results are averaged according to the noise level. The reference models are suffixed with -E-D-A.

SNR (dB) / Features	Inf	20	15	10	5	0	-5	Avg.	R.I. (%)
(a) MFCC Based									
MFCC-E-D-A	89.89	87.24	84.41	78.87	63.78	29.86	-5.82	61.17	-
MFCC-E-T $z = [8,6,5,4,4,3,3,2,2,2,2,2,2]$	93.02	94.15	92.65	88.84	79.22	56.42	19.58	**74.840**	**35.20**
(b) PLPCC Based									
PLPCC-E-D-A	88.99	87.92	84.78	78.97	64.18	32.96	-0.67	62.45	-
PLPCC-E-T $z = [8,6,5,4,4,3,3,3,3,3,2,2,2]$	92.98	94.29	92.89	89.38	79.76	56.86	18.79	**74.99**	**33.40**
(c) LPCC Based									
LPCC-E-D-A	88.13	86.04	83.11	76.64	60.82	29.65	0.02	60.63	-
LPCC-E-T $z = [8,6,5,5,4,4,3,3,2,2,2,2,2]$	91.09	93.26	91.24	86.98	76.45	50.90	10.79	**71.53**	**27.69**

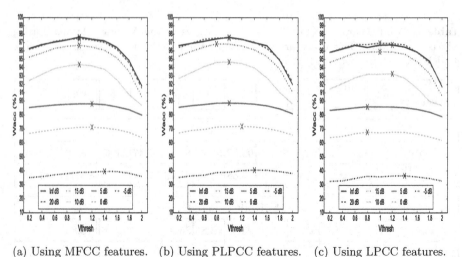

(a) Using MFCC features.   (b) Using PLPCC features.   (c) Using LPCC features.

Fig. 4. Variation of the performance of TFS as a function of V_{thresh} using whole-word HMM. The crosses indicate the maximum word accuracy for each noise level.

define for every phoneme triplets. This was not the case with whole-word HMMs because the state occupancy of a phoneme HMM is usually much shorter.

It can be seen in Fig. 4 that the plots are approximatively convex. The up and down hill-shaped curves support the idea introduced in sections 1 and 4 relative to informative coefficients. The amount of unrelated information added to the frame will be greater than the amount of related information if the concatenated coefficients are too close, or too far, from each other. This phenomenon is reflected in Fig. 4 with an increase in accuracy when V_{thresh} increases, up to some point where it starts to decrease.

Another observation that is worth mentioning in Fig. 4 is the behavior of the best accuracy with respect to the noise level. Apart from LPCC, where it is less clearly identifiable, the maximum result tends to occur at greater V_{thresh} as the noise increases. For example, the best word accuracy appears at $V_{thresh} = 1$ for the least noisy task and at $V_{thresh} = 1.4$ for the noisiest one. Since a greater threshold produces a z that has greater time offsets, our approach seems to act like a noise reduction method by smoothing the signal. Indeed, smoothing a highly noisy signal requires gathering information at a far distance. In this sense, our approach behaves similarly by selecting coefficients that are farther apart.

In summary, our results suggest that frequency-based dynamical features relying on the concatenation of adjacent coefficients helps improve the accuracy, especially for noisy utterances. The results on the Aurora 2 database show that the proposed TFS features achieved a better average word accuracy than the delta features. However, in the context of recognizing clean utterances with whole-word HMMs, our method did not outperform the reference features.

Nonetheless, TFS appears to be a good choice for dynamical features since it performed the best overall, can be learned rapidly from the data and is based on a single specified parameter V_{thresh}.

6 Conclusion

A novel way of improving the dynamics of static speech features was proposed. The issue that was addressed was the susceptibility to noise of derivative operations during the modeling of speech dynamics. The proposed Temporal Feature Selection (TFS) features have shown to improve the robustness of the state of the art delta features in various types of noise. The experimentations have shown that the 3 most standard features, MFCC, PLPCC and LPCC, combined with the TFS features achieved an averaged relative improvement of 20.79% and 32.10% in accuracy for whole-word and phoneme HMMs on the Aurora 2 database.

For further study, we plan to evaluate our approach on the harder problem of large vocabulary continuous speech recognition. Specifically, we will examine to potential of TFS to replace triphone HMMs modeling. Finally, we intend to use deep learning approaches to study the impacts of better feature engineering on the learning process of DNNs.

References

1. Bahl, L., De Souza, P., Gopalakrishnan, P., Nahamoo, D., Picheny, M.: Robust methods for using context-dependent features and models in a continuous speech recognizer. In: 1994 IEEE International Conference on Acoustics, Speech, and Signal Processing, ICASSP 1994, vol. 1, pp. I-533. IEEE (1994)
2. Fukunaga, K.: Introduction to Statistical Pattern Recognition, 2nd edn. Academic Press Professional Inc., San Diego (1990)
3. Furui, S.: Speaker-independent isolated word recognition based on emphasized spectral dynamics. In: IEEE International Conference on Acoustics, Speech, and Signal Processing, ICASSP 1986, vol. 11, pp. 1991–1994. IEEE (1986)
4. Gales, M., Young, S.: The application of hidden markov models in speech recognition. Foundations and Trends in Signal Processing 1(3), 195–304 (2008)
5. Gales, M.J.: Maximum likelihood linear transformations for hmm-based speech recognition. Computer Speech & Language 12(2), 75–98 (1998)
6. Gales, M.J.: Semi-tied covariance matrices for hidden markov models. IEEE Transactions on Speech and Audio Processing 7(3), 272–281 (1999)
7. Gopinath, R.A.: Maximum likelihood modeling with gaussian distributions for classification. In: Proceedings of the 1998 IEEE International Conference on Acoustics, Speech and Signal Processing, vol. 2, pp. 661–664. IEEE (1998)
8. Hossan, M.A., Memon, S., Gregory, M.A.: A novel approach for MFCC feature extraction. In: 2010 4th International Conference on Signal Processing and Communication Systems (ICSPCS), pp. 1–5. IEEE (2010)
9. Jolliffe, I.: Principal component analysis. Springer Series in Statistics, vol. 1. Springer, Berlin (1986)

10. Kumar, K., Kim, C., Stern, R.M.: Delta-spectral cepstral coefficients for robust speech recognition. In: 2011 IEEE International Conference on Acoustics, Speech and Signal Processing (ICASSP), pp. 4784–4787. IEEE (2011)
11. Kumar, N., Andreou, A.G.: Heteroscedastic discriminant analysis and reduced rank HMMs for improved speech recognition. Speech Communication **26**(4), 283–297 (1998)
12. Leggetter, C.J., Woodland, P.C.: Maximum likelihood linear regression for speaker adaptation of continuous density hidden markov models. Computer Speech & Language **9**(2), 171–185 (1995)
13. Lockwood, P., Boudy, J.: Experiments with a nonlinear spectral subtractor (NSS), Hidden Markov models and the projection, for robust speech recognition in cars. Speech Communication **11**(2–3), 215–228 (1992)
14. Oppenheim, A.V., Schafer, R.W., Buck, J.R.: Discrete-time Signal Processing, 2nd edn. Prentice-Hall Inc., Upper Saddle River (1999)
15. Pearce, D., günter Hirsch, H., Gmbh, E.E.D.: The aurora experimental framework for the performance evaluation of speech recognition systems under noisy conditions. In: ISCA ITRW ASR2000, pp. 29–32 (2000)
16. Rath, S.P., Povey, D., Veselỳ, K.: Improved feature processing for deep neural networks. In: Proc. Interspeech (2013)
17. Saon, G., Padmanabhan, M., Gopinath, R., Chen, S.: Maximum likelihood discriminant feature spaces. In: Proceedings of the 2000 IEEE International Conference on Acoustics, Speech, and Signal Processing, ICASSP 2000, vol. 2, pp. II1129–II1132. IEEE (2000)
18. Shrawankar, U., Thakare, V.M.: Techniques for feature extraction in speech recognition system: A comparative study. arXiv:1305.1145 (2013)
19. Trottier, L., Chaib-draa, B., Giguère, P.: Effects of Frequency-Based Inter-frame Dependencies on Automatic Speech Recognition. In: Sokolova, M., van Beek, P. (eds.) Canadian AI. LNCS, vol. 8436, pp. 357–362. Springer, Heidelberg (2014)
20. Weng, Z., Li, L., Guo, D.: Speaker recognition using weighted dynamic MFCC based on GMM. In: 2010 International Conference on Anti-Counterfeiting Security and Identification in Communication (ASID), pp. 285–288. IEEE (2010)
21. Young, S.J., Evermann, G., Gales, M.J.F., Hain, T., Kershaw, D., Moore, G., Odell, J., Ollason, D., Povey, D., Valtchev, V., Woodland, P.C.: The HTK Book, version 3.4. Cambridge University Engineering Department, Cambridge (2006)
22. Yu, D., Seltzer, M.L., Li, J., Huang, J.T., Seide, F.: Feature learning in deep neural networks-studies on speech recognition tasks. arXiv:1301.3605 (2013)
23. Zheng, F., Zhang, G., Song, Z.: Comparison of different implementations of MFCC. Journal of Computer Science and Technology **16**(6), 582–589 (2001)

Simulating Naming Latency Effects

Yannick Marchand[1]([✉]) and Robert Damper[2]

[1] Faculty of Computer Science, Dalhousie University, Halifax, Canada
ymarchan@dal.ca
[2] School of Electronics and Computer Science, University of Southampton,
Southampton, UK
rid@ecs.soton.ac.uk

Abstract. Pronunciation by analogy (PbA) is a data-driven method for converting letters to sound, with potential application to text-to-speech systems. We studied the capability of PbA to account for a broad range of naming latency phenomena in English. These phenomena included the lexicality, regularity, consistency, frequency, and length effects. To simulate these effects, various features of the PbA pronunciation lattice (a data structure that is produced for generating the pronunciation of a spelling pattern) were investigated. These measures included the number of arcs, nodes, pattern matchings and candidate pronunciations. While each of these individual features were able to replicate many of the effects, a measure of complexity that combined the frequency of the words as well as the number of candidates and arcs successfully simulated all of the effects tested.

Keywords: Naming latency · Cognitive model · Reading aloud

1 Introduction

A computerized text-to-speech (TTS) system refers to the ability to play back printed text in a spoken voice. This technology has many important applications in diverse areas including telecommunications-based information services (e.g. reading emails over the telephone), reading machines for blind and non-speaking people, and aids for first and second language learners. Pronunciation by analogy (PbA) is a data-driven technique for the automatic phonemisation of text, first proposed by Dedina and Nusbaum in 1991 in their system PRONOUNCE [12]. Since this time several groups around the world have developed PbA for use in TTS synthesis (e.g. [1,21,27]).

There is an apparent commonality between the computational problem of speech synthesis and the process of human reading. Given this situation, it is surprising that there has been minimal synergy between the two fields. On one hand, both well-established (e.g. [6,8,28]) and emergent (e.g. [25,26,29]) models of reading aloud may have overlooked valuable insights from the speech synthesis community regarding the very challenging computational problem of mapping letters to sounds. On the other hand, machine learning techniques for

© Springer International Publishing Switzerland 2015
D. Barbosa and E. Milios (Eds.): Canadian AI 2015, LNAI 9091, pp. 167–180, 2015.
DOI: 10.1007/978-3-319-18356-5_15

grapheme-to-phoneme conversion (e.g. [4,9,30]) have not been yet investigated to model the cognitive mechanisms of human reading.

Against this background, the present research aims to explore the possibility of PbA to reproduce the results of several seminal psychological studies on naming latencies in healthy English readers that span over three decades. Naming latency corresponds to the time between the presentation of a string of letters and the beginning of its pronunciation.

The remainder of this paper is structured as follows. In the next section, we present an overview of our PbA model. Section 3 outlines and summarizes the results of several important past experiments on naming latency. Section 4 describes the methodology that was conducted in order to simulate the key naming latency effects with PbA. Section 5 presents the results of these simulations in some detail. Finally, section 6 concludes with a discussion of our results.

2   Pronunciation by Analogy

In Pronunciation by Analogy (PbA), an unknown word is pronounced by matching substrings of the input to substrings of known words, hypothesizing a partial pronunciation for each matched substring from the phonological knowledge, and assembling the partial pronunciations. This method consists of four components: the lexical database, the matcher that compares the target input to all the words in the database, the pronunciation lattice (a data structure representing possible pronunciations), and the decision function that selects the 'best' pronunciation among the set of possible ones.

2.1   Lexical Database

The lexicon used in the present work is a manually-aligned version of the *Webster's Pocket Dictionary*. This dictionary contains approximately 20,000 words and was used by Sejnowski and Rosenberg for training their NETtalk neural network for text-to-phoneme conversion [31]. To cater to the cases where a word's spelling has more letters than its pronunciation has phonemes, yet still allow a strict one-to-one correspondence, they introduced a null phoneme symbol. Also, to allow for the (rarer) cases where a word's spelling has fewer letters than its pronunciation has phonemes, they introduced 'new' phonemes, such as /X/ to represent the spelling-to-sound correspondence <x> → /ks/[1] in words like <cortex>.

2.2   Pattern Matching

An input word is matched in turn against all orthographic entries in the lexicon. For a given entry, the process, in the early PRONOUNCE version, starts with

---

[1] In this paper, we use the phoneme symbols of the International Phonetic Alphabet [18] to specify pronunciations.

the input string and the dictionary entry left-aligned. Substrings sharing contiguous, common letters in matching positions are then found and information about these matching substrings and their corresponding aligned phoneme substrings is entered into a pronunciation lattice as described below. One of the two strings is then shifted right by one letter and the process repeated, until the input string and the dictionary entry are right-aligned. (We call this 'partial' pattern matching.)

The implementation with our further developed PbA model [21] features a 'full' pattern matching between input letter string and dictionary entries, as opposed to the 'partial' matching of PRONOUNCE. That is, rather than starting with the two strings left-aligned, we start with the initial letter of the input string $\mathcal{I}$ aligned with the last letter of the dictionary entry $\mathcal{W}$. The matching process terminates not when the two strings are right-aligned, but when the last letter of $\mathcal{I}$ aligns with initial letter of $\mathcal{W}$.

2.3   Pronunciation Lattice

The pronunciation lattice used in the PbA model is a graph data structure used to represent all the matching substrings (and, thereby, the possible pronunciations) of an input word. A node of the lattice represents a matched letter, L_i, at some position, i, in the input. The node is labeled with its position index i and with the phoneme that corresponds to L_i in the matched substring. An arc is placed from node i to node j if there is a matched substring starting with L_i and ending with L_j. The arc is labeled with the phoneme part of the matched substring and its corresponding number of occurrences. Figure 1 shows an example lattice for the pseudoword ('nonword') <shead>.

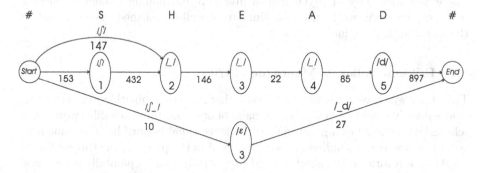

Fig. 1. Pronunciation lattice for the pseudoword <shead>. The delimiter symbol for the space at the beginning and end of the word is #. For simplicity, only a subset of nodes and arcs is shown.

As an illustration of how to interpret the pronunciation lattice, consider the arc between the *Start* node and the node labeled /ɛ/ at position 3. This arc is labeled with the pronunciation /ʃ_/ (where /_/ represents the null phoneme) and with a number of occurrences equals to 10. This means that there were

10 words in the dictionary that matched the pseudoword <shead> from the start of the word up to the letter <e> at position 3 *and* had the pronunciation /ʃ_ɛ.../. These 10 words were: <shed>, <shelf>, <shell>, <shellfish>, <shelter>, <shelve>, <shelves>, <shepherd>, <sheriff> and <sherry>.

2.4   Decision Function

A possible pronunciation for the input corresponds to a complete path through its lattice, from *Start* to *End* nodes, with the output string assembled by concatenating the phoneme labels on the nodes/arcs in the order that they are traversed. (Different paths can, of course, correspond to the same pronunciation.) Scoring of candidate pronunciation uses two heuristics:

1. If there is a unique shortest path, then the corresponding pronunciation is taken as the output.
2. If there are tied shortest paths, then the pronunciation corresponding to the best scoring of these is taken as the output.

The choice of scoring function(s), so as to determine the 'best' candidate pronunciation, is a major consideration in the design and implementation of any PbA model. The scoring function is not utilized in the current simulations and hence it will not be described further. However, the interested reader can refer to [11] for further information.

3   Naming Latency Studies

There is a vast body of psychological literature on naming latency in human readers. Therefore, we focused on simulating well-established effects found in the following key seminal studies.

3.1   Eriksen, Pollak & Montague (1970)

This study was conducted to examine whether readers implicitly speak the word before overtly voicing it [13]. A set of pairs of one- and three-syllables words was selected with the following criteria: (i) the three-syllable word has the same first syllable as the corresponding one-syllable word in the pair; (ii) the three-syllable word is not a form of the shorter word (e.g. <pain> and <painfully> were not allowed); (iii) the two words of each pair have similar frequency. It was shown that three-syllable words had significantly longer naming latencies than the word with only one syllable.

3.2   Forster & Chambers (1973)

A total pool of 75 letter strings were used in their experiment [14]: (i) 15 low frequency words; (ii) 15 high frequency words; (iii) 30 nonwords that were constructed by rearranging the letters of each of the previous words, but preserving pronounceability; (iv) 15 extremely rare, unfamiliar words (i.e., <bice>, <obol>). Two-thirds of the items in each sample were letter strings of four letters whereas the remaining third were six-letter stimuli. It was found that:

1. Naming times for words (excluding unfamiliar words) were shorter than for nonwords (i.e. the lexicality effect);
2. Naming times for high frequency words were shorter than for low frequency words (i.e. the frequency effect);
3. No significant difference in naming time was observed for nonwords and unfamiliar words;
4. The number of letters had a significant effect, with longer words taking longer to name (i.e. the length effect).

3.3   Baron & Strawson (1976)

In this study [2], three types of letter strings were used: (i) 5 lists of 10 regular words that were defined as consistent with "the rules of pronunciation as defined by Veneszky (1970) or by the availability of a number of consistent analogies, so that application of these rules or analogies would yield only one pronunciation" (p. 387); (ii) 5 lists of 10 exception words, i.e., words that violate principles given in (i); (iii) 5 lists of 10 nonsense words that were chosen so as to be pronounced as real words when rules were applied or analogies used. Using the work of [20], frequency of the words were approximately matched. Two main results were found:

1. Regular words were named more quickly than exception words (i.e. the regularity effect);
2. Naming latency for nonsense words was slower than for words.

3.4   Frederiksen & Kroll (1976)

In this work [15], half of the stimulus items were words and half were nonwords. To be pronunceable (i.e., to be pseudowords), the nonwords were derived from the words by randomly changing one vowel, usually the first (e.g. the pseudoword <erbor> was derived from <arbour>. Words were falling in four different frequency classes from the Thorndike and Lorge word count [32]. Also, for each category class, words were selected having four, five, or six letters and either one or two syllables. Their analysis of response latencies indicated that:

1. Words were named more rapidly than pseudowords;
2. There was a significant overall effect of array length;
3. There was a significant effect of word frequency: the higher the frequency class, the faster the words were named.

3.5 Mason (1978)

In experiment 3 of this study [22], a list of 80 word and 80 nonword stimuli was used. Each category of letter strings was composed of two halves of four and six letter stimuli, respectively. All the 80 words were rated high in frequency by the Thorndike and Lorge word count [32]. The findings were the following:

1. Nonwords took longer to name than did words;
2. Six-letter strings took significantly longer to read than did four-letter strings.

3.6 Parkin (1982)

In experiment 3 of this study [23], thirty-two exception words that have a guide to their pronunciation listed in the *Oxford Paperback Dictionary* were selected with corresponding regular words, matched for word frequency and part of speech. This experiment demonstrated that exception words took longer to pronounce than regular words.

3.7 Jared (1997)

Because Jared used them in her study, it is important to define the notion of *consistent* and *inconsistent* words. Traditionally, regular words are held to conform to abstract spelling-to-sound correspondences or rules (like those of [34] and [33]) whereas exception words "break the rules". Glushko introduced a classification that "... may be more appropriate for psychology" (p. 684) of consistent and inconsistent words [16]. Words are deemed consistent or inconsistent according to the consistency of pronunciation of their orthographic neighbours—whether or not they rhyme. Within this new framework, words are classified into: exception, regular and consistent, regular and inconsistent. Most exception words, such as <have>, are inconsistent by the new definition. Although consistent exception words exist, they are mostly "one-of-a-kind" words like <mauve>, <laugh> and <schism> without any orthographic neighbours at all. However, <wave>, which was previously classified as regular, becomes regular and inconsistent. It is regular because all <–ave> monosyllabic words *except* <have> take the pronunciation /–æv/. It is inconsistent because it is an orthographic neighbour of <have>. However, <wade> is regular and consistent since all <–ade> monosyllabic words take the pronunciation /–æd/ without exception.

In the study by Jared, it was demonstrated that spelling-sound consistency not only affects low-frequency words but also high-frequency words [19]. In the first experiment, 80 monosyllabic words were used, half of those were high-frequency and half were low-frequency words. Half the words in each frequency group were inconsistent words and the other half were consistent words. It was demonstrated that:

1. Participants named high-frequency words significantly faster than low-frequency words;
2. Naming latencies were significantly longer for inconsistent words than for consistent words (both for low and high frequency words).

4   Methodology

We examined several indices that contribute to the complexity of the PbA pronunciation lattice as possible predictors of naming latency. These include the numbers of nodes, arcs, pattern matchings and competing candidate pronunciations (i.e. shortest paths as noted earlier). The underlying idea is the more complex the lattice, the longer the processing time is expected to be. The computational simulations were conducted using the PbA method with a 'full' pattern matching between input letter string and dictionary entries such as previously defined in section 2.2.

In an attempt to simulate the different phenomena by the intermediary of a single dependent variable, we also empirically and operationally defined a complexity measure (CM) for a given string to be pronounced, CM(string), by the following equation:

$$CM(String) = \frac{A \times C}{log_b(b + Frequency(String))}$$

A and C are the number of arcs and candidates, respectively. Frequency (String) equals to zero when the string is a nonword otherwise it is equal to the linguistic frequency of the word represented by the string. We used the word frequency list from [20] that is publicly available via the World Wide Web at http://www.psych.rl.ac.uk/ (last accessed 14 January 2015). Note that the simulations were conducted with b=10 and that the statistical significance was assessed with two-tailed tests.

5   Results

5.1   Lexicality Effect

Table 1(a) describes the basic features and the complexity measure of the PbA pronunciation lattice obtained using the stimuli from the four seminal studies that elicited a lexicality effect. All these comparisons were subjected to a paired-sample t-test (when stimuli were matched in pairs in the original studies) or an independent t-test (when the number of items were different or the stimuli did not match in pairs) between all the words and nonwords.

The statistical results are shown in Tables 1(b). They demonstrate that the number of candidates and the complexity measure are excellent predictors of the lexicality effect. Interestingly, the number of arcs is inversely proportional to the naming latency that characterizes this effect. Indeed, the words are read faster in spite of an apparent higher number of arcs.

Also, similar to the study of Forster and Chambers [14], there was no significant difference for nonwords and unfamiliar words for any of the studied features in the simulations.

Table 1. PbA simulation results for the lexicality effect. (a) Means for the basic features of the pronunciation lattices. (b) Paired-sample or independent t-tests between words (W) and nonwords (NW). KEYS—*:p-$value < 0.1$; **:p-$value < 0.05$; ***:p-$value < 0.01$; Sig.: Significance; NS: Not significant.

(a)

Study	Nodes		Arcs		Matchings		Cand.		CM	
	W	NW	W	NW	W	NW	W	NW	W	NW
Forster & Chambers, 73	27.4	27.2	90.3	82.1	7070	6355	1.1	2.9	73.7	258.6
Baron & Strawson, 76	29.9	25.6	102.6	65.1	6803	5188	1.0	2.7	65.9	162.0
Frederiksen & Kroll, 76	28.1	27.8	92.9	80.1	6854	6593	1.1	3.3	89.5	268.0
Mason, 78	29.6	29.5	102.6	89.0	6632	6459	1.0	3.0	55.2	263.5

(b)

Study	Nodes		Arcs		Matchings		Cand.		CM	
	t	Sig.	t	Sig.	t	Sig.	t	Sig.	t	Sig.
Forster & Chambers, 73	0.64	NS	2.46	**	1.87	*	-4.50	***	-4.12	***
Baron & Strawson, 76	2.27	**	3.59	***	1.81	*	-9.56	***	-7.73	***
Frederiksen & Kroll, 76	0.41	NS	3.71	***	0.66	NS	-9.47	***	-7.98	***
Mason, 78	0.13	NS	2.45	**	0.60	NS	-7.72	***	-8.64	***

5.2 Regularity and Consistency Effects

Table 2(a) presents the means of the features of interest using the stimuli of the studies showing a regularity effect whereas Table 3 gives the means for the study of Jared [19] on the consistency effect (i.e. words with consistent spelling to sound are pronounced faster than those that are inconsistent). All these variables were subjected to a paired-sample t-test or an independent t-test between all the regular (or consistent) and irregular (or inconsistent) words.

Table 2. PbA simulation results for the regularity effect. (a) Means for the basic features of the pronunciation lattices. (b) Paired-sample t-tests or independent t-tests between regular and irregular words. KEYS: as for Table 1(b).

(a)

Study	Nodes		Arcs		Matching		Cand.		CM	
	R	I	R	I	R	I	R	I	R	I
Baron & Strawson, 76	29.0	30.8	96.5	108.9	6500	7113	1.1	1.0	68.9	62.8
Parkin, 82	28.6	32.4	103.6	121.4	7588	8060	1.0	1.0	80.3	93.2

(b)

Study	Nodes		Arcs		Matchings		Cand.		CM	
	t	Sig.	t	Sig.	t	Sig.	t	Sig.	t	Sig.
Baron & Strawson, 76	-1.51	NS	-1.92	*	-1.13	NS	0.99	NS	0.91	NS
Parkin, 82	-2.73	**	-2.97	***	-0.71	NS	-1.00	NS	-3.18	***

Table 3. PbA simulation results for the consistency effect found in Jared (1997). Means for the basic features as well as the complexity measure of the pronunciation lattices and paired-sample or independent t-tests between consistent (C) and inconsistent (I) words. KEYS: as for Table 1(b).

	Means		C vs. I	
	C	I	t	Sig.
Nodes	25.0	28.0	-3.05	***
Arcs	81.9	93.9	-2.26	**
Matchings	6789	6486	0.51	NS
Candidates	1.0	1.0	1.00	NS
Complexity measure	54.2	62.8	-1.88	*

The results (see Tables 2(b) and 3) highlight the importance of the number of arcs within the pronunciation lattices to explain the regularity and consistency effects. To a lesser extent, the complexity measure also simulated this phenomenon.

5.3   Frequency Effect

Table 4(a) shows the means of the basic features and the complexity measure in regard to the frequency effect (i.e. high frequency words are named faster than low frequency words). Results of the simulations indicate that none of the basic indices of the pronunciation lattice on their own predict this effect. However, the complexity measure is an excellent predictor, achieving high statistical significance for all three of the studies eliciting this phenomenon (see table 4(b)).

Table 4. PbA simulation results for the frequency effect. (a) Means for the features of the pronunciation lattices. (b) Paired-sample or independent t-tests between high-frequency (HF) and low-frequency (LF). KEYS: as for Table 1(b).

(a)

Study	Nodes		Arcs		Matchings		Cand.		CM	
	HF	LF	HF	LF	HF	LF	HF	LF	HF	LF
Forster & Chambers, 73	26.9	28.0	91.4	89.3	7864	6276	1.0	1.1	44.7	102.7
Frederiksen & Kroll, 76	28.5	27.8	94.7	91.2	6740	6968	1.0	1.3	68.6	110.5
Jared, 97	26.8	26.3	89.2	86.6	6522	6753	1.2	1.0	45.9	71.1

(b)

Study	Nodes		Arcs		Matchings		Cand.		CM	
	t	Sig.	t	Sig.	t	Sig.	t	Sig.	t	Sig.
Forster & Chambers, 73	-0.59	NS	0.23	NS	1.96	*	-1.00	NS	-2.93	**
Frederiksen & Kroll, 76	0.73	NS	0.79	NS	-0.45	NS	-2.46	**	-4.26	***
Jared, 97	1.10	NS	0.98	NS	-1.10	NS	0.96	NS	-10.44	***

5.4 Length Effect

Table 5(a) presents the means of the features obtained using the stimuli of the three studies that demonstrated a length effect between 'short' and 'long' words (see [3] for a review on studies regarding the length effect). Results from tables 5(b) show that all the features with the exception of the number of candidates strongly reproduce the length effect. This is not surprising because longer words likely generate more dense pronunciation lattices.

A one-way analysis of variance with length as factor with three levels (letter strings of 4, 5 and 6 letters) was also conducted on the key features of the pronunciation lattices and the complexity measure for the study of Frederiksen and Kroll [15]. Table 6 shows again that, with the exception of the number of candidates, all the features were demonstrated to be strong predictors of the length effect. However, while the number of nodes, arcs and matchings significantly

Table 5. PbA simulation results for the length effect. (a) Means for the basic features of the pronunciation lattices. (b) Paired-sample or independent t-tests between "short" (S) and "long" (L) items. KEYS: as for Table 1(b).

(a)

Study	Nodes		Arcs		Matchings		Cand.		CM	
	S	L	S	L	S	L	S	L	S	L
Eriksen et al, 70	22.2	50.2	68.4	217.6	5233	12288	1.0	1.4	53.5	253.6
Forster & Chambers, 73	24.2	33.8	71.0	118.8	5485	8929	1.9	2.4	131.1	264.0
Mason, 78	24.2	34.9	72.7	119.1	5520	7572	2.2	1.9	142.1	173.9

(b)

Study	Nodes		Arcs		Matchings		Cand.		CM	
	t	Sig.	t	Sig.	t	Sig.	t	Sig.	t	Sig.
Eriksen et al, 70	-7.79	***	-7.86	***	-6.40	***	-1.31	NS	-3.15	**
Forster & Chambers, 73	-7.95	***	-8.71	***	-6.06	***	-0.98	NS	-2.79	***
Mason, 78	-14.00	***	-10.92	***	-4.97	***	1.25	NS	-1.09	NS

Table 6. PbA simulation results for the length effect found in Frederiksen and Kroll's study (1976). Means for the basic features as well as the complexity measure of the pronunciation lattices and F value between letter strings of different lengths. KEYS: as for Table 1(b).

	Means			Letter string length	
	4	5	6	F	Sig.
Nodes	22.1	28.1	33.6	121.95	***
Arcs	64.5	87.5	107.8	65.75	***
Matchings	4841	6962	8343	33.53	***
Candidates	2.3	2.4	1.8	2.63	*
Complexity measure	145.6	208.8	185.0	2.74	*

explained this phenomenon, the complexity measure only exhibited significant differences ($p < 0.05$) between letter strings of 4 and 5 letters.

6 Conclusion

In this study, we have shown that a method for grapheme-to-phoneme conversion based upon pronunciation by analogy (PbA) was able to successfully reproduce several seminal and robust results on naming latencies, which included:

1. the lexicality effect: nonwords take longer to name than words;
2. the regularity effect: irregular words take longer to name than regular words;
3. the consistency effect: inconsistent words take longer to name than consistent words;
4. the frequency effect: low-frequency words take longer to name than high-frequency words;
5. the length effect: long words take longer to name than short words.

Table 7 summarizes all the results. They indicate that the number of nodes is a poor predictor of the lexicality and frequency effects, failing to reach significance in both cases (at the exception of the Baron & Strawson study [2]). This is not at all surprising, since the number of nodes in a letter string depends fairly directly upon the number of letters and the number of phonemes that these individual letters map to.

The number of arcs seems to be an excellent predictor of the regularity and consistency effects, regular words are seen to have fewer arcs on average than exception words. For the lexicality effect, however, the differences are in the wrong direction. The words consistently have more arcs than the pseudowords

Table 7. PbA simulation results for all the studies under investigation. KEYS: as for Table 1(b).

Effect	Study	Nodes	Arcs	Matchings	Cand.	CM
Lexicality	Forster & Chambers, 73	NS	**	*	***	***
	Baron & Strawson, 76	**	***	*	***	***
	Frederiksen & Kroll, 76	NS	***	NS	***	***
	Mason, 78	NS	**	NS	***	***
Regularity	Baron & Strawson, 76	NS	*	NS	NS	NS
	Parkin, 82	**	***	NS	NS	***
Consistency	Jared, 97	***	**	NS	NS	*
Frequency	Forster & Chambers, 73	NS	NS	*	NS	**
	Frederiksen & Kroll, 76	NS	NS	NS	**	***
	Jared, 97	NS	NS	NS	NS	***
Length	Eriksen et al, 70	***	***	***	NS	**
	Forster & Chambers, 73	***	***	***	NS	***
	Frederiksen & Kroll, 76	***	***	***	**	*
	Mason, 78	***	***	***	NS	NS

despite the fact that words are read faster. Thus, naming latency can not be directly proportional to the number of arcs in the pronunciation lattice.

It can be shown that while each of these individual features were able to replicate many of the effects, the measure of complexity is the only one to simulate all of the effects tested.

In seeking to understand how human readers convert print to sound, intense debate has raged in the cognitive research literature (e.g. [17]) over whether there is a single (connectionist and lexical) route or dual (lexical plus rule-based) routes. The basic notion of dual-route theories [5–8,14,24] is that words are pronounced by lexical look-up whereas non-word letter strings can only be pronounced by abstract letter-to-sound rules. In addition to the fact that PbA was enhanced for use in TTS, it is distinctly different from extant models of reading aloud as it is single-route but, unlike other such models, it is not connectionist in inspiration. Thus, in addition to be one of the best methods for automatic phonemization in text-to-speech synthesis [10], the current results on naming latency encourage us to believe that PbA could rank as a promising alternative computational model of reading aloud.

References

1. Bagshaw, P.C.: Phonemic transcription by analogy in text-to-speech synthesis: Novel word pronunciation and lexicon compression. Computer Speech and Language **12**(2), 119–142 (1998)
2. Baron, J., Strawson, C.: Use of orthographic and word-specific knowledge in reading aloud. Journal of Experimental Psychology: Human Perception and Performance **2**(3), 386–393 (1976)
3. Barton, J.J., Hanif, H.M., Eklinder Björnström, L., Hills, C.: The word-length effect in reading: A review. Cognitive Neuropsychology **31**(5–6), 378–412 (2014)
4. Bisani, M., Ney, H.: Joint-sequence models for grapheme-to-phoneme conversion. Speech Communication **50**(5), 434–451 (2008)
5. Coltheart, M.: Lexical access in simple reading tasks. In: Underwood, G. (ed.) Strategies of Information Processing, pp. 151–216. Academic Press, New York (1978)
6. Coltheart, M., Curtis, B., Atkins, P., Haller, M.: Models of reading aloud: Dual-route and parallel-distributed-processing approaches. Psychological Review **100**(4), 589–608 (1993)
7. Coltheart, M., Patterson, K.E., Marshall, J.C. (eds.): Deep Dyslexia. Routledge and Kegan Paul, London (1980)
8. Coltheart, M., Rastle, K., Perry, C., Langdon, R., Ziegler, J.: DRC: A dual-route cascaded model of visual word recognition and reading aloud. Psychological Review **108**(1), 204–256 (2001)
9. Daelemans, W., van den Bosch, A., Weijters, T.: IGTree: Using trees for compression and classification in lazy learning algorithms. Artificial Intelligence Review **11**(1–5), 407–423 (1997)
10. Damper, R.I., Marchand, Y., Adamson, M.J., Gustafson, K.: Evaluating the pronunciation component of text-to-speech systems for English: A performance comparison of different approaches. Computer Speech and Language **13**(2), 155–176 (1999)

11. Damper, R.I., Marchand, Y.: Information fusion approaches to the automatic pronunciation of print by analogy. Information Fusion **7**(2), 207–220 (2006)
12. Dedina, M.J., Nusbaum, H.C.: Pronounce: A program for pronunciation by analogy. Computer Speech and Language **5**(1), 55–64 (1991)
13. Eriksen, C.W., Pollak, M.D., Montague, W.E.: Implicit speech: mechanism in perceptual encoding. Journal of Experimental Psychology **84**(3), 502–507 (1970)
14. Forster, K.I., Chambers, S.M.: Lexical access and naming time. Journal of Verbal Learning and Verbal Behavior **12**, 627–635 (1973)
15. Frederiksen, J., Kroll, J.: Spelling and sound: Approaches to the internal lexicon. Journal of Experimental Psychology: Human Perception and Performance **2**(3), 361–379 (1976)
16. Glushko, R.J.: The organization and activation of orthographic knowledge in reading aloud. Journal of Experimental Psychology: Human Perception and Performance **5**(4), 674–691 (1979)
17. Humphreys, G.W., Evett, L.J.: Are there independent lexical and non-lexical routes in word processing? An evaluation of the dual route theory of reading. Behavioral and Brain Sciences **8**(4), 689–739 (1985)
18. Handbook of the International Phonetic Association: A Guide to the Use of the International Phonetic Alphabet. Cambridge University Press, Cambridge (1999)
19. Jared, D.: Spelling-sound consistency affects the naming of high-frequency words. Journal of Memory and Language **36**, 505–529 (1997)
20. Kučera, H., Francis, W.N.: Computational Analysis of Present-Day American English. Brown University Press, Providence (1967)
21. Marchand, Y., Damper, R.I.: A multistrategy approach to improving pronunciation by analogy. Computational Linguistics **26**(2), 195–219 (2000)
22. Mason, M.: From print to sound in mature readers as a function of reader ability and two forms of orthographic regularity. Memory & Cognition **6**(5), 568–581 (1978)
23. Parkin, A.J.: Phonological recoding in lexical decision: Effects of spelling-to-sound regularity depend on how regularity is defined. Memory and Cognition **10**(1), 43–53 (1982)
24. Patterson, K.E., Morton, J.: From orthography to phonology: An attempt at an old interpretation. In: Patterson, K.E., Marshall, J.C., Coltheart, M. (eds.) Surface Dyslexia: Neuropsychological and Cognitive Studies of Phonological Reading, pp. 335–359. Lawrence Erlbaum Associates, London (1985)
25. Perry, C., Ziegler, J.C., Zorzi, M.: Nested incremental modeling in the development of computational theories: the CDP+ model of reading aloud. Psychological Review **114**(2), 273 (2007)
26. Perry, C., Ziegler, J.C., Zorzi, M.: Beyond single syllables: Large-scale modeling of reading aloud with the connectionist dual process (CDP++) model. Cognitive Psychology **61**(2), 106–151 (2010)
27. Pirrelli, V., Yvon, F.: The hidden dimension: A paradigmatic view of data-driven NLP. Journal of Experimental and Theoretical Artificial Intelligence **11**(3), 391–408 (1999)
28. Plaut, D.C., McClelland, J.L., Seidenberg, M.S., Patterson, K.E.: Understanding normal and impaired word reading: Computational principles in quasi-regular domains. Psychological Review **103**(1), 56–115 (1996)
29. Pritchard, S.C., Coltheart, M., Palethorpe, S., Castles, A.: Nonword reading: comparing dual-route cascaded and connectionist dual-process models with human data. Journal of Experimental Psychology: Human Perception and Performance **38**(5), 1268 (2012)

30. Rao, K., Peng, F., Sak, H., Beaufays, F.: Grapheme-to-phoneme conversion using long short-term memory recurrent neural networks. In: Proceedings of ICASSP (2015)
31. Sejnowski, T.J., Rosenberg, C.R.: Parallel networks that learn to pronounce English text. Complex Systems 1(1), 145–168 (1987)
32. Thorndike, E.L., Lorge, I.: The Teachers' Word Book of 30,000 Words. Columbia University, NY, Teachers' College (1944)
33. Venezky, R.L.: The Structure of English Orthography. Mouton, The Hague (1970)
34. Wijk, A.: Rules of Pronunciation of the English Language. Oxford University Press, Oxford (1966)

On Assessing the Sentiment of *General* Tweets

Sifei Han[1] and Ramakanth Kavuluru[1,2]([✉])

[1] Department of Computer Science, University of Kentucky, Lexington, KY, USA
[2] Division of Biomedical Informatics, Department of Biostatistics,
University of Kentucky, Lexington, KY, USA
{eric.s.han,ramakanth.kavuluru}@uky.edu

Abstract. With the explosion of publicly accessible social data, sentiment analysis has emerged as an important task with applications in e-commerce, politics, and social sciences. Hence, so far, researchers have largely focused on sentiment analysis of texts involving entities such as products, persons, institutions, and events. However, a significant amount of chatter on microblogging websites may not be directed at a particular entity. On Twitter, users share information on their general state of mind, details about how their day went, their plans for the next day, or just conversational chatter with other users. In this paper, we look into the problem of assessing the sentiment of publicly available *general* stream of tweets. Assessing the sentiment of such tweets helps us assess the overall sentiment being expressed in a geographic location or by a set of users (scoped through some means), which has applications in social sciences, psychology, and health sciences. The only prior effort [1] that addresses this problem assumes equal proportion of positive, negative, and neutral tweets, but a casual observation shows that such a scenario is not realistic. So in our work, we first determine the proportion (with appropriate confidence intervals) of positive/negative/neutral tweets from a set of 1000 randomly curated tweets. Next, adhering to this proportion, we use a combination of an existing dataset [1] with our dataset and conduct experiments to achieve new state-of-the-art results using a large set of features. Our results also demonstrate that methods that work best for tweets containing popular named entities may not work well for general tweets. We also conduct qualitative error analysis and identify future research directions to further improve performance.

1 Introduction

Sentiment analysis (or opinion mining) has gained significant attention from the computer science research community over the last decade due to the rapid growth in e-commerce and the practice of consumers writing online reviews for products and services they have used. Movies, restaurants, hotels, and recently even hospitals and physicians are being reviewed online. Manually aggregating all information available in a large number of textual reviews is impractical. However, discovering different aspects of the product/service that the review is discussing and the corresponding evaluative nature of the review for each of them is computationally challenging given the idiosyncratic and informal nature

© Springer International Publishing Switzerland 2015
D. Barbosa and E. Milios (Eds.): Canadian AI 2015, LNAI 9091, pp. 181–195, 2015.
DOI: 10.1007/978-3-319-18356-5_16

of customer reviews. Sentiment analysis has also been essential in gleaning information from customer surveys that companies routinely conduct. Due to our direct involvement in an ongoing project, we also observe that companies consult researchers to conduct sentiment analysis of emails of their employees to assess personnel morale and to improve organizational behavior and decision making. Recently, in the field of healthcare, researchers have focused on identifying emotions in suicide notes [25] and predicting county level heart disease mortality using Twitter language usage [6].

Due to the short informal nature of messages called *tweets* and the asymmetric network structure, since its introduction in 2006, Twitter has grown into one of the top 10 visited websites in the world with 100 million daily active users who generate over 500 million tweets per day [28]. Instead of going to a product website, Twitter users (henceforth *tweeters*) discuss their opinions and express their sentiments on different topics to their followers and all other users (if they wish). Given a recent study [14] reveals that over 95% of Twitter profiles are public (the default setting), Twitter has become an interesting platform to track sentiment on different topics and to assess the general mood of tweeters at specific locations and times. The informal text also poses challenges through (sometimes intentionally) misspelled words, neologisms, and other short forms that do not occur in dictionaries. Emoticons, abbreviations, user mentions, and hashtags also add to the complexity of analyzing tweet sentiment. We request the readers to refer to a recent survey [15] for details on general approaches to sentiment analysis on Twitter.

Most current efforts that analyze tweet sentiment directly or indirectly focus on tweets that contain popular topics or entities and tend to use datasets that are skewed to contain fewer neutral tweets. The ongoing series of Semantic Evaluation (SemEval) tasks added a Twitter sentiment analysis track in 2013 [19,26] in which the dataset selection was done based on the presence of a popular named entity in the tweet *and* the presence of at least one word with positive or negative sentiment score > 0.3 in SentiWordNet 3.0 [2], a lexical resource that contains positive, negative, and objectivity scores for synsets in WordNet. Although this is justified for tasks that involve analyzing sentiment of tweets that discuss a popular topic or entity, several tweets from the Twitter firehose may not discuss a popular topic. Tweeters might be chatting with others, sharing information on how their day went or their plans for the next few days, or just tweeting about how they feel at the time. Consider the following tweets from our dataset

- I feel so accomplished i had 1 Liter of water ! 1 more to go
- Good thing i have no work today
- Well headed to dorm to nap then open gym to practice
- Started a new diet where I only eat fast food
- Math exam tomorrow is going to kick my butt
- I swear this cold is gonna get the best of me

We noticed that most tweets in our dataset (randomly selected using Twitter streaming API) have the general nature as those in these six examples. Although some of them contain sentiment words, many do not and most tweets are not on any popular topic. This seems to be in line with the original intention of Twitter creators: until November 2009 Twitter had "What are you doing?" as the prompt displayed to the users when they log in; since then this has been changed to "What's happening?" Although this sample tweet list is slightly biased to show sentiment expressing tweets, we notice that many tweets are neutral or objective in nature.

However, to gauge the aggregate sentiment from a geographic location or sentiment expressed by a user group, it is essential to be able to determine the sentiment of all tweets (without any other topic based or sentiment word based selection bias). Such scenarios arise naturally in social sciences, psychology, and health sciences especially in the domain of mental health and substance abuse. For example, researchers might want to analyze the sentiment of tweet streams from users who identify themselves as smokers or vapers (e-cigarette users) or users from a particular area that reports low health rankings (http:// www.countyhealthrankings.org/). Further selection of users can be done based on predicted age group, gender, race, or ethnicity [13, 20, 24]. The results can also be extended to interview or counseling narratives of mental health patients by aggregating sentiment expressed in each sentence. However, except for the lone effort by Agarwal et al. [1], we are not aware of attempts to determine sentiment of a set of *general* tweets[1] collected using the Twitter streaming API. Even in their effort, Agarwal et al. assume equal proportion for the three sentiment classes, which our manual analysis shows is not realistic. So in our current effort

1. We manually estimate the proportion of [positive : negative : neutral] tweets to be 29% (26–32%) : 18% (16–21%) : 53% (50–56%) from a sample of 1000 randomly selected tweets selected from a set of 20 million tweets collected through Twitter streaming API in 2013. We also estimate that only 10% (7–13%) of the tweets have named entities in them. The 95% confidence intervals of the proportions calculated using Wilson score [31] are shown in parentheses.
2. Adhering to this estimated class proportion, we combine the dataset used by Agarwal et al. [1] with our dataset to create a larger dataset and conduct experiments with a broad set of features to identify a combination of features that offers the best performance (macro average of positive and negative sentiment F-scores). We also show that our best model improves over Agarwal et al.'s results on their original dataset with equal class proportions. Furthermore, we also show that a system comparable to the top performer [17] in SemEval tasks may not suffice for our general tweets, warranting identification of high performing feature subsets.
3. We analyze the confusion matrix and manually identify causes for certain types of errors and future research directions to improve sentiment analysis of general tweets.

---

[1] Note that some of these tweets may contain named entities or popular topics but we are not prescreening those that contain such themes

2   Background and Related Work

Sentiment analysis has emerged as an important sub-discipline within natural language processing research in computer science. Given it is very difficult for human users to exhaustively read and understand large numbers of potentially subjective narratives, automated methods have gained prominence over the past decade pursued first as document level classification tasks [21,27] and subsequently as sentence level [9], and recently as phrase level [19,32] tasks. Unsupervised approaches (e.g., [27]) that take advantage of sentiment lexicons, supervised approaches (e.g., [22]) that employ statistical learning, and semi-supervised approaches (e.g., [33]) that automatically generate training data have evolved as different alternatives that are currently being used in a hybrid fashion to obtain state-of-the-art results. Purely lexicon based approaches suffer from low recall and statistical learning approaches that rely only on tweet content and labeled data often offer low precision, especially with smaller training datasets; this has been mitigated to some extent with the advent of Internet crowd sourcing opportunities such as Amazon Mechanical Turk for generating large training datasets. Although manually building high coverage sentiment lexicons is impractical, automated approaches to induce them have resulted in significant performance gains [11]. This has proven especially useful for Twitter data given its 140 character limit and the extremely informal nature of communication, due to which popular hand-built lexicons were found insufficient.

One of the first notable attempts in sentiment analysis for tweets was by Go et al. [8] who used supervised learning and emoticon based distance supervision to acquire training data. Researchers have also focused on target dependent sentiment classification [4,10] where the sentiment is associated with a target concept. From 2013, a shared task [19] on Twitter sentiment analysis has been added to the annual SemEval workshop. Researchers at NRC-Canada entered the best performer [17] in the 2013 SemEval task. They designed a sophisticated hybrid sentiment analysis system that incorporates both hand-built and automatically constructed sentiment lexicons as features, besides using the conventional ngram features, in a supervised learning framework. Recently, they improved upon their results [11] by generating separate lexicons for affirmative and negated contexts. Although these efforts significantly advance the state-of-the-art in tweet sentiment analysis, they all use datasets that have been curated to contain popular topics during the collection period. Named entities or event names such as iPhone, Gaddafi, AT&T, Kindle, and Japan Earthquake are used and in the case of SemEval tasks, additionally, presence of sentiment expressing words is also required, thus inherently skewing the dataset to subjective tweets [19]. It is not clear whether methods that produce the best results on these datasets work best for general tweets as indicated in Section 1, where we discuss applications of sentiment analysis of general tweets.

Agarwal et al. [1] curated a set of random tweets collected using Twitter streaming API and conducted supervised learning experiments with a broad set of features also incorporating some sentiment lexicons. As they point out, their effort is the first in looking at such tweets without any pre-screening constraints and to our knowledge is the only such attempt. They introduce a new tree kernel representation of tweets and show that this representation performs on par with traditional approaches that involve content based features and sentiment features including emoticons, parts of speech, lexicon based prior polarity scores, and presence of hashtags, user mentions, and URLs. However, they assume that the positive, negative, and neutral classes are equal in proportion, which our analysis shows is not realistic. Hence, in our current effort we first estimate the proportion of the three classes, build a representative dataset, and conduct experiments to identify feature combinations that achieve best performance.

3 Datasets and Performance Measures

We sampled tweets using Twitter streaming API periodically in 2013 and collected over 20 million English tweets from which we curated two randomly selected tweet datasets one with 10,000 tweets and another with 1000 tweets. We used the larger dataset to conduct automated analysis of different characteristics of general tweets and the smaller dataset to manually determine the distribution of positive, negative, and neutral class distributions.

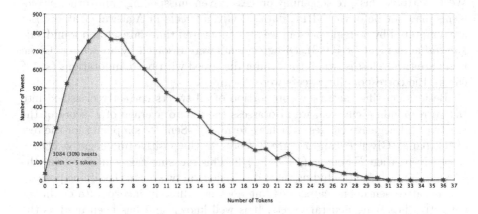

Fig. 1. Distributions of tweets with different numbers of tokens from a dataset of 10,000 randomly selected tweets

In Figure 1 we plot the number of tokens in a tweet on the x-axis and the corresponding number of tweets in the larger dataset of 10,000 tweets. We consider each user mention, emoticon, URL, and hashtag as an individual token. As we can see from the figure, 30.8% (29.9–31.7%) of the tweets have fewer than 6 tokens. Based on this dataset, we also observe that 12.5% (11.9–13.2%) of general tweets have URLs, 36.1% (35.2–37.1%) contain user mentions, 13.1% (12.5–13.8%) use hashtags, and 5.3% (4.9–5.8%) have emoticons in them, where the ranges in parentheses represent the 95% confidence intervals computed based on Wilson score [31].

We have two annotators independently perform a three way classification of smaller 1000 tweet dataset instances into positive, negative, or neutral classes. The two annotators were given general instructions on classifying the tweets and were later asked to discuss and resolve disagreements through discussion. We had *moderate* agreement ($\kappa = 0.54$) based on the general rule of thumb [12] on observer agreement for categorical data. Based on the consolidated judgments we estimate the class proportion ratio [positive : negative : neutral] to be 29% (26–32%) : 18% (16–21%) : 53% (50–56%) with 95% confidence intervals shown in parentheses. Although the proportions are different, the proportion of positive tweets is larger than that of negative tweets even in the SemEval datasets [19]. Adhering to our estimated proportion, we combine our dataset with the dataset used in Agarwal et al. [1] to build a consolidated larger dataset of 3523 tweets with 1844 neutral, 1011 positive, and 668 negative tweets. Since neutral tweets are the majority, we first merge neutral tweets from both datasets, and randomly select positive and negative tweets according to our estimated proportion. Since the dataset in [1] has non-English tweets, we use Natural Language Toolkit (NLTK [3]) 'words' corpus' English word subset to first automatically filter tweets when at least 40% of words in the tweet are English words; these filtered tweets are subsequently manually filtered to obtain only English tweets. We undertook this pre-screening process, given misspellings and other neologisms may not be in the NLTK dataset. Given tweets in Agarwal et al. dataset are annotated by a single person, we also annotated that dataset and found *substantial* agreement ($\kappa = 0.8$). Disagreements were resolved by an arbiter. Given there are three classes, when the arbiter disagreed with both annotations ($< 1\%$ cases), the corresponding tweets were discarded.

Besides classifier accuracy (proportion of all tweets correctly classified into the corresponding classes), we also assess the macro average of F-score of the positive and negative classes, which we term as F-Sent for simplicity for the rest of this paper. Given F_+ and F_- are F-scores for the positive and negative classes respectively, then F-Sent $= (F_+ + F_-)/2$. This measure takes into account the FPs and FNs caused (including those due to neutral tweets) in classifying positive and negative sentiment categories but does not directly incorporate credit for correctly classifying neutral tweets. It is well known and has been used as the main measure in the SemEval [19] Twitter sentiment analysis tasks.

4   Supervised Classification Framework

We follow the hybrid approach of employing sentiment lexicons as features in the supervised framework while also using the conventional content based feature (e.g., n-grams) and Twitter specific features (e.g., emoticons and hashtags). Our main classifier is the well known linear support vector machine from the LIBLINEAR [7] library made available in the scikit-learn [23] machine learning framework. We use automatic class weighting supported through the classifier and use the default one-vs-rest approach for three way classification of sentiment categories. Free text is pre-processed in general to minimize the noise in the feature space for text classification. For our tweet dataset we fully replicate the approach used by Agarwal et al. [1, Section 4] by replacing tweet targets (user mentions using the @ symbol) and URLs with specialized tokens since specific user mentions and URLs tokens often do not constitute meaningful features. Emoticons are replaced by their polarity based on the emoticon polarity dictionary built by Agarwal et al. [1]. We replace negation words (e.g., not, no, never, n't, cannot) with a single "NOT" token and expand popular slang acronyms (e.g., rofl, lol) to full forms. We incorporate a large set of features used by Agarwal et al. [1] and Kiritechnko et al. [11] and introduce new lexico-syntactic features that combine sentiment expressing words with their parts of speech and dependency edges involving them.

Lexical features: We use word unigrams and bigrams (henceforth called just ngrams) as the base features with feature weighting based on Naives Bayes (NB) scores [29] computed using the training data. We refer to this way of using NB scores as input to an SVM classifier as NBSVM as introduced by Wang and Manning [29]. The numbers of tokens with all capitalized letters, hashtags, elongated words[2], contiguous sequences of question marks, exclamation marks, or a combination of both are also included in the feature list.

Syntactic features: We incorporate numbers of each part-of-speech (POS) tag type as a feature. Since sentiment lexicons often record word polarity scores without specifying the word POS (except the MPQA subjectivity lexicon), it is difficult to compensate for the polarity scores that might be incorrectly considered as features. For example, consider the sentences "I like watching movies" and "It smells like popcorn". Sentiment lexicons might have a high positive score for the word 'like' but it does not apply for the second sentence. So we include a new lexico-syntactic binary feature <w>-POS(w) where 'w' is a sentiment expressing word found in sentiment lexicons and POS(w) is its part-of-speech as observed in an input tweet. For such words, based on the dependency parse [5] of a tweet, we also introduce another lexico-syntactic binary feature <w>-<g/d>-<dtype>, where 'dtype' is the type of a dependency relation involving 'w' and 'g/d' is determined based on whether the relation has 'w' as a governor (g) or dependent (d). For example running Stanford parser on "it smells like popcorn" generates dependencies `prep(smells, like)` and `pobj(like, popcorn)` which give the features `like-d-prep` and `like-g-pobj`. This is to capture the effects of syntactic relations involving sentiment words on the overall tweet sentiment.

Sentiment lexicon based features: A sentiment lexicon typically has a list of words with the corresponding polarity expressed simply as a binary positive or negative categorization. More recent lexicons, especially those curated automatically, assign numerical scores that indicate the polarity of the word where a positive (negative) value typically indicates a positive (negative) polarity and the

---

[2] We do not consider elongated versions for ngrams and shorten such tokens as in [1]. However, we look at the original tweet text to determine the number of such words.

magnitude of the value corresponds to strength of the sentiment. For our experiments we use the hand-built Bing Liu lexicon [9], MPQA subjectivity lexicon [32], and the NRC-Canada Emotion Lexicon [18]. For these lexicons, since numerical scores are not explicitly provided, we choose appropriate integer scores based on the polarity and any corresponding strength/intensity information available following the approach by Kiritchenko et al. [11]. We also use an automatically created sentiment lexicon, the Hashtag Sentiment Lexicon (HSLex), constructed by researchers at NRC-Canada using a dataset of tweets with a few hashtagged emotion words. We use their latest version [11] of this lexicon where they generate different scores for affirmative and negated contexts. Given these different lexicons the actual features are as follows:

1. For each lexicon used, the total score of all sentiment expressing ngrams that occur in the tweet, the total score of only positive (negative) ngrams, the maximum score among positive (negative) ngrams, score of the last token in the tweet, and all these scores computed separately for unigrams and bigrams. The scores for ngrams in a negated context (as identified in [22]) within a tweet are obtained from the negated context lexicon for the automatically created HSLex lexicon [11].
2. For each lexicon used, the total numbers of sentiment expressing ngrams, negation words, positive ngrams, negative ngrams, and all these counts computed separately for unigrams and bigrams. We also include the numbers of positive (negative) emoticons and also their presence (so binary) as the last token of the tweet.
3. For each lexicon used, similar to how scores and counts for ngrams are computed (items 1 and 2 in this list), we also incorporate as features, sentiment (aggregated) scores and counts for different parts of speech that occur in a tweet. This is based on the link between the unigrams in the tweet and the associated POS tags and the presence of such unigrams in the lexicons.

5   Experiments, Results, and Discussion

We split our dataset into 80% training and 20% test sets using stratified sampling with class proportions maintained according to the distribution in the full dataset. We ran 5-fold cross validation hundred times (using distinct shuffles) on the training dataset to identify the best feature combination among all features described in Section 4. The best combination chosen was the one that had the maximum average F-Sent score over those 100 iterations. Given the large number of features, for computational tractability of considering all possible combinations, we divided all features into ten distinct groups: 1. four groups from the lexicon based features (corresponding to the list at the end of Section 4 with separate score and count groups from the third item); 2. three groups from syntactic features (POS tag counts and the two new lexico-syntactic features as singleton groups); and 3. three groups from lexical features (ngrams, NBSVM weighting, and all Twitter specific features such as all-caps words and elongated words as one group). The best feature combination based on our experiments is the union of all features excluding the following features: POS tag counts, emoticon counts, lexical features such as elongated or all-caps words, and the lexico-syntactic feature that joins a sentiment word with its POS tag. Using this best feature combination, we used cross validation and grid search to identify the best regularization parameter and tolerance value for stopping criteria for the SVM classifier. We finally trained using the best feature combination and parameter settings and ran our model on the test set to obtain accuracy of 70.70% and F-Sent of 62.87%. However, since this is based on a single 80:20 split of our dataset, we repeated our experiments over hundred distinct 80:20 splits and obtain results shown in Table 1 which shows a mean F-Sent of 62.28 with a 95% confidence interval of 61.82–62.75%.

Table 1. Average accuracy (Acc.) and F-Sent over 100 distinct 80%-20% train-test splits when using all features, the best feature combination with feature group ablated performances

Features	Train Stats		Test Stats		
	Acc.	F-Sent	Acc.	F-Sent	95% CI F-Sent
Best Combination	71.56%	64.48%	70.47%	62.28%	61.82–62.75%
– Dependency Feature	71.50%	64.44%	70.45%	62.17%	61.75–62.60%
– NBSVM	69.64%	61.76%	70.13%	62.49%	62.02–62.95%
– Ngram+NBSVM	67.01%	58.60%	67.33%	58.99%	58.57–59.40%
– Lexicon Features	65.29%	53.23%	65.56%	53.43%	52.90–53.96%
All features	68.52%	60.11%	69.05%	60.87%	60.42–61.32%
NRC-Lite	68.52%	58.09%	68.57%	59.39%	58.97–59.82%

In Table 1, rows 2–5 indicate the performance if we remove feature classes from the best combination. The biggest drop in performance is obtained when lexicon features removed resulting in a 9% drop in F-Sent and 6% drop in accuracy. As noted by Kiritchenko et al. [11], the drop due to ngrams ablation is significantly less (row 4) compared to removing lexicon features. Dropping the NBSVM weighting causes 2.72% loss in F-Sent in training but shows a negligible increase in performance over the best combination test average. Although test accuracy drops when ablating NBSVM weighting, it is also negligible in contrast with the corresponding drop of nearly 2% in training. The drop in performance due to the dependency based feature is also not significant. These results for NBSVM weighting and dependency features could be due to the sparsity of token frequencies and (word, dependency type) pair frequencies, respectively, and need further investigation with a larger dataset. The penultimate row of Table 1 shows that identifying the best combination results in 1.41% improvement in test F-Sent score and over 3% improvement for the training F-Sent score. We also experimented with a system comparable to that used by NRC-Canada researchers [11] which we call NRC-Lite since we removed certain features, specifically, word cluster scores, ngrams that are longer than 2 tokens (given the sparsity), non-contiguous ngrams, and lexicon features from the Sentiment140 lexicon [11]. From the final row in Table 1, we notice that our best performer test F-Sent is 2.91% higher than that for NRC-Lite and the gains are over 6% for training F-Sent. However, many of our features are from [11] and identifying the best feature combination might have been the key in obtaining higher performance.

Although most current approaches directly model sentiment analysis into a direct three way classification problem, there is a more intuitive two stage approach where a binary classifier first distinguishes tweets that carry sentiment from objective/neutral tweets. For such tweets identified through the first stage model, a second binary classifier identifies whether the tweet expresses positive or negative sentiment. We curated separate subsets of our dataset for the corresponding classifiers in these two stages and experimentally identified the best feature combination for both types of classifiers in exactly the same way as we did for the direct three way classification. Finally, over hundred different 80-20% train-test splits of our data, we obtained a mean test F-sent of 60.25% (59.79–60.71%) and mean accuracy of 69.93% (69.61–70.25%) with 95% confidence intervals show in parentheses. Compared with our best results (first row of Table 1), we notice a drop of 2% in F-Sent and 1% in accuracy. This has been our experience with the two-stage approach even in other text classification domains that have hierarchical class structures. Given neutral tweets cause the most errors (more later in Section 6), we believe that the first stage classifier propagates errors to the second stage to an extent that limits the overall performance of the approach.

Table 2. Average performance measures with our best combination based on 5-fold cross-validation using 100 distinct shuffles of the original Agarwal et al. [1] dataset with equal class proportions

Measures	Agarwal et al.	Our Best Combination	
		Mean	95% CI
Accuracy	60.50%	62.85%	62.78–62.92%
F-Sent	60.23%	64.68%	64.61–64.76%
F_+	59.41%	64.07%	63.98–64.16%
F_-	61.04%	65.30%	65.21–65.39%
$F_{neutral}$	60.15%	59.74%	59.65–59.82%

We conducted additional experiments to see how our best feature combination performs on the original general tweet dataset by Agarwal et al. [1] with equal proportions for the three classes (1709 tweets per class). We followed their approach of five-fold cross validation and obtained results as shown in Table 2, which indicates an improvement of over 4% in F-Sent and 2% in accuracy. Since our features are geared towards identifying tweets with polarity, we notice a significant increase in F_+ and F_- and a negligible drop in $F_{neutral}$.

6   Qualitative Error Analysis

In Table 3, we display the test and training error confusion matrices where the rows represent ground truth and columns are predicted classes. A glance at them shows that in both scenarios most errors involve neutral tweets. To be precise, 86% of test errors and 89% of training errors are caused due to neutral tweets. This is the main motivation for our effort and this observation strongly backs our belief that datasets with a realistic distribution of the three classes should be used without pre-screening bias.

Table 3. Confusion Matrices for Training and Test Datasets

T\P	+	-	N	T\P	+	-	N
+	142	13	47	+	542	38	230
-	16	71	46	-	45	304	186
N	48	36	284	N	166	132	1177

(a) Test Matrix	(b) Training Matrix

Given this situation, we manually analyzed a few misclassified tweets. Since we did not impose a constraint on the length of the tweets, we found several examples of short tweets that have been misclassified. Consider the negative tweet, I'm not fine, misclassified as a positive tweet. The main clue is the negation word followed by the word 'fine'. However, in the automatically created HSLex [11], we find a score of 0.832 for fine_NEG. While this might not be the main reason, there is not much additional information to rely on for the learner for such short tweets. We believe a specific customization of the features and classification framework for short tweets might be essential. Although our best combination did not include elongated words and other tweet specific lexical features, based on our manual analysis, we believe these features might play a crucial role for shorter tweets. Based on our observation in Figure 1, since 30% of tweets are short tweets with fewer than 6 tokens, we believe this to be an interesting research direction. However, our initial experiments on building two separate classifiers for short (≤ 5 tokens) and long (> 5 tokens) tweets did not result in overall performance improvements, potentially due to the very small size of the training dataset for short tweets. However, we noticed that the percentage of neutral tweets increases by 10% in short tweets compared to the full dataset. This further confirms that a more involved customization for short tweets is essential.

Consider this positive tweet misclassified as a neutral tweet: @jenna_bandi ohhhh my lorddddd your a lifesaver. The bigram "ohhh my" has a positive score of 1.64 but "your a" has a negative score of -1.05 in HSLex. Due to a missing space between 'life' and 'saver', we missed important evidence given both words have positive scores. Splitting up potential bigrams into constituent unigrams (in addition to retaining the original token) might provide more evidence toward the correct sentiment of the tweet. Tweets that start out positive (negative) but end up conveying a more negative (positive) sentiment latter might need special handling. Consider this negative tweet misclassified as neutral: `@vixxybabyy hopefully a GED !!! But even that might not happen for this one`. Researchers have had success [30] by simply splitting the tweet in the middle (in terms of word count) and treating tokens in both halves as having a separate feature type. We will employ this approach in our future work.

We end with an example of a neutral tweet misclassified as a positive tweet: `This ice is supppppa cold, but then again it is ice`. This tweet showcases the complexity of identifying neutral tweets. Even if the slang word is correctly identified as 'super', the overall sentiment might still be positive given 'super' and 'super cold' both have positive sentiment scores in HSLex.

7   Conclusion

Most current efforts in sentiment analysis of Twitter data are focused on datasets that are biased toward tweets that contain popular named entities and sentiment expressing words. While there is merit to this focus, it is also important to consider general tweets most of which (90% according to our estimate) do not contain named entities and are essentially conversational chatter about tweeters' daily activities and personal situations or their general mood. Sentiment analysis of such tweets can help study the sentiment expressed via Twitter by different groups of tweeters based on demographics (age, race/ethnicity, gender, location) and additional attributes (e.g., smokers, vapers) across time and to correlate this information [16] with additional locational information (county health rankings, urban/rural indices). To our knowledge, there is only one earlier effort that looks at general tweets by Agarwal et al. [1], although the authors assume equal proportion of positive, negative, and neutral tweets.

In this paper, we first estimated the proportion of the three classes using manual annotation of a random sample of 1000 tweets selected from over 20 million tweets collected in 2013. Our analysis showed that class proportions are skewed and that more than half of the tweets are neutral justifying additional efforts for general tweet sentiment analysis. Based on the estimated proportion, we constructed a new dataset and conducted experiments using well known features, including those derived from sentiment lexicons. We also introduced additional lexico-syntactic features based on part-of-speech tags and dependency parses for sentiment expressing words. Unlike prior efforts, we identified best feature combinations based on repeated cross validation experiments on different shuffles of the training data. We demonstrated that our best feature combination provides statistically significant performance improvements over using all features as indicated by non-overlapping 95% confidence intervals (last column of the first and penultimate rows in Table 1), which justifies our approach of identifying feature subsets instead of simply using a large set of features. Our feature ablation experiments demonstrated that the lexicon based features contribute the most to the performance of our models, corroborating the findings of other researchers [11]. Additionally, we also showed that models based on our best feature combination outperform prior approaches on the original equal proportioned dataset of general tweets by Agarwal et al. [1]. At the time of this writing, our current effort is the first to study the distribution of the sentiment classes in general tweets and the associated sentiment analysis of such tweets based on a dataset constructed according to the estimated class proportions.

We conducted qualitative error analysis of our results and identified important future research directions. Besides improvements identified in Section 6, we believe that a larger dataset might be more suitable for further research in assessing the sentiment of general tweets from the Twitter stream, especially in building separate classifiers for short and long tweets. Given the presence of a large number of neutral tweets, it might also be more desirable to employ more than two annotators through an online crowd sourcing approach.

Acknowledgments. Many thanks to anonymous reviewers for their detailed comments that greatly helped improve the quality of this paper. We are also grateful to Apoorv Agarwal for providing the dataset used in his prior efforts. The project described in this paper was supported by the National Center for Advancing Translational Sciences (UL1TR000117). The content is solely the responsibility of the authors and does not necessarily represent the official views of the NIH.

References

1. Agarwal, A., Xie, B., Vovsha, I., Rambow, O., Passonneau, R.: Sentiment analysis of twitter data. In: Proceedings of the Workshop on Languages in Social Media, pp. 30–38. Association for Computational Linguistics (2011)
2. Baccianella, S., Esuli, A., Sebastiani, F.: SentiWordNet 3.0: An enhanced lexical resource for sentiment analysis and opinion mining. In: LREC, vol. 10, pp. 2200–2204 (2010)
3. Bird, S., Klein, E., Loper, E.: Natural Language Processing with Python. O'Reilly Media (2009)
4. Chen, L., Wang, W., Nagarajan, M., Wang, S., Sheth, A.P.: Extracting diverse sentiment expressions with target-dependent polarity from twitter. In: Proceedings of the Sixth International Conference on Weblogs and Social Media, ICWSM, pp. 50–57 (2012)
5. de Marneffe, M., MacCartney, B., Manning, C.: Generating typed dependency parses from phrase structure parses. In: Proceedings of the Fifth International Conference on Language Resources and Evaluation (LREC), pp. 449–454 (2006)
6. Eichstaedt, J.C., Schwartz, H.A., Kern, M.L., Park, G., Labarthe, D.R., Merchant, R.M., Jha, S., Agrawal, M., Dziurzynski, L.A., Sap, M., et al.: Psychological language on twitter predicts county-level heart disease mortality. Psychological Science **26**(2), 159–169 (2015)
7. Fan, R.-E., Chang, K.-W., Hsieh, C.-J., Wang, X.-R., Lin, C.-J.: Liblinear: A library for large linear classification. Journal of Machine Learning Research **9**, 1871–1874 (2008)
8. Go, A., Bhayani, R., Huang, L.: Twitter sentiment classification using distant supervision. Technical report, Dept. of Computer Science, Stanford Univ. (2009)

9. Hu, M., Liu, B.: Mining and summarizing customer reviews. In: Proceedings of the Tenth ACM SIGKDD International Conference on Knowledge Discovery and Data Mining, pp. 168–177. ACM (2004)

10. Jiang, L., Yu, M., Zhou, M., Liu, X., Zhao, T.: Target-dependent twitter sentiment classification. In: Proceedings of the 49th Annual Meeting of the Association for Computational Linguistics: Human Language Technologies, vol. 1, pp. 151–160. Association for Computational Linguistics (2011)

11. Kiritchenko, S., Zhu, X., Mohammad, S.M.: Sentiment analysis of short informal texts. Journal of Artificial Intelligence Research, 723–762 (2014)

12. Landis, J., Koch, G.: The measurement of observer agreement for categorical data. Biometrics **33**(1), 159–174 (1977)

13. Liu, W., Ruths, D.: What's in a name? using first names as features for gender inference in twitter. In: Proceedings of the AAAI Spring Symposium: Analyzing Microtext, pp. 10–16 (2013)

14. Liu, Y., Kliman-Silver, C., Mislove, A.: The tweets they are a-changin': Evolution of twitter users and behavior. In: Proceedings of the Eighth AAAI International Conference on Weblogs and Social Media (ICWSM) (2014)

15. Martínez-Cámara, E., Martín-Valdivia, M.T., Ureña-López, L.A., Montejo-Ráez, A.R.: Sentiment analysis in twitter. Natural Language Engineering **20**(01), 1–28 (2014)

16. Mitchell, L., Frank, M.R., Harris, K.D., Dodds, P.S., Danforth, C.M.: The geography of happiness: Connecting twitter sentiment and expression, demographics, and objective characteristics of place. PloS One **8**(5), e64417 (2013)

17. Mohammad, S.M., Kiritchenko, S., Zhu, X.: NRC-Canada: Building the state-of-the-art in sentiment analysis of tweets. In: Proceedings of the Annual SemEval Workshop, pp. 321–327 (2013)

18. Mohammad, S.M, Turney, P.D.: Emotions evoked by common words and phrases: Using mechanical turk to create an emotion lexicon. In: Proceedings of the NAACL HLT 2010 Workshop on Computational Approaches to Analysis and Generation of Emotion in Text, pp. 26–34. Association for Computational Linguistics (2010)

19. Nakov, P., Kozareva, Z., Ritter, A., Rosenthal, S., Stoyanov, V., Wilson, T.: Semeval-2013 task 2: Sentiment analysis in twitter. In: Proc. SemEval (2013)

20. Nguyen, D., Gravel, R., Trieschnigg, D., Meder, T.: How old do you think i am? a study of language and age in twitter. In: Proceedings of the Seventh International AAAI Conference on Weblogs and Social Media (ICWSM), pp. 439–448 (2013)

21. Pang, B., Lee, L.: A sentimental education: Sentiment analysis using subjectivity summarization based on minimum cuts. In: Proceedings of the 42nd Annual Meeting on Association for Computational Linguistics, p. 271. Association for Computational Linguistics (2004)

22. Pang, B., Lee, L., Vaithyanathan, S.: Thumbs up?: sentiment classification using machine learning techniques. In: Proceedings of the ACL 2002 Conference on Empirical Methods in Natural Language Processing, vol. 10, pp. 79–86. Association for Computational Linguistics (2002)

23. Pedregosa, F., Varoquaux, G., Gramfort, A., Michel, V., Thirion, B., Grisel, O., Blondel, M., Prettenhofer, P., Weiss, R., Dubourg, V., Vanderplas, J., Passos, A., Cournapeau, D., Brucher, M., Perrot, M., Duchesnay, E.: Scikit-learn: Machine learning in Python. Journal of Machine Learning Research **12**, 2825–2830 (2011)

24. Pennacchiotti, M., Popescu, A.-M.: A machine learning approach to twitter user classification. In: Proceedings of the Fifth International AAAI Conference on Weblogs and Social Media (ICWSM), pp. 281–288 (2011)

25. Pestian, J.P., Matykiewicz, P., Linn-Gust, M., South, B., Uzuner, O., Wiebe, J., Cohen, K.B., Hurdle, J., Brew, C.: Sentiment analysis of suicide notes: A shared task. Biomedical Informatics Insights **5**(suppl. 1), 3 (2012)
26. Rosenthal, S., Nakov, P., Ritter, A., Stoyanov, V.: Semeval-2014 task 9: Sentiment analysis in twitter. In: Proc. SemEval (2014)
27. Turney, P.D.: Thumbs up or thumbs down?: semantic orientation applied to unsupervised classification of reviews. In: Proceedings of the 40th Annual Meeting on Association for Computational Linguistics, pp. 417–424. Association for Computational Linguistics (2002)
28. Twitter, Inc., Registration with United States securities and exchanges commission (2013). http://www.sec.gov/Archives/edgar/data/1418091/000119312513390321/d564001ds1.htm
29. Wang, S., Manning, C.D.: Baselines and bigrams: Simple, good sentiment and topic classification. In: Proceedings of the 50th Annual Meeting of the Association for Computational Linguistics: Short Papers, vo. 2, pp. 90–94. Association for Computational Linguistics (2012)
30. Wang, W., Chen, L., Thirunarayan, K., Sheth, A.P.: Harnessing Twitter "big data" for automatic emotion identification. In: 2012 International Conference on Social Computing (SocialCom), pp. 587–592. IEEE (2012)
31. Wilson, E.B.: Probable inference, the law of succession, and statistical inference. Journal of the American Statistical Association **22**(158), 209–212 (1927)
32. Wilson, T., Wiebe, J., Hoffmann, P.: Recognizing contextual polarity in phrase-level sentiment analysis. In: Proceedings of the Conference on Human Language Technology and Empirical Methods in Natural Language Processing, pp. 347–354. Association for Computational Linguistics (2005)
33. Yu, N., Kubler, S.: Semi-supervised learning for opinion detection. In: 2010 IEEE/WIC/ACM International Conference on Web Intelligence and Intelligent Agent Technology (WI-IAT), vol. 3, pp. 249–252. IEEE (2010)

Real-Time Sentiment-Based Anomaly Detection in Twitter Data Streams

Khantil Patel$^{(\boxtimes)}$, Orland Hoeber, and Howard J. Hamilton

Department of Computer Science, University of Regina, Regina, Canada
{patel26k,orland.hoeber,howard.hamilton}@uregina.ca

Abstract. We propose an approach for real-time sentiment-based anomaly detection (RSAD) in Twitter data streams. Sentiment classification is used to split the data into independent streams (positive, neutral, and negative), which are then analyzed for anomalous spikes in the number of tweets. Four approaches for evaluating the data streams are studied, along with the parameters that adjust their sensitivity. Results from an evaluation show the effectiveness of a probabilistic exponentially weighted moving average (PEWMA) coupled with a sliding window that uses median absolute deviation (MAD).

Keywords: Data stream mining · Anomaly detection · Twitter · Social media analytics · Sentiment classification · Real-time processing

1 Introduction

Time-series data streams have become a popular way of characterizing the data generated by real-time applications with a temporal attribute. Since such data can introduce new patterns very quickly, *data stream mining* has drawn interest from many researchers, with a focus on developing anomaly detection techniques that are both computationally efficient and memory efficient [4,5,10]. Anomaly detection in time-series data streams is challenging in three aspects [11]: (1) the dynamic nature of the data streams may result in changes in the data distributions over time (called *concept drift*); (2) storing the data for further analysis is not feasible given the high-velocity and unbounded length of the data; and (3) the analysis must happen sufficiently quickly to be able to operate in real-time.

Twitter has become a popular micro-blogging platform where millions of users express their opinions on a wide range of topics on a daily basis via *tweets*, producing large amounts of data every second that can be modelled as time-series data streams and analyzed for anomalies. Twitter allows real-time collection of streams of tweets related to any specified topic keywords, hash tags, or user names through their *public streams* service [13]. This easy access to the data has enabled researchers to study and propose a broad range of techniques, including visual analytics [6,8], sentiment analysis [2,12], and anomaly detection [5].

The work in this paper is motivated by the challenge of providing users with timely information about different opinions relevant to topics of interest without

© Springer International Publishing Switzerland 2015
D. Barbosa and E. Milios (Eds.): Canadian AI 2015, LNAI 9091, pp. 196–203, 2015.
DOI: 10.1007/978-3-319-18356-5_17

requiring continual observation. In order to derive public opinions, tweets can be subjected to sentiment classification, resulting in a labelling of individual tweets as positive, neutral, or negative [2]. A visual analytics based approach has been used in our prior work to discover and analyze the temporally changing sentiment of tweets posted by fans in response to micro-events occurring during a multi-day sporting event [6]. However, with this approach, in order to discover emerging micro-events that are causing significant increases in positive, neutral, or negative tweets, one must analyze data continuously. The goal of the research described in this paper is to detect, in real-time, anomalies in Twitter sentiment data streams, thereby providing alerts concerning the change to the analysts and enabling them to conduct further analysis immediately.

Real-time sentiment-based anomaly detection (RSAD) starts by classifying the tweets and aggregating them in temporal bins of a fixed interval (e.g., 15 minutes). Candidate anomalies are detected based on their deviation from the distribution of recent data; these are then compared to other previously seen anomalies within a sliding window to identify legitimate anomalies. This approach is resilient in the face of concept drift, makes use of an incrementally updatable model, and is efficient enough to handle high-velocity data streams.

2 Methodology

We consider a data point to be an *anomaly* if it deviates sufficiently from nearby data points or a specified group of data points in the past. We define a *candidate anomaly* to be a data point that deviates from the most recent data points. Moreover, if this candidate anomaly deviates from the group of other previously detected candidate anomalies in some limited timeframe, we consider it a *legitimate anomaly*. In the remainder of this section, we explain the approach used in RSAD to detect these types of anomalies.

2.1 Pre-processing

A unique feature of RSAD is the detection of anomalies within pre-classified data streams. The rationale for this is to allow for the independent detection of anomalous increases in tweets that are *positive, neutral,* or *negative* in nature. From the perspective of anomaly detection, we can consider the classification process as a pre-processing step. We use an online sentiment analysis service called Sentiment140 [12], which was designed specifically to address the short and cryptic nature of English language tweets.

In order to turn the streams of tweets into time-series data, we aggregate them over a pre-determined interval of time (e.g., 15 minutes). The granularity of this binning affects the sensitivity to small-scale vs. large-scale anomalies and can be set based on an expectation of the velocity patterns of the tweets for the given query. Since the goal is to analyze these data streams based only on the tweet frequency, once the classification and temporal binning are performed, the actual contents of the tweets are discarded. All that remains is the number of positive, neutral, and negative tweets that were seen in each time period. These frequency counts serve as the data points for the anomaly detection stage.

2.2   Candidate Anomaly Detection

To detect the candidate anomalies from among the local context of data points, we consider a deviation-based approach using two possible methods for determining the average of the previously seen data points: exponentially weighted moving average (EWMA) [9] and probabilistic exponentially weighted moving average (PEWMA) [3]. Although each of these approaches have been used to detect outliers in streaming data in two separate contexts [3,8], it is not clear which is more appropriate for the RDAS approach and Twitter data.

An anomaly score of a data point d_t is calculated to represent its deviation from the mean of the data points in its neighbourhood. In the streaming context, the *local neighbours* of an newly arrived data point are the ones that recently arrived because the following data points have not yet been received. The candidate anomaly score (CAS) is evaluated using following formula:

$$CAS(d_t) = \frac{|d_t - \mu_{c(t-1)}|}{\mu_{c(t-1)}} \tag{1}$$

where t is the time of current bin, and $\mu_{c(t-1)}$ is an exponentially weighted moving average of previous data points, as explained shortly. CAS was adapted from the A-ODDS technique [10], in which the neighbourhood density of each data point is determined using a probability density estimator and then the anomaly score of a data point is computed in terms of the relative distance between its neighbourhood density and the average neighbourhood density of recent data points. In the A-ODDS approach, the neighbourhood consists of a set of data points up to radius r on both sides of the data point. In our work, CAS is the relative distance of d_t to a weighted average of previous points.

If the CAS of the current data point is near zero; the point is close to the other data points. If the CAS of the current data point is a large value, then it is significantly larger or smaller than the other data points. In order to label the current data point as a candidate anomaly, the CAS should be larger than the standard deviation of the previously seen data points by some factor. The threshold condition for a data point d_t to be so labeled is given as:

$$CAS(d_t) > \tau_c * \sigma_{c(t-1)} \tag{2}$$

where $\sigma_{c(t-1)}$ is the standard deviation of the data points, and τ_c is a threshold factor for candidate anomalies. τ_c can be set by a domain expert according to the particular features of the data stream. A lower value of τ_c increases sensitively to clustered anomalies; a higher one increases sensitivity to dispersed anomalies.

With each new data point, it is necessary to update $\mu_{c(t)}$ and $\sigma_{c(t)}$. A naïve approach would be to maintain all the data points in the N previous steps, and use the standard formulation to evaluate $\mu_{c(t)}$ and $\sigma_{c(t)}$. However, with a weighted average, the oldest data point contributes too little to the measures to justify keeping N data points at all times. Another approach is to use an exponentially weighted moving average (EWMA) [9] and calculate $\mu_{c(t)}$ and $\sigma_{c(t)}$ incrementally for a new data point d_t according to the equations:

$$\mu_{c(t)} = \alpha_{EWMA} * \mu_{c(t-1)} + (1 - \alpha_{EWMA}) * d_t \qquad (3)$$

$$\sigma_{c(t)} = \alpha_{EWMA} * \sigma_{c(t-1)} + (1 - \alpha_{EWMA}) * |d_t - \mu_{c(t-1)}| \qquad (4)$$

Here $0 < \alpha_{EWMA} < 1$ is the decay weighting factor. The α_{EWMA} parameter controls the weight distribution between the new data point d_t and the old mean $\mu_{c(t-1)}$; a value of 0 implies no weight on the history, while a value of 1 implies all weight on the history. An inherent assumption with the EWMA approach is that the mean is changing at most gradually with respect to the exponential weighting parameter α_{EWMA}. Thus, a significant change in d_t will result in a significant change in $\mu_{c(t)}$ and a large increase in $\sigma_{c(t)}$.

To increase resiliency against such changes in d_t, the value of weighting parameter α_{EWMA} can be dynamically adjusted. More precisely, if d_t changes to a greater degree with respect to recent data, then a higher weight (α_{EWMA} close to 1) should be given to the recent data points; otherwise more weight should be given to d_t. Probabilistic EWMA (PEWMA) [3] adjusts the weighting parameter based on the probability of the occurrence of the value of the current data point. The probabilistic weighting parameter is given as $\alpha_{PEWMA} = \alpha_{EWMA}(1 - \beta P_t)$, where P_t is the probability of occurrence of d_t and β is the weight placed on P_t. The parameter α_{EWMA} is multiplied by $(1 - \beta P_t)$ to reduce the influence of an abrupt change in d_t on the moving average.

The probability density estimator equation with the standard normal distribution for P_t is given as $P_t = \frac{1}{\sqrt{2\pi}} \exp\left(-Z_t^2/2\right)$.

While evaluating P_t for the current data point d_t, it may happen that $P_t \to 0$, if $\sigma_{c(t-1)} \to \infty$. To avoid such situations, normalization is applied to the input data points to obtain a zero-mean and unit standard deviation random variable $Z_t = (d_t - \mu_c)/\sigma_c$. The factor $\frac{1}{\sqrt{2\pi}}$ is the constant height and it is selected to normalize P_t such that $0 < P_t < \frac{1}{\sqrt{2\pi}}$. The drawback of considering the standard normal distribution is that for larger value of d_t, $P_t \to 0$. However, our approach does not require that the deviation of d_t be large, as long as it is sufficiently deviated from the underlying data distribution. By adjusting equation 3, with the probabilistic weighting factor [3], we get:

$$\mu_{c(t)} = \alpha_{PEWMA} * \mu_{c(t-1)} + (1 - \alpha_{PEWMA}) * d_t \qquad (5)$$

2.3   Legitimate Anomaly Detection

To detect whether a candidate anomaly should be considered a legitimate anomaly, we use a one-sided sliding window of length W (e.g., 6 days). In contrast to the conventional method of keeping all past data points in the sliding window, we keep only the data points identified as candidate anomalies. We consider a window-based deviation approach using two possible methods for determining the deviation of the data points in the window: standard deviation (STD), based on the simple arithmetic mean, and median absolute deviation (MAD), based on the median. While each approach has been used to detect outliers in static time series data, it is not clear which is better for a sliding window containing only candidate anomalies.

To determine whether a candidate anomaly should be considered as legitimate, the legitimate anomaly score (LAS) is calculated. The LAS of a data point represents its deviation from the mean of the candidate anomalies in the window. For the data point d_t, LAS is computed using the following equation:

$$LAS(d_t) = \frac{|d_t - \mu_{w(t-1)}|}{\mu_{w(t-1)}} \tag{6}$$

where $\mu_{w(t-1)}$ is the (unweighted) mean of the candidate anomalies in the window up to and including time $t-1$. LAS gives the relative distance of d_t to this mean.

The significance of LAS is similar to that of CAS in equation 1. Thus, in order to label a data point d_t as a legitimate anomaly, LAS should be sufficiently large. The cutoff condition is given as:

$$LAS(d_t) > \tau_l * \sigma_{w(t-1)} \tag{7}$$

where $\sigma_{w(t-1)}$ is the standard deviation (STD) estimated from all candidate anomalies in the window up to and including time $t-1$ and τ_l is a threshold factor for legitimate anomalies.

Each time another candidate anomaly is encountered, we update the mean $\mu_{w(t)}$ and standard deviation $\sigma_{w(t)}$ with respect to the sliding window. Since only candidate anomalies are kept in the window, the number of data points is relatively small. In such cases, the standard deviation measure is strongly affected by presence of extreme values [7]. As a result, statistical measures that are robust against extreme anomalies, such as median and median absolute deviation (MAD), are recommended [7]. The median $\hat{\mu}_{w(t)}$ is the median of all previously detected candidate anomalies in the window. The median absolute deviation (MAD) $\hat{\sigma}_{w(t)}$ is the median of the absolute deviations from the median of the candidate anomalies.

3   Preliminary Experimental Evaluation

In order to evaluate and compare the different alternatives for identifying candidate and legitimate anomalies, we performed anomaly detection experiments using Twitter data streams collected during 2013 Le Tour de France cycling races [6]. This event was held from June 29 - July 21, 2013, and is the premier race in professional cycling. The data set contains 449,077 English-language tweets retrieved from the Twitter public stream that were posted using the official hash tag ("#tdf") during the event period. Since the event is no longer live, for the purposes of this evaluation, we simulated an artificial data stream using these tweets. Given the features of this data set, the aggregation period was set to 15 minutes, and the sliding window length was set to 6 days.

The combination of the two models that can be applied to detect candidate anomalies (EWMA and PEWMA) and the two models that summarize the statistical properties of the sliding window (STD and MAD) result in four approaches to be evaluated. We named the approaches EWMA-STD, EWMA-MAD, PEWMA-STD, and PEWMA-MAD. The threshold parameters for the candidate and legitimate anomaly detection steps (τ_c and τ_l, respectively) were independently manipulated in the range $[1, 5]$. The decay factors for EWMA and PEWMA were fixed at $\alpha_{EWMA} = 0.97$ and $\alpha_{PEWMA} = 0.99$ respectively, which are optimal minimum mean square error parameters in many settings [3].

For the experiments, we leveraged an open source, real-time distributed stream processing framework, called *Apache Storm* [1]. The four approaches were implemented in the Storm framework independently and then evaluated with the input of the simulated data stream. In the absence of classification labels indicating known anomalies in the tweets data stream, we worked with domain experts to assess the false positives and false negatives identified in the data. For each experimental setting, precision and recall were calculated, along with the F-score. Furthermore, since our goal was to discover an approach that works well across all three sentiment classes, we averaged the F-score over the positive, neutral, and negative data streams for each experimental setting.

3.1 Results and Analysis

The results of the experiments in the manipulation of the threshold parameters τ_c and τ_l are provided in Figure 1. First, observe that the STD approach (Figures 1a and 1c) is sensitive to the value of τ_l. As this parameter was increased, the F-score decreased because the method for determining if a candidate anomaly is considered a legitimate anomaly became more strict. Although this increase resulted in high precision (those that met this criteria were clearly anomalies), the recall was adversely affected, with many actual anomalies not being detected. This pattern held for both the EWMA and PEWMA approaches.

For the EWMA-MAD approach, a similar pattern of the F-score decreasing as τ_l increases holds (Figure 1b). Furthermore, as τ_c increases, there is also a general pattern of the F-score decreasing. Extreme anomalies have a significant impact on the mean value, and make it difficult to discover additional anomalies in the local context when the threshold value is high. The result was high precision and low recall. The PEWMA-MAD approach (Figure 1d) was more resilient to the settings of the parameters than the other three approaches. We hypothesize that PEWMA is more effective than EWMA at calculating the average value in the local context, which caused the candidate anomalies in the sliding window to also be more representative of the true anomalies, and MAD is more effective than STD at choosing the legitimate anomalies. The highest F-score achieved across all 100 experimental settings was 0.80 (PEWMA-MAD, $\tau_c = 4$, $\tau_l = 4$). While the other methods approached this value for certain settings, given the

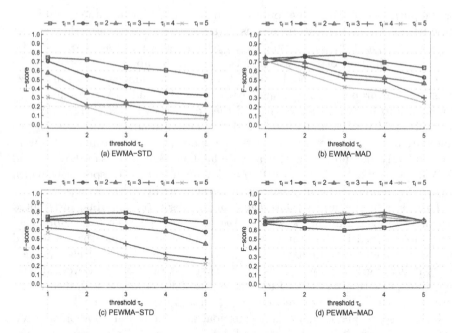

Fig. 1. Candidate anomalies identified as legitimate using STD and MAD, estimated with mean and the median respectively. (Tweets aggregated at 15 minutes interval)

resilience of PEWMA-MAD to the threshold parameters, we conclude that it is the superior approach for this data set.

3.2   Real-Time Performance

In terms of computational complexity, our candidate anomaly detection method is linear in the number of data points (bins of data) due to the incremental nature of calculating the EWMA and PEWMA. Our approach for deciding whether a candidate anomaly is a legitimate anomaly loops over all of the candidate anomalies within the current window. As such, this step has a complexity of $O(n)$, where n is the maximum number of potential candidate anomalies. Given a window size of 6 days and an aggregation interval of 15 minutes, the worst-case value for n is 576. Clearly, with these settings, the approach can be considered to run in real-time. Even with an extremely high velocity data stream, as long as the aggregation interval is kept in the minute-range, the approach will be able to keep up on a sufficiently fast computer system.

4   Conclusion and Future Work

In this paper we highlighted the problem of real-time detection of changes in the sentiment in Twitter data steams. We proposed the RSAD approach for efficiently detecting anomalies in presence of temporal drift. In the experimental evaluation of the four approaches for detecting anomalies in the 2013 Le Tour de France data set, we found that the PEWMA-MAD approach was accurate and resilient to the settings of threshold parameters. Future work will evaluate the RSAD approach over multiple datasets and compare it to other approaches from the literature. Furthermore, we wish to expand this approach to identify cyclical patterns in the data, in order to exclude them from being detected as anomalies.

References

1. Apache Storm. https://storm.apache.org/ (accessed January 1, 2015)
2. Bifet, A., Frank, E.: Sentiment knowledge discovery in twitter streaming data. In: Pfahringer, B., Holmes, G., Hoffmann, A. (eds.) DS 2010. LNCS, vol. 6332, pp. 1–15. Springer, Heidelberg (2010)
3. Carter, K.M., Streilein, W.W.: Probabilistic reasoning for streaming anomaly detection. In: Proceedings of the Statistical Signal Processing Workshop (SSP), pp. 377–380. IEEE (2012)
4. Gupta, M., Gao, J., Aggarwal, C.C., Han, J.: Outlier detection for temporal data: A survey. IEEE Transactions on Knowledge and Data Engineering **26**(9), 2250–2267 (2014)
5. Guzman, J., Poblete, B.: On-line relevant anomaly detection in the twitter stream: an efficient bursty keyword detection model. In: Proceedings of the ACM SIGKDD Workshop on Outlier Detection and Description, pp. 31–39. ACM (2013)
6. Hoeber, O., Hoeber, L., Wood, L., Snelgrove, R., Hugel, I., Wagner, D.: Visual twitter analytics: exploring fan and organizer sentiment during Le Tour de France. In: Proceedings of the VIS Workshop on Sports Data Visualization, pp. 1–7. IEEE (2013)
7. Leys, C., Ley, C., Klein, O., Bernard, P., Licata, L.: Detecting outliers: Do not use standard deviation around the mean, use absolute deviation around the median. Journal of Experimental Social Psychology **49**(4), 764–766 (2013)
8. Marcus, A., Bernstein, M.S., Badar, O., Karger, D.R., Madden, S., Miller, R.C.: Twitinfo: aggregating and visualizing microblogs for event exploration. In: Proceedings of the SIGCHI Conference on Human Factors in Computing Systems, pp. 227–236. ACM (2011)
9. Münz, G., Carle, G.: Application of forecasting techniques and control charts for traffic anomaly detection. In: Proceedings of the ITC Specialist Seminar on Network Usage and Traffic. Logos Verlag (2008)
10. Sadik, S., Gruenwald, L.: Online outlier detection for data streams. In: Proceedings of the Symposium on International Database Engineering & Applications, pp. 88–96. ACM (2011)
11. Sadik, S., Gruenwald, L.: Research issues in outlier detection for data streams. SIGKDD Explor. Newsl. **15**(1), 33–40 (2014)
12. Sentiment140. http://www.sentiment140.com/ (accessed December 10, 2014)
13. Twitter Public Streams. https://dev.twitter.com/streaming/public (accessed December 10, 2014)

Sentiment and Factual Transitions in Online Medical Forums

Victoria Bobicev[1], Marina Sokolova[2,3]([✉]), and Michael Oakes[4]

[1] Technical University of Moldova, Chisinau, Moldova
vika@rol.md
[2] Institute for Big Data Analytics, Halifax, Canada
[3] Faculty of Medicine and Faculty of Engineering, University of Ottawa,
Ottawa, Canada
sokolova@uottawa.ca
[4] Research Group in Computational Linguistics, University of Wolverhampton,
Wolverhampton, UK
Michael.Oakes@wlv.ac.uk

Abstract. This work studies sentiment and factual transitions on an online medical forum where users correspond in English. We work with discussions dedicated to reproductive technologies, an emotionally-charged issue. In several learning problems, we demonstrate that multi-class sentiment classification significantly improves when messages are represented by affective terms combined with sentiment and factual transition information (paired t-test, P=0.0011).

1 Introduction

Online public forums discuss personal experience and often convey the sentiments and emotions of the forum participants. Personal sentiments expressed in the posted messages[1] set interaction patterns among the members of online communities and have a strong influence on the public mood [5,15]. Several studies found that shared online emotions can improve personal well-being and empower patients in their battle against an illness [6,12].

Sentiment transition has become a popular topic in sentiment analysis following maturity of online communities and availability of relevant data. However, most of such studies work with only positive and negative polarity and analyze propagation of positive and negative sentiments [9,12]. In this study, conducted on forums where users communicate in English, we work with the sentiment categories of *encouragement, gratitude, confusion* and factual categories of *facts, endorsement* and analyze how these categories change in consecutive messages; for example:

We thank anonymous reviewers for thorough and helpful comments. This work has been supported by NSERC Discovery grant.

[1] Terms "messages" and "posts" are equivalent in this work.

© Springer International Publishing Switzerland 2015
D. Barbosa and E. Milios (Eds.): Canadian AI 2015, LNAI 9091, pp. 204–211, 2015.
DOI: 10.1007/978-3-319-18356-5_18

im spotting more brown today... its darker brown and more of it. I'm terrified and so sad. Had to work all day and I was just fighting back the tears all day. I don't know how to do this anymore and put on that fake smile every day and act like I'm just a happy person when my heart is breaking on the inside. **confusion**

Have you tested yet? I know a lot of people get this brown spotting, and even red, and are still pregnant. It's so hard going through this, I hope it works out for the best for you. **encouragement**

The paper is organized as follows: in section 2 we describe related work, followed by data analysis in section 3, then we present our sentiment and factual transitions in section 4, multi-class classification results in section 5, and our discussion and future work in section 6.

2   Related Work

Emergence of infoveillance and infodemiology stipulated analysis of health-related affects and emotions displayed on Twitter [3,4,11] and public forums dedicated to health [8,12].

Studies of sentiment propagation in social networks is an emerging area of sentiment analysis [1,5,12]. It has been shown that accuracy of sentiment classification in consecutive messages strongly depends on the topic of debates and can vary from 53% to 69%, with 60% accuracy obtained on health care discussions [1]. Results obtained for three online communities have shown that the original polarity is preserved in the follow-up posts: negative posts mainly trigger negative follow-ups, positive posts - positive follow-ups, and neutrality brings forth neutrality [5]. In [12], the authors studied positive and negative sentiment dynamics on Cancer Survivors Network. They sought propagation of the sentiment expressed in the initial message of a discussion and found that an estimated 75%-85% of the forum participants change their sentiment in a positive direction through online interactions with other community members. We, on other hand, analyze transitions happening between several sentiment and factual categories.

Sentiment propagation between two Twitter interlocutors has been studied in several works, [10] being the most cited. In it, the authors selected sequences of tweets (aka chains) based on positive and negative emoticons. These chains were analyzed for the propagation of 6 emotions: *love, joy, surprise, anger, sadness,* and *fear*. Our current study is different in 3 aspects: we study discussions carried out by many participants, the text of forum messages is quite different from tweets, and we analyze sentiment and factual categories.

The use of general and domain-specific affective lexicons in health-related studies were compared by [8] and [13]. Both studies showed a preference for domain-specific lexicons over general lexicons. [8] applied SentiWordNet, Subjectivity Lexicon and a polarized domain lexicon to classify drug reviews into positive, negative and neutral. Combination of the three lexicons produced the best classification of positive reviews (F-score$_{pos}$ = 0.62) and the best classification of negative reviews (F-score$_{neg}$ = 0.32); the authors did not report the overall classification results. [13] adapted Pointwise Mutual Information to build a domain lexicon HealthAffect. In 6-class sentiment classification, HealthAffect produced an F-score = 0.52 whereas WordNetAffect produced an F-score = 0.30. [1] classified difference in opinions within discussion threads. They represented text by unigrams, Linguistic Inquiry and Word Count (LIWC), and the MPQA dictionary. For HealthCare discussions the best accuracy 60.64% was achieved when the text was represented by LIWC. The cited works did not consider how ambiguous labels should be re-solved. Our study, on other hand, suggests a disambiguation procedure to resolve annotation conflicts.

3   Data Set and Annotation Process

We work with data obtained from the In Vitro Fertilization (IVF) website[2]. The data was introduced in [13] and became available on demand. On the IVF forum, 95% of the participants are women. We analyze discussions from the IVF Ages 35+ sub-forum[3]. Discussions with < 10 posts usually consisted of a question and factual replies. Discussions with > 20 posts usually ventured outside the initial topic. As a result, we kept 80 discussions, with 10 − 20 posts each; the average was 17 posts per discussion. Those discussions gave us 1322 posts authored by 359 participants. We considered the pool of authors and the number of messages are sufficiently large and heterogenous to be representative of the forum discussions and sentiments appearing in them.

We applied the annotation process by Wiebe et al.[14] where annotators were asked to identify private state expressions in context, instead of judging separate words and phrases. The process has been implemented in two steps [13]. In the first step, annotators were asked to mark messages in an unsupervised manner (i.e., we did not provide them with pre-set sentiment categories). The result was a selection of 35 types of sentiments, including worry, concern, doubt, impatience, support, hope, happiness, enthusiasm, thankfulness. In the next step, we gradually generalized the sentiments into three groups: confusion, encouragement, gratitude. After that, each post was **independently** assigned by two annotators with one dominant sentiment. The annotators reached a strong agreement with Fleiss Kappa = 0.737. If annotators assigned a message to different categories, the message was labeled *ambiguous*. We considered *confusion* as a non-positive label whereas *encouragement* and *gratitude* were considered as positive labels.

As expected, not all posts were predominantly sentimental. Participants speak about treatment outcomes, share details about medical procedures and clinics, and discuss symptoms and conditions. Messages presenting only factual information were marked as *facts*. Some posts contained factual information followed by emotional utterances. Those utterances almost always conveyed support and implicit approval ("hope, this helps", "I wish you all the best", "good luck"). We called this category of messages *endorsement*. Our set of categories *encouragement, gratitude, confusion, facts, endorsement* was a trade-off between

---

[2] www.ivf.ca/forums
[3] http://ivf.ca/forums/forum/166-ivf-ages-35

the two requirements: it had to be adequate to cover the emotional characteristics of the messages; it should should be small enough to allow the automated learning of the categories.

4   Sentiment and Factual Transitions in the Forum Discussions

Our annotation process gave the following empirical results: *facts* were found to be the most frequent category, with 433 messages, followed by *encouragement* – 310 messages, *endorsement* – 162, *gratitude* – 124, *confusion*, with 117 messages, was found to be the least frequent category. There were 176 *ambiguous* messages on which annotators did not agree on the label. We observed that despite a large number of participants and time delay messages were related within discussions: every posted message answered to one or several previous messages, and in many cases did not diverge from the discussed topic.

We collected information about changes of sentiment and factual categories during discussions. We built a matrix A where an element a_{i_j} shows occurrence of a message with category i (rows) being followed by a message with category j (columns) (the left side of Table 1). The row "start" indicates categories of the first messages of the discussions; the column "end" indicates categories of the last messages of the discussions. Note that barring 0s for *encouragement* and *gratitude* in the start row, the other elements of matrix A are non-zero. This means that there exists at least one transition between every pair of categories, i.e., the transitions form a strongly connected balanced digraph, where for each node its indegree (i.e., the number of head points) is equal to its outdegree (i.e., the number of tail points).

Table 1. Transitions of the sentiment and factual categories in discussions. For each category, the greatest transition probability is in **bold**, the smallest is in *italics*.

prev	The transition matrix of sentiments							The matrix of sentiment transition probabilities						
	ambig	confus	encour	endors	facts	gratit	end	ambig	confus	encour	endors	facts	gratit	end
start	21	45	0	1	13	0	0	0.27	0.57	0.00	0.02	0.17	0.00	0.00
ambig	39	8	44	19	51	7	13	0.22	0.05	0.25	0.11	**0.29**	*0.04*	0.01
confus	14	23	31	20	39	3	1	0.11	0.18	0.24	0.16	**0.30**	*0.03*	0.01
encour	34	7	112	31	47	41	20	0.12	*0.03*	**0.39**	0.11	0.17	0.15	0.07
endors	9	7	28	33	58	16	6	0.06	*0.05*	0.18	0.22	**0.37**	0.11	0.04
facts	45	7	59	46	197	40	31	0.11	*0.02*	0.14	0.11	**0.47**	0.10	0.08
gratit	13	9	36	12	28	17	9	0.11	*0.08*	**0.30**	0.10	0.23	0.14	0.08

Based on the transition matrix A, we built a stochastic matrix P by dividing each value by the row total. This matrix of probabilities forms a first-order Markov model, where the probability of each category of message depends on the nature of the previous message (the right side of Table 1). For each category, we mark in **bold** the greatest probability of the next category and in *italics* – the smallest probability of the next category (shown on Figure 1). Figure 2 shows the most likely categories at the beginning and the end of discussions.

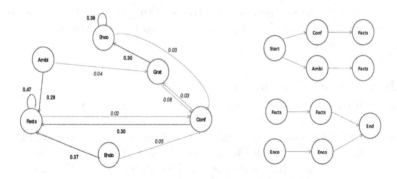

Fig. 1. Greatest and smallest transition probabilities for each category

Fig. 2. Most likely beginning and end of discussions

The transition matrixes show that two categories tend to reinforce themselves: *encouragement* is most frequently followed by *encouragement*, and *facts* – by *facts*. On other hand, *confusion* and *gratitude* trigger an emotional change in discussions: *confusion* is most often followed by *facts*, and *gratitude* – by *encouragement*. A message of *confusion* is very rarely followed by one of *gratitude* or appears at the end of the discussion.

5   Machine Learning Applications

To assess usability of the set of categories in automated studies of data, we solve four multi-class classification problems:

6-class classification where 1322 posts are classified into *confusion, encouragement, endorsement, gratitude, facts, ambiguous*

5-class classification where the *ambiguous* class is removed and the remaining 1146 posts are classified into the other 5 classes.

4-class classification where 1322 posts are grouped as follow: *facts* and *endorsement* classes make up a *(factual)* class, *encouragement* and *gratitude* classes become a *positive* class, and the *confusion* and *ambiguous* classes remain.

3-class classification where 176 *ambiguous* messages are removed, and the remaining 1146 messages are classified as *positive, confusion* or *factual* as in 4-class classification.

We applied SVM (linear kernel, c = 1) and used 10-fold cross-validation to compute multi-class Precision (Pr), Recall (R), and F-score (F). For the benchmark comparison with other sentiment analysis methods, we represented our data by SentiWordNet; this representation had 4 039 attributes. We tested three types of message representation:

1. sentiment transition, where messages are represented by two attributes whose values are categories assigned by each annotator to the previous message, two attributes whose values are categories assigned by each annotator to the next message, and three attributes showing whether the previous, current and next messages are first, middle, or last ones – 7 attributes altogether;
2. affective words, where we represent messages through a domain-specific lexicon HealthAffect (Section 2); this representation had 1 193 attributes;
3. combination of the two representations, with 1 200 attributes.

Table 2 reports the results. In terms of Fscore, our results are comparable or better than those reported in previous multi-classification studies [1,13]. In terms of the lexicon comparison, the results are two-fold: for 6- and 5- class classification, multi-categorical HealthAffect provided for better accuracy than polarity-oriented SentiWordNet; on 4- and 3-class classification where the emphasis was on differentiation between positive and negative sentiments, SentiWordNet provided for better accuracy than HealthAffect.

Table 2. Results of multi-class classification problems. The benchmark results are in *italics*. For each problem, the best P, R and F are in **bold**. The baseline: 6-class – F=0.162, 5-class – F=0.207, 4-class – F=0.281, and 3-class – F=0.356.

Representation	# attr.	6-class			5-class			4-class			3-class		
		P	R	F	P	R	F	P	R	F	P	R	F
SentiWordNet	4039	*0.385*	*0.405*	*0.387*	*0.485*	*0.485*	*0.478*	*0.502*	*0.525*	*0.509*	*0.632*	*0.635*	*0.631*
Sentiment transit.	7	0.395	0.408	0.356	0.441	0.458	0.414	0.475	0.528	0.489	0.620	0.620	0.605
HealthAffect	1193	0.402	0.416	0.405	0.496	0.494	0.491	0.501	0.517	0.505	0.628	0.620	0.620
HealthAffect and sentiment transit.	1200	**0.431**	**0.430**	**0.423**	**0.512**	**0.513**	**0.511**	**0.541**	**0.549**	**0.549**	**0.663**	**0.653**	**0.660**

The obtained results support our hypothesis that information on sentiment and factual transitions is critical for reliable message classification. Although the transitions were represented by 7 attributes, they provided for the results which were not statistically different from the results obtained on the 1 193 affective terms (paired t-test, P=0.1152). On the other hand, the combination of the transition attributes and affective terms provided for more significant results than affective terms (paired t-test, P=0.0011).

This analysis shows that a Markov model of sentiment and factual transitions can be used to find the most likely class of message when the classifier was unable to do this definitively. Conditional Random Fields (CRF) are another type of classifier which takes into account that the classes of different members of a sequence or pattern might depend on each other [2].

6 Discussion and Future Work

In this work, we have focused on sentiment and factual transitions in public discussions on online medical forums. We have identified *encouragement, gratitude, confusion* sentiments that represent emotions expressed by the forum participants and two factual categories of *facts, endorsement*. We have shown that there is at least one transition between each pair of categories. Solving a series of multi-class sentiment classification problems, we demonstrated that 7 attributes of sentiments transitions provided for the results which were not statistically different from the results obtained on the 1 193 affective terms (paired t-test, P=0.1152). At the same time, adding sentiment transitions to affective terms in message representations statistically improved the results (paired t-test, P=0.0011).

Our results on sentiment and factual transitions can be used in an emerging field of social mining, especially to domains where participants post emotionally-charged messages (e.g., ecology, climate change). A Markov model can help in disambiguation when the annotators disagree on the message sentiments or when automatic classifiers cannot give a definitive class decision. This can be done by analogy with the CLAWS part-of-speech disambiguator. Although the annotators reached a strong agreement in assigning sentiments to the messages, 176 messages, or 13.3% of data, were assigned different sentiments, thus labeled as *ambiguous*. To resolve the disagreement problem, we can evoke probabilities from Table 1. In cases where there are many different annotations or classification labels for a long sequence of messages there will be too many possible sequences to find the most probable by trying them all out in turn, so a "short cut" like the Viterbi algorithm would be used, which would only consider certain "most promising" paths.

Although study of affective lexicons was not the main goal of this work, we have noticed that a smaller number of HealthAffect terms provided for comparable or better results than a considerably large number of SentiWordNet terms. We suggest further exploration of what set of the affective terms can better represent messages in learning experiments. Application of Conditional Random Fields is another venue of future work.

References

1. Anand, P., Walker, M., Abbott, R., Fox Tree, J.E., Bowmani, R., Minor, M.: Cats rule and dogs drool!: classifying stance in online debate. In: The 2nd Workshop on Computational Approaches to Subjectivity and Sentiment Analysis, ACL-HLT 2011, pp. 1–9 (2011)
2. Blunsom, P., Cohn, T.: Discriminative word alignments with conditional random fields. In: Proceedings of COLING/ACL 2006, pp. 65–72 (2006)
3. Bobicev, V., Sokolova, M., Jafer, Y., Schramm, D.: Learning sentiments from tweets with personal health information. In: Kosseim, L., Inkpen, D. (eds.) Canadian AI 2012. LNCS, vol. 7310, pp. 37–48. Springer, Heidelberg (2012)
4. Chew, C., Eysenbach, G.: Pandemics in the Age of Twitter: Content Analysis of Tweets during the 2009 H1N1 Outbreak. PLoS One (2010)
5. Chmiel, A., Sienkiewicz, J., Thelwall, M., Paltoglou, G., Buckley, K., Kappas, A., Holyst, J.: Collective Emotions Online and their Influence on Community Life. PLoS One (2011)

6. Coulson, N.: Receiving social support online: an analysis of a computer-mediated support group for individuals living with irritable bowel syndrome. CyberPsychology and Behavior **8**(6), 580–584 (2005)
7. Culotta, A.: Lightweight methods to estimate influenza rates and alcohol sales volume from Twitter messages. Language Resources and Evaluation **47**, 217–238 (2013)
8. Goeuriot, L., Na, J., Kyaing, W., Khoo, C., Chang, Y., Theng, Y., Kim, J.: Sentiment lexicons for health-related opinion mining. In: The 2nd ACM SIGHIT International Health Informatics Symposium, pp. 219–225 (2012)
9. Hassan, A., Abu-Jbara, A., Radev, D.: Detecting subgroups in online discussions by modeling positive and negative relations among participants. In: Proceedings of EMNLP, pp. 59–70 (2012)
10. Kim, S., Bak, J., Jo, Y., Oh, A.: Do you feel what i feel? social aspects of emotions in twitter conversations. In: Proceedings of ICWSM, pp. 495–498 (2012)
11. Paul, M., Dredze, M.: You are what you tweet: analyzing twitter for public health. In: International Conference on WSM (2011)
12. Qiu, B., Zhao, K., Mitra, P., Wu, D., Caragea, C., Yen, J., Greer, G., Portier, K.: Get online support, feel better-sentiment analysis and dynamics in an online cancer survivor community. In: Privacy, Security, Risk and Trust (PASSAT) and 2011 IEEE Third International Conference on Social Computing (SocialCom), pp. 274–281 (2011)
13. Sokolova, M., Bobicev, V.: What sentiments can be found in medical forums?. In: Recent Advances in Natural Language Processing, pp. 633–639 (2013)
14. Wiebe, J., Wilson, T., Cardie, C.: Annotating Expressions of Opinions and Emotions in Language. Language Resources and Evaluation **39**(2–3), 165–210 (2005)
15. Zafarani, R., Cole, W.D., Liu, H.: Sentiment propagation in social networks: a case study in livejournal. In: Chai, S.-K., Salerno, J.J., Mabry, P.L. (eds.) SBP 2010. LNCS, vol. 6007, pp. 413–420. Springer, Heidelberg (2010)

Abstractive Meeting Summarization
as a Markov Decision Process

Gabriel Murray[(✉)]

University of the Fraser Valley, Abbotsford, BC, Canada
gabriel.murray@ufv.ca
http://www.ufv.ca/cis/gabriel-murray/

Abstract. The task of abstractive summarization is formulated as a Markov Decision Process. Value Iteration is used to determine the optimal policy for natural language generation. While the approach is general, in this work we apply the system to the problem of automatically summarizing meeting conversations. The generated abstracts are superior to generated extracts according to intrinsic measures.

Keywords: Abstractive summarization · Natural language generation · Markov decision process · Value iteration

1 Introduction

This paper represents ongoing work on the problem of automatic abstractive summarization of meeting conversations. Here we focus on the task of natural language generation (NLG), and formulate the task as a Markov Decision Process (MDP). A variant of Value Iteration is used to determine the optimal policy for text generation, based on combining information from n-gram language models and combined term-weighting metrics. We show that the short generated abstracts are superior to text extracts according to intrinsic measures.

2 Related Work

The overall research task is one of automatic summarization [1]. While our novel approach is general and thus applicable to many domains, we apply it here to the domain of meeting conversations, and Carenini et al. [2] provide an overview of techniques for summarizing conversational data. Renals et al. [3] survey various work that has been done analyzing meeting interactions, including automatic summarization. Most work on automatic summarization in general, including meeting summarization, has been *extractive* in nature. In the context of meetings, this approach translates to finding the sentences that are the most salient, and concatenating them to form the minutes of the meeting [4,5]. There has been a surge of research on abstractive meeting summarization over the past five years [6–9], in part because extrinsic studies show that end-users rate extractive summaries very poorly in comparison with generated abstracts, particularly if the meeting transcripts are filled with errors from a speech recognition component [10].

© Springer International Publishing Switzerland 2015
D. Barbosa and E. Milios (Eds.): Canadian AI 2015, LNAI 9091, pp. 212–219, 2015.
DOI: 10.1007/978-3-319-18356-5_19

3   Abstractive Summarization as an MDP

In this section we describe the architecture of the MDP system, as well as its inputs.

3.1   Abstractive Community Detection

The MDP summarization approach requires an upstream component for *abstractive community detection*, where meeting sentences are clustered into (potentially overlapping) groups such that each group will be realized by a single abstractive summary sentence. We use the following abstractive community detection approaches for comparison: 1) The system of Murray et al. [11] that uses a supervised logistic regression classifer to build a sentence graph, followed by the CONGA graph community detection algorithm [12]. 2) A simple unsupervised approach that uses k-means clustering to build a sentence graph, followed by divisive clustering to the desired number of clusters. 3) Human gold-standard sentence communities.[1]

3.2   The Summarization MDP

In the summarization MDP approach, states correspond to the unique word types that occur in the sentence cluster. In each state, the available actions correspond to selecting the next word to generate, so that in a given state there are as many available actions as there are possible next words. Figure 1 illustrates this with a simple toy example, where the sentence cluster consists of only two sentences: *The cat sat* and *It sat there*. Each of the five words {the, cat, sat, it, there} is associated with a state at each time-step. Arrows correspond to actions that can be selected to generate a new word, so that an arrow from some word w_i to another word w_j can be thought of as an invocation of an action $generate(w_j)$.[2] The leftmost state is the START state and the rightmost state is the STOP state. It is possible to transition to the STOP state at any step after the first step. Once in the STOP state, there are no further available actions. With this toy example, the resulting generated sentence can be between one and three words long, e.g. *Cat, Cat sat, Cat sat there*, etc.

For a given cluster of sentences, we can *unroll* the MDP state structure for a particular number of time-steps, also known as a horizon h. In the toy example of Figure 1, the horizon is $h = 3$. In practice, for a given cluster of sentences we set h to be equal to the average length of the sentences in the cluster. This constrains the length of the abstractive sentence we will generate to be between 1 and h words.

---

[1] The terms *community* and *cluster* are used interchangeably.

[2] Since there is a state for each word at each time-step, it would be more proper to use state notation such as w_{it} for the ith word at the tth time-step.

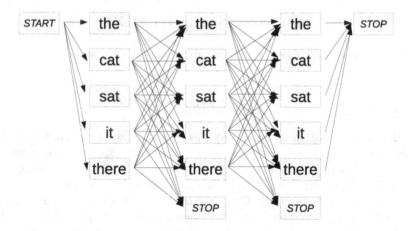

Fig. 1. The MDP State Structure for a Toy Example Where the Sentence Cluster Consists of Two Sentences: *The cat sat* and *It sat there.* The horizon here is $h = 3$.

3.3 Value Iteration

We desire to learn the best action to take in each step; that is, if we are in a state corresponding to some word w_1, what is the best word w_2 to generate next? More formally, if π is a policy that maps states to actions, we want to learn the optimal policy π^* that gives the highest expected value when following the policy. Value Iteration allows us to iteratively estimate the expected value of being in some state s and following the optimal policy. At each iteration k, we apply the following update:

$$V_k[s] = \max_a \sum_{s'} P(s'|s,a)\,(R(s,a,s') + \gamma V_{k-1}[s'])$$

In general, $P(s'|s,a)$ is the probability of winding up in state s' after carrying out action a in state s, while $R(s,a,s')$ is the *immediate reward* of carrying out that action and ending up in state s'. The term $\gamma V_{k-1}[s']$ is the discounted expected value of state s' from the previous iteration. In these experiments, the discount factor is set to $\gamma = 0.9$.

We now interpret the first two terms in the context of the abstractive summarization problem. The first term $P(s'|s,a)$ is simply a bigram language probability that could be rewritten as $P(w_j|w_i, generate(w_j))$. In fact, the bigram probabilities are interpolated from two language models, one estimated from just the sentences in the cluster and one estimated from the entire meeting.

The second term $R(s,a,s')$ could be rewritten as $R(w_i, generate(w_j), w_j)$ and is simply the sum of the term-weight scores for w_i and w_j, with the term-weighting scheme being *Gain* [13] multiplied by inverse document frequency (*idf*). That is, if the term-weight scheme tw is defined as

$$tw(w) = Gain(w) * idf(w)$$

then

$$R(w_i, generate(w_j), w_j) = tw(w_i) + tw(w_j)$$

For STOP states, the immediate reward is set to 1, favouring the generation of shorter sentences. A lower reward would favour generating longer sentences.

The third term $\gamma V_{k-1}[s']$ is simply the discounted value of state s' from the previous iteration of the algorithm. In these experiments, the discount factor is set to $\gamma = 0.95$.

Note that the above term $P(w_j|w_i, generate(w_j))$ indicates that an action $generate(w_j)$ is non-deterministic: it only results in moving to the state w_j at the following timestep with a probability equal to the estimated bigram probability. Otherwise no word is generated and the state is not changed. This non-determinism is only present during Value Iteration to ensure that we are working with proper probabilities that sum to 1. Once the optimal policy is learned and we generate our abstractive text, described below in Section 3.5, generating a word is fully deterministic. With that in mind, we can fully expand the update rule that is applied to each state on each iteration, until convergence:

$$V_k[w_i] = \max_{generate(w_j)} P(w_j|w_i, generate(w_j))(R(w_i, generate(w_j), w_j) + \gamma V_{k-1}[w_j])$$
$$+ (1 - P(w_j|w_i, generate(w_j)))(0 + \gamma V_{k-1}[w_i])$$

After Value Iteration has completed, we can determine π^* by selecting the action with the highest expected value in each state.

3.4 State Thinning

In early experiments, after Value Iteration was completed it was observed that for some words the optimal action varied little across the time-steps. That is, for some word w_i, the optimal action $\pi^*(w_i)$ is the same at time-step 2 or 3 as it is at time-step 15 or 16. If we end up generating a word more than once in a summary sentence, this can lead to disfluencies such as *I was following the following the following the path*. To mitigate this problem, during Value Iteration we periodically do *state thinning*. On every fifth iteration, we begin at the START state and follow the currently estimated optimal policy. If any word (excluding stopwords) is generated more than once, we retain only the state corresponding to the first generation of that word, and remove all other states in the optimal sequence that correspond to subsequent generations of that word. While this has the benefit of reducing disfluencies such as the example above, it typically causes Value Iteration to need additional iterations before converging.

3.5 Generation

Once π^* is learned for a particular cluster of sentences, we generate the abstractive sentence for that cluster by beginning at the START state and following

$\pi^*(START)$, generating each word in sequence until we reach a STOP state. A flexibility of this MDP approach is that we could also manually specify that the sentence needs to begin with a particular sequence $\{w_1, w_2, \cdots, w_n\}$ and then start generating the remainder of the sentence according to $\pi^*(w_n)$. As mentioned above, the generation step is fully deterministic: if $\pi^*(w_i) = generate(w_j)$ then we generate w_j with certainty.

Figure 2 reproduces the toy example from Figure 1, simulating the completion of value iteration and the utilization of the resulting optimal policy to generate an abstractive sentence *Cat sat there*. For clarity, we have expanded the notation to include time-step information for each state, so that $sat_{t=2}$ refers to a state at time-step 2 corresponding to the word token *cat*.

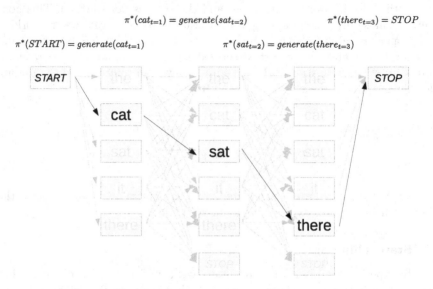

Fig. 2. Following the Optimal Policy to Generate the Sentence *Cat Sat There*

4   Corpora and Evaluation

For these experiments, we use the AMI [14] meeting corpus, which includes gold-standard abstractive and extractive summaries, with many-to-many links between the summary types. This many-to-many mapping between extractive and abstractive summaries provides us with gold-standard abstractive communities, which we evaluate alongside the two automatic abstractive community detection approaches.

For evaluation, we use ROUGE [15], which compares machine summaries with human gold-standard summaries, using n-gram and skip n-gram overlap. Specifically, we use ROUGE-1 (unigram overlap), ROUGE-2 (bigram overlap), and ROUGE-SU4 (skip bigram overlap). Our gold-standard summaries are human-authored abstracts rather than human-selected extracts. For all machine summaries, the summary length was limited to 100 words.

5 Sample Summary

Below is a sample summary generated using the automatic abstractive approach (with the supervised + CONGA abstractive community detection system as input). This is a 100-word summary of meeting IS1009d from the AMI corpus, in which the group finalized their design for a remote control. They discussed the buttons it would contain, the colours, materials, look-and-feel, and the snail-shell shape of the remote.

```
1. I think I'm pretty happy with the mute buttons.
2. Maybe two mute buttons.
3. The on-off button is that the colours very attractive.
4. I think voice recognition is our big selling point.
5. It should have nine channel buttons, a next button,
volume control features, colour, contrast, sharpness, etcetera.
6. I think shape is one thing.
7. It's blue in colour buttons, it is a snail shape, so it is soft
rubber.
8. Colour, i think im pretty happy with the financing, there was
our voice recogniser.
9. And then we have to interface push buttons.
10. And all the buttons will have a shell shape.
```

Some sentences represent entire extracted sentences from the meeting transcript, while others are compressed versions of meeting sentences. A few summary sentences (e.g. 3, 7, 8, 9) are synthesized from several meeting sentences, with varying quality.

6 Results

The ROUGE scores for all systems are shown in Table 1. There are three sets of results for the abstractive MDP system, corresponding to the three abstractive community detection systems used as input, described earlier in Section 3.1. We also compare with two extractive systems: a *greedy* system that simply selects the highest-scoring dialogue act units in the meeting, and an *exemplar* system that first does abstractive community detection but then just selects the highest-scoring sentence from each community, rather than doing abstractive summarization. For all systems that require automatic community detection / clustering, the number of clusters for these experiments is set to 18, which is the average number of sentences in the human abstract summaries for this corpus.

The overall finding is that all of the abstractive MDP-based systems are superior to the extractive systems on all ROUGE metrics. The differences between the three abstractive systems are relatively small, though the community detection approach of Murray et al. [11] yields abstractive summary results that are comparable to using human gold-standard clusters. Note that the ROUGE scores shown in Table 1 seem low for all systems. This is due to two reasons: because we are creating very short (100-word summaries), leading to high precision but very low recall, and because our ROUGE model summaries are human *abstracts* rather than human *extracts*.

Table 1. ROUGE F-Scores (w/ confidence intervals) for Abstractive Summaries

System	ROUGE-1	ROUGE-2	ROUGE-SU4
EXTRACTIVE			
Greedy	0.136 (0.118-0.155)	0.021 (0.015-0.028)	0.040 (0.032-0.047)
Exemplar	0.121 (0.107-0.134)	0.022 (0.014-0.031)	0.031 (0.025-0.037)
ABSTRACTIVE			
(MDP Systems)			
Unsupervised	0.154 (0.133-0.174)	0.028 (0.021-0.036)	0.044 (0.036-0.052)
Supervised	**0.186** (0.167-0.205)	0.033 (0.025-0.040)	**0.056** (0.049-0.064)
Human Clust.	0.185 (0.168-0.207)	**0.034** (0.025-0.044)	0.053 (0.046-0.063)

7   Conclusion

The experimental results show the superiority of the abstractive approach compared with two extractive summarization methods, according to the ROUGE metrics. For generating very concise meeting summaries, this is a promising approach that combines both compression and abstractive synthesis, as demonstrated by the sample summary in Section 5. However, that sample output also demonstrates that summary sentences that are derived by synthesizing multiple meeting sentences can be highly ungrammatical or nonsensical. A promising idea is to combine this bottom-up MDP approach with a top-down template-filling summarization system. As described earlier, a flexibility of the MDP approach is that we could add a constraint that the sentence must begin with a particular sequence $\{w_1, w_2, \cdots, w_n\}$, e.g. based on templates learned from training corpora, and then generate the remainder of the sentence according to $\pi^*(w_n)$.

We have formulated the problem of abstractive summarization as an MDP, and applied the system to meeting conversations. The generated abstractive summaries are ranked higher than extractive summaries according to intrinsic measures. The approach is general, and future work will apply this fully unsupervised MDP summarization system to other domains.

References

1. Nenkova, A., McKeown, K.: Automatic summarization. Foundations and Trends in Information Retrieval **5**(2–3), 103–233 (2011)
2. Carenini, G., Murray, G., Ng, R.: Methods for Mining and Summarizing Text Conversations, 1st edn. Morgan Claypool, San Rafael (2011)
3. Renals, S., Bourlard, H., Carletta, J., Popescu-Belis, A.: Multimodal Signal Processing: Human Interactions in Meetings, 1st edn. Cambridge University Press, New York (2012)

4. Xie, S., Favre, B., Hakkani-Tür, D., Liu, Y.: Leveraging sentence weights in a concept-based optimization framework for extractive meeting summarization. In: Proc. of Interspeech 2009, Brighton, England (2009)
5. Gillick, D., Riedhammer, K., Favre, B., Hakkani-Tür, D.: A global optimization framework for meeting summarization. In: Proc. of ICASSP 2009, Taipei, Taiwan (2009)
6. Mehdad, Y., Carenini, G., Tompa, F., Ng, R.: Abstractive meeting summarization with entailment and fusion. In: Proc. of ENLG 2013, Sofia, Bulgaria, pp. 136–146 (2013)
7. Liu, F., Liu, Y.: Towards abstractive speech summarization: Exploring unsupervised and supervised approaches for spoken utterance compression. IEEE Transactions on Audio, Speech and Language Processing $21(7)$, 1469–1480 (2013)
8. Wang, L., Cardie, C.: Domain-independent abstract generation for focused meeting summarization. In: Proc. of ACL 2013, Sofia, Bulgaria, pp. 1395–1405 (2013)
9. Mehdad, Y., Carenini, G., Ng, R.: Abstractive summarization of spoken and written conversations based on phrasal queries. In: Proc. of ACL 2014, Baltimore, MD, USA, pp. 1220–1230 (2014)
10. Murray, G., Carenini, G., Ng, R.: The impact of asr on abstractive vs. extractive meeting summaries. In: Proc. of Interspeech 2010, Tokyo, Japan, pp. 1688–1691 (2010)
11. Murray, G., Carenini, G., Ng, R.: Using the omega index for evaluating abstractive community detection. In: Proceedings of the NAACL 2012 Workshop on Evaluation Metrics and System Comparison for Automatic Summarization, Montreal, Canada, pp. 10–18 (2012)
12. Gregory, S.: An algorithm to find overlapping community structure in networks. In: Proc. of ECML/PKDD 2007, Warsaw, Poland (2007)
13. Papineni, K.: Why inverse document frequency?. In: Proc. of NAACL 2001, pp. 1–8 (2001)
14. Carletta, J.E., et al.: The AMI meeting corpus: a pre-announcement. In: Renals, S., Bengio, S. (eds.) MLMI 2005. LNCS, vol. 3869, pp. 28–39. Springer, Heidelberg (2006)
15. Lin, C.Y., Hovy, E.H.: Automatic evaluation of summaries using n-gram co-occurrence statistics. In: Proc. of HLT-NAACL 2003, Edmonton, Calgary, Canada, pp. 71–78 (2003)

Data Mining and Machine Learning

Rule Extraction from Random Forest: The RF+HC Methods

Morteza Mashayekhi[(✉)] and Robin Gras

School of Computer Science, University of Windsor, Windsor, ON, Canada
{mashaye,rgras}@uwindsor.ca
http://www.uwindsor.ca

Abstract. Random forest (RF) is a tree-based learning method, which exhibits a high ability to generalize on real data sets. Nevertheless, a possible limitation of RF is that it generates a forest consisting of many trees and rules, thus it is viewed as a black box model. In this paper, the RF+HC methods for rule extraction from RF are proposed. Once the RF is built, a hill climbing algorithm is used to search for a rule set such that it reduces the number of rules dramatically, which significantly improves comprehensibility of the underlying model built by RF. The proposed methods are evaluated on eighteen UCI and four microarray data sets. Our experimental results show that the proposed methods outperform one of the state-of-the-art methods in terms of scalability and comprehensibility while preserving the same level of accuracy.

Keywords: Rule extraction · Random forest · Hill climbing

1 Introduction

Random forest (RF) is an ensemble learning method for both classification and regression that constructs and integrates multiple decision trees at training step using bootstrapping. Additionally, it aggregates the outputs of all trees via plurality voting in order to classify a new input. It has few parameters to tune and it is robust against overfitting. It runs efficiently on large data sets and can handle thousands of input variables. Moreover, RF has an effective method for estimating missing data, and has some mechanisms to deal with unbalanced data sets [4]. In some applications, RF outperforms well-known classifiers such as support vector machines (SVMs) and neural networks (NNs) [5,18]. Despite good performance of RF in different domains, its major drawback is that, it generates a 'black box'model in the sense that it does not have the ability to explain and interpret the model in an understandable form [23,27] given that it generates a vast number of propositional if-then rules. As a result, ensemble predictors such as RF are very rarely used in domains where making transparent models is mandatory, such as predicting clinical outcomes [23]. In order to bear this limitation, the hypothesis generated by RF should be transformed into a more comprehensible representation.

© Springer International Publishing Switzerland 2015
D. Barbosa and E. Milios (Eds.): Canadian AI 2015, LNAI 9091, pp. 223–237, 2015.
DOI: 10.1007/978-3-319-18356-5_20

To obtain a comprehensible model which is simpler to interpret, accuracy is often sacrificed. This fact is normally referred to as the 'accuracy vs. comprehensibility tradeoff'. The importance of accuracy or comprehensibility is completely related to the application. One way to obtain a transparent model is to induce rules directly from the training set or to build a decision tree. However, another option is to take advantage of the good performance of the existing opaque models such as SVMs, RF, or NNs and generate rules based on them. This process is called rule extraction (RE), which is aimed at providing explanations for the predictive models' outputs. There are two different rule extraction methods based on an opaque model: decompositional and pedagogical [11]. Decompositional methods extract rules at the level of individual units of the prediction model such as neurons in neural networks, and therefore rely on the model's architecture. In contrast, in pedagogical approaches, the predictive model is only used to produce predictions.

In previous years, a high number of rule extraction methods using trained NNs and SVMs have been published (see [11] for a good survey). Nevertheless, in the case of the RF model, few research projects have been conducted. In this paper, the RF+HC methods for the interpretation of the RF model are proposed. The proposed methods can be treated as a decompositional rule extraction approach given that we employed all the generated rules by RF, which are dependent on the number of trees and also the tree structures in the RF.

This paper is organized as follows: the background description including foundation of the RF followed by a discussion of related research projects are explained in section 2. In section 3, the RF+HC methods are introduced. Experimental results of our methods applied to several data sets and comparisons are described in section 4. Finally, we present our conclusions along with possible future directions for our work.

2   Background

2.1   Random Forest

The RF is an ensemble learning method such that successive trees do not depend on the previous ones and each tree is constructed independently using a bootstrap sample of the data set. At the end, a majority voting procedure is used for making predictions. In addition, each node is split using the best feature among a subset of features (m) randomly chosen at that node. Parameter m is usually equal to $0.5 \times sqrt(n)$, $sqrt(n)$, or $2 \times sqrt(n)$ where n is the number of features. Error estimations are performed on a subset of data which are not included in the bootstrap sample at each bootstrap iteration (This subset is called out-of bag or OOB). RF can also estimate the importance of a feature by permutation of the values associated with a feature and comparing of the average OOB error before and after the permutation over all trees. However, it does not consider the dependency between features.

RF deserves to be considered as one of the important prediction methods because it demonstrates a high prediction accuracy and it can be used for clustering and feature selection applications as well [4,7,22]. Moreover, estimating the out-of-bag error often eliminates the need for cross-validation. More importantly, as it generates a multitude of propositional if-then rules, which is the most widespread rule type in RE domain, it has a very high potential to provide clear explanations and interpretations of its underlying model.

2.2 Related Work

One of the projects focusing on this topic was conducted by Zhang et al. [27] to search for the smallest RF. Although their method is not a rule extraction strategy, it seeks out a sub-forest that can achieve the accuracy of a large RF. They used three different measures in order to determine the importance of trees in terms of their predictive power. The experimental results demonstrate that such a sub-forest with performance as good as a large forest exists. Latinne et al. [13] attempted to reduce the number of trees in RF using the McNemar test of significance on the prediction outputs of the trees. Similarly, others tried to select an optimal sub-set of decision trees in RF [2,16,26]. These methods are not really rule extraction methods and mostly concentrate on reducing the number of decision trees in the RF or in a similar ensemble method such as bagging.

There are also some other methods to increase comprehensibility of an ensemble or RF by compacting them into one decision tree. For example, a single decision tree was used to approximate an ensemble of decision trees [24]. In this method, class distributions were estimated from the ensemble in order to determine the tests to be used in the new tree. A similar method was employed to approximate the RF with just one decision tree [12]. The aim was to generate a weaker but transparent model using combinations of regular training data and test data initially labeled by the RF.

Other methods with different approaches were proposed to select an optimal set of rules generated by RF [9,17]. More specifically, Liu et al. [14,15] used RF as an ensemble of rules and proposed a joint rule extraction and feature selection method (CRF) based on 1-norm regularized RF, using sparse encoding and solving an optimization problem applying linear programming method.

3   RF + HC Methods

RE can be expressed as an optimization problem [8] and one solution of this problem is to apply heuristic search methods. These methods overcome the complexity of finding the best rule set, which is an NP-hard problem.

In this section, we present our algorithm (Algorithm 1) to extract comprehensible rules from RF as follows. The algorithm consists of four parts: In the first part, RF is constructed and all the rules in the forest are extracted into the *Rs* set. The second part of the algorithm computes the score of all rules based on the *RsCoverage*, a sparse matrix that shows which rules cover each sample and its corresponding class label. Afterwards, the scores are assigned to the rules in order to control the rule selection process, which can be based on different factors such as accuracy and rule coverage. We used equation (1) that has been shown to be a promising fitness function [20]:

Algorithm 1. RF+HC

```
Input: trainSet, testSet, iniRuleNo, treeNo
Step 1: // Construct Random Forest
    RF = trainRF( trainSet, treeNo );
    Rs = getAllTerminalNodes (RF);
Step 2: // Compute rules coverage
    m = size( trainSet );
    n = size(Rs);
    RsCoverage=zeros( m, n);
    foreach sample in trainSet {
        foreach rule in Rs
            if match(rule, sample)
                RsCoverage (sample, rule) = class;
    }
    RScore = ruleScore (RsCoverage);
Step 3: // Repeat the HC method to obtain best rules
    iniRs = getRuleSet(RScore, n, iniRuleNo);
    impRs = iniRs; bestRs=iniRs;
    for i=1 to MaxIteration {
        impRs = HeuristicSearch (impRs, RScore);
        if Acc_{impRs} >Acc_{bestRs}
            bestRs = impRs;
        impRs = getRuleSet( RScore, n, iniRuleNo);
    }
Step 4: // Calculate the accuracy on test set
    calcPerformance(testSet,bestRs);
```

$$ruleScore_1 = \frac{cc - ic}{cc + ic} + \frac{cc}{ic + k} \qquad (1)$$

In this formula, cc (correct classification) is the number of training samples that are covered and correctly classified by the rule. Variable ic (incorrect classification) is the number of incorrectly classified training samples that are covered by the rule. Finally, k is a predefined positive constant value. In our case $k=4$, though other values can be used as it is mostly to avoid the denominator becomes zero and there is no significant change in the results by modifying k). This scoring function ensures the retention of the rules with higher classification accuracy and higher coverage and to remove the noisy rules. Obviously, other fitness measures can be used instead. One possibility would be to employ the rule score based on metrics such as number of features in the extracted rule set and number of antecedents to increase the quality of rules in terms of comprehensibility. In the third step of the algorithm, a fitness proportionate selection method is used $iniRuleNo$ times to generate an initial rule set ($iniRs$) with a probability to select a rule proportional to its score. In order to search the RF rules space, we used the random-restart stochastic hill climbing method, which gives a local optimum point of the search space based on the random start locations.

Any other search methods such as simulated annealing or genetic algorithm can be applied instead of HeuristicSearch function in the algorithm. We repeated the search with a predefined maximum number of iterations (*MaxIteration*), each time with a new initial rule set. This can compensate some of the deficiencies in hill climbing due to the randomized and incomplete search strategy [21]. The hill climbing algorithm, searches for the best neighbor, the one with the highest score, of the current location based on equation (1) in the search space and by changing (adding/removing) one rule to the current rule set. For adding/removing a rule, we used the same fitness proportionate selection procedure that was employed for producing the *iniRs*. The hill climbing score function was defined based only on the overall accuracy because the scoring schema of the second step already took into account both rule coverage and rule accuracy. If the new movement in the rule set space improves the score value, that change is retained. Otherwise it is discarded and then another neighbor in the rule space is sought. We repeat this step for a pre-defined maximum number of iterations (*MaxIteration*). Finally, in the fourth step, we apply the best extracted rule set on the test set to evaluate the generalization ability of the extracted rules.

One of the RF characteristics is that there is no pruning while it is constructed. Therefore, we expect to have long rules (with a large number of antecedents) in the rule set as well as in the extracted rule set using the proposed algorithm. Having long rules damages the interpretability of the model and thus it should be considered in the applications for which the interpretation of the rules is important. Therefore, we proposed the second algorithm, which is basically similar to Algorithm 1 except that a modified version of the rule score function (i.e. equation (2)) was used, where rl shows rule length or number of antecedents. We called the new method RF+HC_CMPR. In the RF+HC_CMPR method more generalized rules (shorter length rules with higher accuracy) have higher priority than the more specialized rules (the longer rules with lower accuracy) based on the following equation:

$$ruleScore_2 = ruleScore_1 + \frac{cc}{rl} \tag{2}$$

The inputs of the proposed methods are the training/test sets, initial number of rules (*iniRuleNo*) and the number of trees in the RF (see Algorithm 1). Variable *iniRuleNo* adjusts the tradeoff between accuracy and comprehensibility. In cases where prediction ability is important, higher values are used and in cases where the interpretation of the underlying model is important lower values should be used. For the implementation, we used Matlab as the same as the source code available for the CRF method.

4   Experiments and Discussion

To compare our proposed methods with other methods, we also applied CRF [14,15] and RF on 22 different data sets. Different criteria have been proposed to evaluate a RE algorithm [11]. For instance, accuracy is defined as the ability of extracted rules to predict unseen test sets. Another major factor is comprehensibility, which is not easy to measure due to the subjective nature of this concept. There are different factors that are used to determine comprehensibility such as, the number of rules and the average number of antecedents. Another desirable characteristic of a RE method is its potential to be applicable to a wide range of applications. If a RE algorithm is applicable to data sets with a large number of samples, features, or classes then it is said to be scalable. This scalability notion includes time and algorithm complexity.

In our work, we measured the average accuracy of 10 times 3-fold cross-validation (by randomizing the data set for every repetition) for evaluating accuracy, as it gives more accurate results in compare to one time k-fold cross-validation. This measure demonstrates the prediction and generalization ability of the extracted rules. Majority voting is used to classify a sample when more than one rule covers a sample. We assumed a default rule such that the samples not covered by any of the extracted rules are simply assigned to the high frequency class in the dataset. In the RF+HC methods, due to their stochastic nature, we repeated the whole procedure 10 times and computed the average results along with their standard deviations. For the CRF method, 10 different values for the lambda parameter (which indicates the tradeoff between the number of rules and accuracy) were used. To determine these values, we did a few pilot runs with each data set separately. To determine the best lambda, a cross-validation step is incorporated in the CRF method such that it selects the lambda value, which gives the minimum error for cross-validation. In order to show the comprehensibility of the methods, we considered the number of rules, maximum rule length, and total number of antecedents in the extracted rule set. For the CRF method, these values are related to the lambda parameter value, which gives the lowest cross-validation error. On the other hand, those of RF+HC methods are related to 10 repetitions of the process. All the values are rounded to the closest integer value.

Scalability is one of the most important evaluation metrics often overlooked in most of the RE methods such as the CRF method. We measured the computational time as a metric to evaluate the scalability. To have a fair comparison, we used 10 different lambdas in the CRF method and we divided the required time to find the best lambda by 10. This means that we only considered the time for cross-validation using the best lambda plus the time for training and test steps. We considered the cross-validation time because it is an important part of the CRF method , which finds the best lambda in each iteration of the algorithm. At each iteration, that includes cross-validation, optimization, and feature selection, the features in the extracted rules are kept and the rest are removed. In the case of RF+HC, we repeated each experiment 10 times and then divided the overall time by 10. We also considered the hill climbing repetition time (*MaxIteration*) in order to calculate the computation time.

As the input of the proposed algorithms, we should specify the initial number of rules (*iniRuleNo*) for each data set. We used 500 decision trees to build RF with $m=sqrt(n)$, which is a default value mostly used in the literature, where (m) is the number of features randomly chosen at that node for splitting. In the random-restart hill climbing, we repeated hill climbing from 10 initial rule sets. We took *MaxIteration = 500* in all of our experiments. Higher values provide hill climbing with more opportunities for likely improving the rule set score, although it did not happen in our case. For comparing the proposed methods with CRF in terms of performance, comprehensibility, and computation time complexity, we used Wilcoxon and Friedman tests as suggested in [6].

4.1 Data Sets

We used 22 data sets with various characteristics in terms of the number of features, the number of samples, and the number of classes to observe how the performance of the proposed methods varies depending on the data set type. Eighteen data sets were taken from UCI machine learning repository [3] and another four data sets (Golub [10], Colon [1], Nutt [19], Veer [25]), which are gene expression microarray data sets. The extreme cases are Veer with 24188 features, Magic with 19020 samples, and Yeast and Cardio with 10 classes (see Table 1).

Table 1. Data sets along with their characteristics

Data set	Feature#	Class#	Sample#
Breast Cancer	9	2	699
Magic	10	2	19020
Musk Clean1	166	2	476
Wine	13	2	178
Wine Quality	11	6	1599
Iris	4	3	150
Yeast	8	10	1485
Cardiography	20	10	1726
Balance Scale	4	3	625
Cmc	9	3	1473
Glass	9	6	214
Haberman	3	2	306
Iono	34	2	351
Segmentation	19	7	210
Tae	5	3	151
Zoo	16	7	101
Ecoli	7	8	336
Spam	57	2	4601
Golub	5147	2	72
Colon	2000	2	62
Glimo Nutt	12625	2	50
Veer	24188	2	77

4.2   Accuracy and Generalization Ability

On average, both the RF+HC and RF+HC_CMPR methods gave almost the same level of accuracy as the CRF method with marginal differences (Table 2). Moreover, all three methods obtained 96% of the RF accuracy for the whole data sets on average. For some datasets, they demonstrated higher accuracy than RF such as for Tae, Cmc, and Golub with RF+HC or Tae and Clean with CRF method. A similar result was observed in [28] when the authors used a NN ensemble to extract the rules, observing higher accuracy for extracted rules than for the underlying model.

The generalization ability of RF+HC is due to the selection of the high score rules in RF and it is also due to some level of stochasticity, which results in assigning odds to the rules with low scores in the training set, but they may be important for unseen data. Comparing the accuracy of CRF method with the proposed methods revealed that the null hypothesis with $\alpha = 0.05$ cannot be rejected with $z = 0.41$ (CRF vs. FR+HC) and $z = 0.42$ (CRF vs. RF+HC_CMPR), while the critical z value is -1.96 in Wilcoxon test. Therefore, the difference is not significant, which proves that two methods are equivalent in terms of accuracy.

Table 2. Percentage accuracy of the RF+HC, RF+HC_CMPR, CRF, and RF methods on the selected data sets along with the standard deviations in parenthesis

Data set	RF+HC	RF+HC_CMPR	CRF	RF
Cancer	96.18 (0.32)	96.23 (1.56)	95.71 (1.01)	96.65 (1.75)
Magic	85.37 (0.46)	85.6 (0.28)	83.65 (1.3)	88.12 (0.3)
Clean	81.34 (3.25)	83.17 (4.3)	88.45 (1.55)	88.68 (2.18)
Wine	92.07 (3.29)	95.93 (1.8)	91.93 (5.91)	98.99 (0.9)
Wineqlty	65.13 (1.93)	62 (1.8)	62.79 (0.57)	68.59 (3.47)
Iris	93.36 (2.4)	94.12 (3.25)	94.4 (2.61)	96.40 (1.67)
Yeast	59.98 (1.5)	61.3 (0.7)	55.02 (2.75)	62.02 (1)
Cardio	81.74 (0.82)	82 (0.6)	84.01 (0.84)	85.67 (2.19)
BalancS	84.48 (0.52)	83.75 (2.36)	82.87 (2.86)	87.24 (1.6)
Cmc	52.87 (0.99)	52.6 (2.5)	49.42 (3.65)	52.46 (2.57)
Glass	74.33 (2.7)	73.75 (7.3)	72.77 (2.15)	78.02 (7.51)
Haber	67.69 (2.1)	69.14 (1.7)	70.2 (4.42)	73.92 (4.2)
Iono	90.14 (3.53)	91.9 (3.3)	91.45 (1.6)	93.16 (1.9)
Segment	87.54 (1.86)	89.97 (2.4)	88.86 (3.7)	93.14 (2.1)
Tae	57.60 (3.46)	53.45 (4.1)	62.29 (4.8)	55.60 (1)
Zoo	91.33 (9.6)	92.96 (5.87)	93.94 (8.2)	97.02 (2)
Ecoli	84.2 (3.11)	79.9 (4)	86.67 (11.54)	86.96 (1.74)
Spam	94.04 (0.71)	94.33 (0.5)	94.2 (1.05)	95.24 (0.3)
Golub	93.00 (6.7)	87.25 (7.6)	86.11 (9.62)	92.5 (4.5)
Colon	74.76 (5.26)	76.1 (3.9)	82.46 (17.94)	75.00 (11.85)
Glimo	64.11 (4.26)	66.3 (7.36)	54.9 (8.99)	71.69 (14.47)
Veer	58.27 (7.88)	63.11 (8)	60.97 (8.99)	66.43 (13.76)

4.3   Comprehensibility

Although a feature selection phase was incorporated in the CRF method, our methods were superior in the number of extracted rules in all the data sets except the Golub data set (Table 3). The number of rules extracted by RF+HC or RF+HC_CMPR in average are 0.6% of the total number of rules in RF while that of CRF is 11.66%, which demonstrates very good improvement compared to RF and CRF . The proposed methods significantly reduced the number of rules in comparison to CRF ($z = -4.06$) and as a result improved the comprehensibility. However, the difference in terms of rule numbers for the two proposed methods was not significant ($z = -1.89$). There is one dataset, i.e. Golub, for which CRF extracted only one rule. In such cases, the extracted rule is related to one class and it can only explain that class. However, there is no information and interpretation regarding the other class(es). Therefore, we believe that this type of rule set is not fully comprehensible as it cannot describe the underlying model completely. We found an issue in the implementation of the CRF method, which will affect the results. When the number of rules is reported, only the rules with the weights greater than a threshold (in this case 10e-6) are considered. However, all the extracted rules are used to do prediction for the test set which is not correct. The CRF results in Table 3 corresponds to the correct number of rules.

We used the modified version of the rule score function (i.e. equation (2)) in order to give higher priority to the more generalized rules. Table 3 shows the comparison between the original algorithm and RF+HC_CMPR. The results showed that RF+HC_CMPR have a stronger impact on the maximum rule length and also on the total number of antecedents (42% and 18% decrease respectively) in the rule set in comparison with RF+HC. In addition, we observed no significant change in the accuracy. These results indicate that RF+HC_CMPR improves the comprehensibility significantly ($z = -4.16$).

Comparing the CRF method with the two proposed methods using Wilcoxon test (critical z=-1.96) indicates that RF+HC had a significant lower maximum rule length ($z = -3.13$) and also number of antecedents ($z = -4.07$) in compare to CRF. RF+HC_CMPR was superior in all data sets in terms of maximum rule length ($z = -4.09$) and number of antecedents ($z = -4.07$) except for the maximum rule length for Golub.

One important aspect of comprehensibility is the number of rules extracted from an underlying model. However, we have to consider the importance of the tradeoff between accuracy and comprehensibility. The extracted rules should not only be concise but also have good performance on unseen samples. This is, in fact, the main objective of rule extraction. Therefore, a good rule extraction method should consider two facts simultaneously: comprehensibility and generalization ability, although it should be adjustable based on the application. For example, for the Magic dataset, RF generates 608155 rules with approximately 88% accuracy. This number of rules shows the complexity of the model for this dataset. RF+HC methods extract only about 0.4% of the RF rules and give about 85% accuracy for this data set. We still can generate fewer rules by decreasing *iniRuleNo*, although it will reduce accuracy. Therefore, what needs to be considered in order to have a fair judgment is the combination of the number of rules and accuracy. The results we have presented in this paper correspond to the smallest number of rules in order to achieve a level of accuracy as close as possible to the level of accuracy for RF. We provided the samples of extracted rules for two data sets in the Table 4.

Table 3. Each cell shows 'Number of extracted rules / Maximum length of rule / Total number of antecedents'in each method. The values in bold show the best results.

Data set	RF+HC	RF+HC_CMPR	CRF	RF
Cancer	36/8/159	**33/6/129**	463/9/1940	12075/13/65869
Magic	2604/8/8186	**2597/3/5697**	3182/37/50668	608155/58/8514170
Clean	83/15/586	**78/10/473**	104/18/947	18392/20/150309
Wine	16/8/64	**14/5/55**	176/7/619	7590/10/26784
Wineqlty	**1258**/21/12301	1259/**12/10526**	2282/24/22256	138889/30/1757860
Iris	13/6/39	**11/5/28**	43/5/145	4202/9/13222
Yeast	**1037**/25/13460	1303/**13/11621**	1836/27/18430	126936/32/1469328
Cardio	1609/20/15720	**1606/11/12951**	2121/20/19003	126412/22/1150839
BalancS	88/9/471	**83/5/339**	360/11/1768	19764/13/124447
Cmc	332/16/2390	**322/10/1818**	2025/19/14695	74257/22/754197
Glass	88/13/398	**59/8/335**	10050/12/30662	16530/16/115932
Haber	28/13/165	**25/8/140**	410/16/2417	19697/18/142512
Iono	41/11/193	**36/7/145**	155/12/784	10641/14/57312
Segment	**42**/10/267	54/**6/175**	11134/13/24065	9905/12/59837
Tae	91/13/495	**76/8/359**	177/13/997	14437/16/93530
Zoo	16/6/66	**15/4/51**	185/7/608	4954/9/17615
Ecoli	**138**/11/762	141/**7/649**	8900/14/29421	16761/16/105260
Spam	476/34/5076	**473/21/4228**	1154/41/14852	118878/44/1455859
Golub	9/3/18	6/2/10	**1/3/3**	2322/4/4939
Colon	**17/4/46**	19/3/39	27/5/85	2620/6/8154
Glimo	**9/3/23**	12/2/20	17/4/47	1953/4/4716
Veer	18/4/45	**17/3/33**	39/5/128	3254/6/8513

4.4 Complexity and Scalability

We found a significant difference in terms of computational time between our methods and CRF ($z = -4.07$). For all data sets, the RF+HC methods were superior to CRF with the exception of the Iris data set, which had only a one-second difference (Table 5). More specifically, in some cases with large numbers of classes such as Yeast, Glass, Ecoli, and Segment, our methods were 136, 310, 518, and 842 times faster than CRF respectively. We observed the same circumstance for data sets with a large number of samples such as Magic, Spam, and CMC such that RF+HC and RF+HC_CMPR were 13, 18, and 130 times faster than CRF. On average, the overhead time for the proposed methods and CRF method was 1.12, and 11.8 times respectively relative to RF time.

Moreover, we observed more computational time for CRF when there was a larger number of classes (Table 5) because the CRF method considers c classifiers

Table 4. Sample of rules extracted by the RF+HC_CMPR method from Iris and Golub data sets (The features in the data sets are shown by "V" and a subscripted number. The consequence of each rule is specified by a class label, for example, "Class 1". The value in the parenthesis is the score of the rule based on equation 2. Acc. is test set accuracy).

Iris (Acc. 98%)
$V_4 \leq 0.80$: Class 1 (38.50)
$V_3 \leq 2.70$: Class 1 (38.50)
$V_3 \leq 2.60$: Class 1 (38.50)
$V_3 \leq 4.85$ & $V_3 > 2.70$: Class 2 (20.88)
$V_2 \leq 3.05$ & $V_3 \leq 4.75$ & $V_3 > 2.45$ & $V_2 > 2.55$: Class 2 (10.50)
$V_4 > 0.80$ & $V_2 > 2.95$ & $V_4 \leq 1.70$ & $V_3 \leq 5.15$: Class 2 (5.50)
$V_4 > 1.60$: Class 3 (43.50)
Golub (Acc. 100%)
$V_{1727} > 1570.00$: Class 1 (35.73)
$V_{4572} > 1116.50$ & $V_{737} \leq 526.50$: Class 1 (18.25)
$V_{3607} \leq 13177.00$ & $V_{4005} > 44.50$: Class 1 (18.81)
$V_{4969} \leq 540.50$: Class 2 (21.00)
$V_{4648} > 489.00$: Class 2 (16.46)
$V_{4929} > 5863.00$: Class 2 (12.25)
$V_{1556} > 2699.00$ & $V_{3595} \leq 2939.50$: Class 2 (10.67)
$V_{1394} \leq 63.50$ & $V_{3776} \leq 211.00$: Class 2 (9.31)
$V_{4594} \leq 530.50$: Class 2 (6.67)

(c is number of classes) and finds a weight vector for each class. When there are a relatively large number of samples and a large number of classes simultaneously, the CRF method has an even worse performance. In addition, a large number of features can increase the computational time as CRF has a repeating feature selection step. However, in RF+HC methods, the overhead time on top of RF in RF+HC method has a strong linear correlation with the number of samples in the data sets ($R^2 = 0.994$).

4.5 Overall Comparison and Major Contributions

The major contributions of the proposed methods in comparison to RF are that they refine RF in selecting the most valuable rules, which leads to a huge decrement in the number of rules i.e. 0.6% of the random forest rules, while at the same time attaining 96% of the RF accuracy with a reasonable overhead time on top of RF time. In addition, both methods improved the comprehensibility in comparison with CRF while retaining the same accuracy. RF+HC decreased the number of rules, the maximum rule length, and the total number of antecedents by 27%, 16%, and 49% respectively in average. RF+HC_CMPR also reduced them by 25%, 50%, and 59%. The RF+HC methods decreased the computational time in 21 of the 22 data sets. Moreover, for the data sets with a large

Table 5. Computational time for RF+HC, RF+HC_CMPR, CRF, and RF in second

Dataset	RF+HC	RF+HC_ CMPR	CRF	RF
Cancer	16	16	36	5
Magic	1409	1425	19338	1050
Clean	34	34	118	26
Wine	4	5	13	1
Wineqlty	52	56	5317	17
Iris	4	9	3	1
Yeast	46	49	6276	15
Cardio	80	83	6410	31
BalancS	17	17	233	4
Cmc	36	36	4696	14
Glass	5	5	1551	1
Haber	10	10	15	2
Iono	9	9	24	3
Segment	7	7	5900	2
Tae	5	5	14	1
Zoo	3	3	14	1
Ecoli	9	9	4669	2
Spam	236	239	4479	166
Golub	230	230	253	228
Colon	56	56	62	54
Glimo	633	633	720	631
Veer	3165	3165	3558	3162

number of samples and/or a large number of classes, they were much faster (up to about 800 times) in terms of the computational time. Table 6 summarizes the overall comparisons of RF+HC and RF+HC_CMPR with the CRF method. The numbers in the table specify the average rank of each method for Friedman test computed for the mentioned criteria in the table, where lower value demonstrates the better method. The Friedman test showed significant difference between the average ranks and the mean rank for each criterion. However, the difference was marginal for the accuracy as it was also confirmed by the Wilcoxon test. These results show that our proposed methods are better than the CRF in terms of

Table 6. Comparison summary for different methods. The values are the average rank with the standard deviation in the parenthesis.

	RF+HC	RF+HC_CMPR	CRF
Accuracy	2.23 (0.81)	1.73 (0.7)	2.05 (0.9)
Rule#	1.77 (0.53)	1.32 (0.48)	2.91 (0.43)
Time	1.34 (0.24)	1.7 (0.37)	2.95 (0.21)
MaxCond	2.11 (0.26)	1.02 (0.11)	2.86 (0.35)
Cond#	2 (0)	1 (0)	3 (0)

number of rules, computational time, maximum rule length, and also number of antecedents while they keep level of accuracy as the same as CRF method.

5 Conclusions and Future Works

In this paper, we introduced new rule extraction methods from RF. Experimental results showed that these methods are superior to the CRF method in terms of comprehensibility while keeping the same level of accuracy. In addition, our methods are much more scalable than the state-of-the-art method, CRF and they can be applied more generally and on data sets with various characteristics.

This work can be extended in several different directions in future research. We plan to compare the proposed methods with other related methods, especially the ones described in [2,17,26]. Another possible direction would be improving the rule score and fitness function based on other metrics such as number of features in the extracted rule set and number of antecedents to increase the quality of rules in terms of comprehensibility. Yet another direction is to examine other heuristic search methods such as simulated annealing, tabu search, and genetic algorithms in order to find better sets of rules than those obtained with hill climbing.

Acknowledgments. This research was supported by the CRC grant 950-2-3617 and NSERC grant ORGPIN 341854. We greatly appreciate Brian MacPherson for his comments on this paper.

References

1. Alon, U., Barkai, N., Notterman, D.A., Gish, K., Ybarra, S., Mack, D., Levine, A.J.: Broad patterns of gene expression revealed by clustering analysis of tumor and normal colon tissues probed by oligonucleotide arrays. Proceedings of the National Academy of Sciences **96**(12), 6745–6750 (1999)
2. Bernard, S., Heutte, L., Adam, S.: On the selection of decision trees in random forests. In: International Joint Conference on Neural Networks, IJCNN 2009, pp. 302–307. IEEE (2009)
3. Blake, C., Keogh, E., Merz, C.J.: Uci repository of machine learning data bases MLRepository. html (1998). www.ics.uci.edu/mlearn
4. Breiman, L.: Random forests. Machine Learning **45**(1), 5–32 (2001)
5. Caruana, R., Niculescu-Mizil, A.: An empirical comparison of supervised learning algorithms. In: Proceedings of the 23rd International Conference on Machine Learning, ICML 2006, pp. 161–168. ACM (2006)
6. Demšar, J.: Statistical comparisons of classifiers over multiple data sets. J. Mach. Learn. Res. **7**, 1–30 (2006)

7. Díaz-Uriarte, R., Andres, S.A.D.: Gene selection and classification of microarray data using random forest. BMC Bioinformatics **7**(1), 3 (2006)
8. Friedman, J.H., Fisher, N.I.: Bump hunting in high-dimensional data. Statistics and Computing **9**(2), 123–143 (1999)
9. Friedman, J.H., Popescu, B.E.: Predictive learning via rule ensembles. The Annals of Applied Statistics, 916–954 (2008)
10. Golub, T.R., Slonim, D.K., Tamayo, P., Huard, C., Gaasenbeek, M., Mesirov, J.P., Coller, H., Loh, M.L., Downing, J.R., Caligiuri, M.A., et al.: Molecular classification of cancer: class discovery and class prediction by gene expression monitoring. Science **286**(5439), 531–537 (1999)
11. Huysmans, J., Baesens, B., Vanthienen, J.: Using rule extraction to improve the comprehensibility of predictive models. DTEW-KBI_0612, 1–55 (2006)
12. Johansson, U., Sonstrod, C., Lofstrom, T.: One tree to explain them all. In: 2011 IEEE Congress on Evolutionary Computation (CEC), pp. 1444–1451. IEEE (2011)
13. Latinne, P., Debeir, O., Decaestecker, C.: Limiting the number of trees in random forests. In: Kittler, J., Roli, F. (eds.) MCS 2001. LNCS, vol. 2096, pp. 178–187. Springer, Heidelberg (2001)
14. Liu, S., Patel, R.Y., Daga, P.R., Liu, H., Fu, G., Doerksen, R., Chen, Y., Wilkins, D.: Multi-class joint rule extraction and feature selection for biological data. In: 2011 IEEE International Conference on Bioinformatics and Biomedicine (BIBM), pp. 476–481. IEEE (2011)
15. Liu, S., Patel, R.Y., Daga, P.R., Liu, H., Fu, G., Doerksen, R.J., Chen, Y., Wilkins, D.E.: Combined rule extraction and feature elimination in supervised classification. IEEE Transactions on NanoBioscience **11**(3), 228–236 (2012)
16. Martinez-Muoz, G., Hernández-Lobato, D., Suárez, A.: An analysis of ensemble pruning techniques based on ordered aggregation. IEEE Transactions on Pattern Analysis and Machine Intelligence **31**(2), 245–259 (2009)
17. Meinshausen, N.: Node harvest. The Annals of Applied Statistics, 2049–2072 (2010)
18. Näppi, J.J., Regge, D., Yoshida, H.: Comparative performance of random forest and support vector machine classifiers for detection of colorectal lesions in ct colonography. In: Yoshida, H., Sakas, G., Linguraru, M.G. (eds.) Abdominal Imaging. LNCS, vol. 7029, pp. 27–34. Springer, Heidelberg (2012)
19. Nutt, C.L., Mani, D.R., Betensky, R.A., Pablo Tamayo, J., Cairncross, G., Ladd, C., Pohl, U., Hartmann, C., McLaughlin, M.E., Batchelor, T.T., et al.: Gene expression-based classification of malignant gliomas correlates better with survival than histological classification. Cancer Research **63**(7), 1602–1607 (2003)
20. Sarkar, B.K., Sana, S.S., Chaudhuri, K.: A genetic algorithm-based rule extraction system. Applied Soft Computing **12**(1), 238–254 (2012)
21. Selman, B., Gomes, C.P.: Hill-climbing search. Encyclopedia of Cognitive Science (2006)
22. Shi, T., Horvath, S.: Unsupervised learning with random forest predictors. Journal of Computational and Graphical Statistics **15**(1) (2006)
23. Song, L., Langfelder, P., Horvath, S.: Random generalized linear model: a highly accurate and interpretable ensemble predictor. BMC Bioinformatics **14**(1), 5 (2013)
24. Van Assche, A., Blockeel, H.: Seeing the forest through the trees: learning a comprehensible model from an ensemble. In: Kok, J.N., Koronacki, J., Lopez de Mantaras, R., Matwin, S., Mladenič, D., Skowron, A. (eds.) ECML 2007. LNCS (LNAI), vol. 4701, pp. 418–429. Springer, Heidelberg (2007)

25. Veer, L.J., Dai, H., Vijver, J.V.D., He, Y.D., Hart, A.A.M., Mao, M., Peterse, H.L., Kooy, K., Marton, M.J., Witteveen, A.T., et al.: Gene expression profiling predicts clinical outcome of breast cancer. Nature **415**(6871), 530–536 (2002)
26. Yang, F., Wei-hang, L., Luo, L., Li, T.: Margin optimization based pruning for random forest. Neurocomputing **94**, 54–63 (2012)
27. Zhang, H., Wang, M.: Search for the smallest random forest. Statistics and its Interface **2**(3), 381 (2009)
28. Zhou, Z.-H., Jiang, Y., Chen, S.-F.: Extracting symbolic rules from trained neural network ensembles. Ai Communications **16**(1), 3–15 (2003)

A Density-Penalized Distance Measure
for Clustering

Behrouz Haji Soleimani[1]([⊠]), Stan Matwin[1,2], and Erico N. De Souza[1]

[1] Faculty of Computer Science, Dalhousie University, Halifx, NS, Canada
{behrouz.hajisoleimani,erico.souza}@dal.ca
[2] Institute for Computer Science, Polish Academy of Sciences, Warsaw, Poland
stan@cs.dal.ca

Abstract. Measure of similarity between objects plays an important
role in clustering. Most of the clustering methods use Euclidean metric
as a measure of distance. However, due to the limitations of the partitioning clustering methods, another family of clustering algorithms called
density-based methods has been developed. This paper introduces a new
distance measure that equips the distance function with a density-aware
component. This distance measure, called Density-Penalized Distance
(DPD), is a regularized Euclidean distance that adds a penalty term to
Euclidean distance based on the difference between the densities around
the two points. The intuition behind the idea is that if the densities
around two points differ from each other, they are less likely to belong
to same cluster. A new point density estimation method, an analysis on
the computational complexity of the algorithm in addition to theoretical analysis of the distance function properties are also provided in this
work. Experiments were conducted in five different clustering algorithms
and the results of DPD are compared with that obtained by using three
other standard distance measures. Nine different UCI datasets were used
for evaluation. The results show that the performance of DPD is significantly better or at least comparable to the classical distance measures.

1 Introduction

The aim of the clustering is to determine the structure of the data to find different
groups of objects where each group contains similar objects. So, every clustering
algorithm needs a measure to calculate the similarity between different objects.
Distance and similarity are often used interchangeably. Closer their distance is,
more similar the objects are. Many standard distance measures exist and can be
used in clustering algorithms including Manhattan, Euclidean, and Mahalanobis
distances. However, Euclidean distance is the most intuitive and best-known
distance measure and is widely used in machine learning including clustering
algorithms. Some of the clustering algorithms, namely partitioning methods or
centroid-based methods such as K-means [1], use a representative for each cluster and assign points to their nearest cluster. These methods implicitly assume
spherical shaped convex clusters with equal radius because the distance function that they use does not take into account anything about the distribution

© Springer International Publishing Switzerland 2015
D. Barbosa and E. Milios (Eds.): Canadian AI 2015, LNAI 9091, pp. 238–249, 2015.
DOI: 10.1007/978-3-319-18356-5_21

of data or the radius of clusters. Therefore, they have trouble when the clusters have different variances. Model-based clustering methods (e.g. EM clustering [2]) make certain assumptions on the distribution of data and they take into account the variance of clusters but they still cannot find arbitrary shape clusters. Density-based clustering methods (e.g. DBSCAN [3], GDBSCAN [4] and OPTICS [5]) were developed to address these weaknesses. They try to find connected dense components in the data, so they can find arbitrary shape clusters. However, they have problem in finding overlapping clusters. Moreover, they will not work well with clusters of different density, because they usually use a global density threshold. In fact, when we have different levels of density in the data, both distance-based and density-based clustering methods would not perform well. Due to these drawbacks, we tried to combine the notion of density inside the distance function, but rather than building a new clustering algorithm, we combine the two concepts as a new distance function.

There are families of distance metric learning methods in which the metric learning is embedded into the clustering algorithm. Spectral clustering methods are among this category. They use the eigenvectors of the similarity matrix to transform the data into a new space in which the clustering can be done more accurately. Adaptive Metric Learning (AML) [6] and Nonlinear Adaptive Metric Learning (NAML) [7] try to improve the clustering result by a data transformation and improve the data transformation by the result of clustering iteratively. In fact, they formulate the problem as an optimization procedure in which the transformation matrix and the cluster indicator matrix are optimized in a quasi-static manner. The method stipulates that the cluster indicator matrix should be found using the kernel k-means method. Consequently, the metric learning is integrated with the clustering algorithm. [8] proposed a circular shift invariant k-means algorithm using a new distance measure. [9] introduces a point symmetry-based distance measure which is also integrated in the clustering algorithm. However, our proposed distance measure is structurally different and independent of the clustering algorithm.

Another category of unsupervised distance metric learning methods are called manifold learning methods. They try to learn an underlying low-dimensional manifold in such a way that geometric relationships between the data points are preserved. Some of the well-known methods in this category are ISOMAP [10], Locally Linear Embedding (LLE) [11], Laplacian Eigenmap [12] and Locality Preserving Projection (LPP) [13]. ISOMAP is a global method and tries to preserve the geodesic inter-point distances. LLE and Laplacian Eigenmap are local methods and try to preserve the local neighborhood structure. LPP is an extension of Laplacian Eigenmap and finds linear projective maps that preserve the neighborhood structure optimally. But what if the data does not rely on a lower dimensional manifold? This category of methods are also known as nonlinear dimensionality reduction methods, and they are different from our work. Even though each distance measure provides a metric space and it can be seen as a data transformation technique, the purpose of the dimensionality reduction methods are different.

This work proposes a new distance metric, a regularized Euclidean distance called Density-Penalized Distance (DPD). DPD takes into account the density level around the points and adds a penalty to the Euclidean distance. The DPD is evaluated using different types of clustering algorithms including K-means [1], K-medoids [14], CLARANS [15], one density-based clustering (DBSCAN [3]) and one unnormalized Spectral clustering algorithm [16]. By adding a density regularization part to the Euclidean distance, DPD can be easily used in clustering algorithms to address some of their weaknesses, mainly the presence of clusters with different densities. This work also provides theoretical results that show DPD satisfies the axioms of a distance function. In a set of extensive experiments, we show that our proposed distance function outperforms a number of known distance measures and never underperforms them. Statistical tests show that DPD performs significantly better than standard distance metrics most of the time.

The paper is organized as follows: In section 2 the proposed distance function is explained in details. In section 3 theoretical proof that the proposed dissimilarity measure satisfies the required properties of a distance function is provided. Graphical representation of different metric spaces obtained from different distance functions are provided in section 4. Computational complexity of the method is analyzed and discussed in section 5. The experimental platform, distance metrics, clustering algorithms, performance evaluation measures and statistical tests are explained in section 6. And finally, the results and comparison between different distance functions are presented in section 7.

2   Proposed Distance Measure

The proposed distance function is a density-regularized distance measure and it equips the distance function with a density-aware component. Therefore, first a measure of density should be defined in order to calculate the density around each point. There are several standard techniques for density estimation which have the general form of $\frac{k}{Vn}$ where n is the total number of points, V is usually a fixed volume around the point and k is the number of points inside that volume. One of the well-known density estimation methods is the Parzen window density estimation which is also known as kernel density estimation [17][18]. It is a non-parametric probability density function estimation. For estimating the density at point x, it places a window function at that point and counts the number of points inside the window weighted by their distance to x. The general equation for kernel density estimation is:

$$P(x) = \frac{1}{n} \sum_{i=1}^{n} \frac{1}{h^d} K(\frac{x - x_i}{h}) \qquad (1)$$

where d is the number of dimensions, $K(.)$ is the window function or kernel function in d-dimensional space and $h > 0$ is the bandwidth parameter of the kernel or the window size, so h^d would be an approximation of the volume.

$K(.)$ is usually chosen to be a Gaussian kernel in which case h would be the standard deviation of the Gaussian.

Since the main contribution of the work is to equip the distance function with a density-aware component, one can use any arbitrary method of density estimation. However, not all density estimation methods when added to the distance function will satisfy the properties of the distance. Therefore, a new point density estimation method is also proposed in this work. This point density estimation method is based on the distances to the nearest neighbors. Intuitively, the distance between a data point and its nearest neighbors has an inverse relationship with its density.

Let $\mathbf{x}_i = [x_{i1}x_{i2}...x_{id}]^T$ be a point in $\Re^d$ space and r_{ij} be the distance between point $\mathbf{x}_i$ to its j-th nearest neighbor. Let we define $R_k(\mathbf{x}_i) = [r_{i1}r_{i2}...r_{ik}]^T$ to be the vector of the distances between point $\mathbf{x}_i$ and its k nearest neighbors. The norm of this vector can be used as a sparseness measure around point $\mathbf{x}_i$ that is inversely related to its density.

Once a measure of density around points is defined, the regularization term can be defined and added to the Euclidean distance. The regularization term penalizes the distance of the two points based on the difference between the densities around the two points. The general form of the density-penalized distance (DPD) measure can be written as follows:

$$DPD_k(\mathbf{x}_i, \mathbf{x}_j) = d(\mathbf{x}_i, \mathbf{x}_j) + \lambda \|R_k(\mathbf{x}_i) - R_k(\mathbf{x}_j)\|_p \qquad (2)$$

where $d(\mathbf{x}_i, \mathbf{x}_j)$ is the Euclidean distance between $\mathbf{x}_i$ and $\mathbf{x}_j$, $R_k(\mathbf{x}_i)$ is the vector of distances between $\mathbf{x}_i$ and its k nearest neighbors and $\lambda > 0$ is the regularization parameter to balance the Euclidean distance and the penalty term.

If the density around $\mathbf{x}_i$ and $\mathbf{x}_j$ are similar, the penalty term would be close to zero and the density-penalized distance between $\mathbf{x}_i$ and $\mathbf{x}_j$ would be close to their Euclidean distance. The larger the difference between densities, the more penalty is added, and the density-regularized distance between the points grows. In fact, this regularization term tries to push clusters of different densities away from each other. Our distance function implicitly assumes a fairly uniform density for each of the clusters and it tries to preserve the distribution of data inside clusters. At the same time, it tries to reshape the global structure of clusters by pushing them away from each other.

Our proposed distance when used in the clustering context assumes uniform density inside each cluster but not in the entire data. This means that different clusters can have different densities. Only the points that belong to the same cluster are encouraged to have similar density which is a fair assumption. Most of the density-based clustering methods make stronger assumptions on the density. For instance, DBSCAN [3] assumes uniform density in all clusters by using a global threshold (ϵ) for detecting the core points.

Regarding the parameters of the distance function, considering the fact that L_p-norms have the property: $\|x\|_1 \geq \|x\|_2 \geq ... \geq \|x\|_\infty$, we suggest that $\lambda = \frac{1}{\sqrt[p]{k}}$ where k is the size of nearest neighbors distance vector and p comes from the p-norm that we are using. This way, λ is adaptive and it depends on the average number of terms that are effectively contributing to the value of norm. Experiments show that such adaptive λ results in a good distance, balancing Euclidean distance and density penalty term, and its performance is not significantly different from the best λ that is found by tuning. Consequently, the method is not sensitive to the value of k and no parameter tuning is needed as we used L_2 regularization with a fixed k in all of our experiments. Empirically, the performance of the distance function is the same with any value of k in a reasonable range $5 \leq k \leq 20$. So, the density-penalized distance measure can be written as:

$$DPD_k(\mathbf{x}_i, \mathbf{x}_j) = \|\mathbf{x}_i - \mathbf{x}_j\|_2 + \frac{1}{\sqrt[p]{k}}\|R_k(\mathbf{x}_i) - R_k(\mathbf{x}_j)\|_p \tag{3}$$

3 Properties of the Distance Function

In this section we will provide a proof that our density-penalized distance measure has all the required properties of a distance function. A metric d on set X is a function that maps $X \times X$ to a real number, $d : X \times X \to \Re$. Each metric is required to satisfy four properties: non-negativity, identity of indiscernible, symmetry and triangle inequality.

- Non-negativity: $DPD_k(\mathbf{x}_i, \mathbf{x}_j) \geq 0$. $DPD_k(\mathbf{x}_i, \mathbf{x}_j)$ is the weighted sum of two L_p-norms. Therefore, $DPD_k(\mathbf{x}_i, \mathbf{x}_j) \geq 0$.
- Identity of indiscernible: $DPD_k(\mathbf{x}_i, \mathbf{x}_j) = 0$ if and only if $\mathbf{x}_i = \mathbf{x}_j$. Obviously when $\mathbf{x}_i = \mathbf{x}_j \Rightarrow \|\mathbf{x}_i - \mathbf{x}_j\|_2 = 0$. Additionally, $\mathbf{x}_i = \mathbf{x}_j \Rightarrow R_k(\mathbf{x}_i) = R_k(\mathbf{x}_j) \Rightarrow \|R_k(\mathbf{x}_i) - R_k(\mathbf{x}_j)\|_p = 0$. Therefore, $DPD_k(\mathbf{x}_i, \mathbf{x}_j) = 0$. The backward direction is also obvious. When $DPD_k(\mathbf{x}_i, \mathbf{x}_j) = 0$ both norms have to be zero that implies $\mathbf{x}_i = \mathbf{x}_j$.
- Symmetry: $DPD_k(\mathbf{x}_i, \mathbf{x}_j) = DPD_k(\mathbf{x}_j, \mathbf{x}_i)$. L_p-distance is symmetric $\|A - B\|_p = \|B - A\|_p$. Therefore, sum of two L_p-distances is also symmetric. $DPD_k(\mathbf{x}_i, \mathbf{x}_j) = DPD_k(\mathbf{x}_j, \mathbf{x}_i)$.
- Triangle inequality: $DPD_k(\mathbf{x}_i, \mathbf{x}_l) \leq DPD_k(\mathbf{x}_i, \mathbf{x}_j) + DPD_k(\mathbf{x}_j, \mathbf{x}_l)$. If V is a complex vector space, a norm is defined as a function $f : V \to \Re$. Each norm has to satisfy the triangle inequality which can be stated as $f(x + y) \leq f(x) + f(y), \quad \forall x, y \in V$. Consequently, L_p-norms have this property $\|A + B\|_p \leq \|A\|_p + \|B\|_p$

$$\|\mathbf{x}_i - \mathbf{x}_l\|_2 = \|\mathbf{x}_i - \mathbf{x}_j + \mathbf{x}_j - \mathbf{x}_l\|_2 \leq \|\mathbf{x}_i - \mathbf{x}_j\|_2 + \|\mathbf{x}_j - \mathbf{x}_l\|_2 \tag{4}$$

$$\|R_k(\mathbf{x}_i) - R_k(\mathbf{x}_l)\|_p = \|R_k(\mathbf{x}_i) - R_k(\mathbf{x}_j) + R_k(\mathbf{x}_j) - R_k(\mathbf{x}_l)\|_p$$
$$\leq \|R_k(\mathbf{x}_i) - R_k(\mathbf{x}_j)\|_p + \|R_k(\mathbf{x}_j) - R_k(\mathbf{x}_l)\|_p \tag{5}$$

By multiplying equation 5 by a positive λ and adding it to equation 4 we will have:

$$\|\mathbf{x}_i - \mathbf{x}_l\|_2 + \lambda\|R_k(\mathbf{x}_i) - R_k(\mathbf{x}_l)\|_p \leq$$
$$\|\mathbf{x}_i - \mathbf{x}_j\|_2 + \lambda\|R_k(\mathbf{x}_i) - R_k(\mathbf{x}_j)\|_p + \|\mathbf{x}_j - \mathbf{x}_l\|_2 + \lambda\|R_k(\mathbf{x}_j) - R_k(\mathbf{x}_l)\|_p \tag{6}$$

$$DPD_k(\mathbf{x}_i, \mathbf{x}_l) \leq DPD_k(\mathbf{x}_i, \mathbf{x}_j) + DPD_k(\mathbf{x}_j, \mathbf{x}_l) \tag{7}$$

4   Visualization of the Metric Space

So far, we have introduced a new distance measure and proved its properties. Each set equipped with a distance function which gives the distance between elements of the set, provides a metric space. When we work with numerical data, the set is the $\Re^d$ space and each data point is a member of this set $x_i \in \Re^d, i = 1, 2, ..., n$. Euclidean distance which is a special case of Minkowski distance where $p = 2$ is the most intuitive distance measure. Consequently, we usually use the Euclidean space for illustration purposes. But when we use a different distance function, the distribution of data would be different from that of Euclidean space and we need a new representation based on the new metric space. In this section, we provide a visualization for different metric spaces each of which equipped with a distance function. We will see the effect of distance function on the distribution of data and how it affects the results of a clustering method.

At first, the pair-wise distances between all the points are calculated and a matrix $D_{n \times n}$ of the pair-wise distances is obtained where n is the number of data points. Afterwards we need to find a new representation of data in $\Re^{d'}$ space and locate our data in the new metric space in such a way that it satisfies the distance matrix D. We used a classic multidimensional scaling method [19] to embed the points in a two dimensional space.

Figure 1a shows a dataset consisting of three Gaussian clusters which are distributed in a two dimensional space. Figure 1f shows the results of k-means algorithm on the original data using Euclidean distance. As it can be seen, Euclidean distance is not good for finding the optimal boundary when we have clusters with different variance or density. Figure 1 visualizes different metric spaces in pairs and in each pair the distribution of data in that metric space at the top and the results of k-means clustering using that distance function at the bottom. Figures 1b, 1c, 1d and 1e illustrate the distribution of the data with their true labels in metric spaces obtained from Manhattan distance, Chebyshev distance, Mahalanobis distance and our proposed density-penalized distance, respectively. Figures 1g, 1h, 1i and 1j display the results of k-means algorithm in the new metric spaces obtained from Manhattan distance, Chebyshev distance, Mahalanobis distance and our proposed DPD, respectively. As it can be seen from the figures, the distance function has a strong influence on the result of the clustering. In the presence of clusters with different densities, DPD makes the boundary between the clusters more clear by pushing the clusters away from each other.

5   Computational Complexity

The proposed distance measure can be either used inside the clustering algorithm instead of Euclidean distance (i.e. online usage) or it can be used to restructure the data before running the clustering algorithm (i.e. offline usage). Let n be the number of points, k be the number of nearest neighbors, d be the number of dimensions, c be the number of clusters and i be the number of iterations

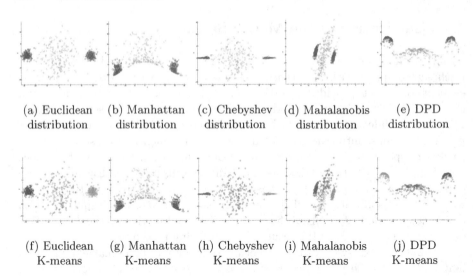

(a) Euclidean distribution (b) Manhattan distribution (c) Chebyshev distribution (d) Mahalanobis distribution (e) DPD distribution

(f) Euclidean K-means (g) Manhattan K-means (h) Chebyshev K-means (i) Mahalanobis K-means (j) DPD K-means

Fig. 1. Visualization of the different metric spaces

needed in the algorithm. In offline usage, pairwise Euclidean distances are needed and can be found in $O(dn^2)$. Afterwards, the k nearest neighbors for all the points can be found using a linear scan in $O(kn^2)$. Then the pairwise density regularized part can be calculated in $O(kn^2)$ and it can be added to the Euclidean distances to form the final DPD distance matrix in $O(n^2)$. Therefore, in this case the total computational complexity for transforming the space would be $O(dn^2 + kn^2 + kn^2 + n^2) = O((d+k)n^2)$. This is quadratic in terms of number of data points $O(n^2)$ and it will be added to the computational complexity of the clustering algorithm. The offline approach is a bit costly because the full distance matrix is needed for the transformation.

In the online approach the cost will depend on the clustering algorithm. For instance, lets consider k-means clustering with $O(idcn)$ runtime. At each iteration of the algorithm we have to calculate the distances between the centroids and the points. The Euclidean distances can be calculated in $O(cnd)$. But for DPD we also need the point density estimates for the centroids and all the points. k nearest neighbors of $(c+n)$ points among n points and their corresponding distances can be calculated using fast nearest neighbor search methods [20][21] in $O(dk(c+n)\log n)$. Then the density regularized part can be calculated in $O(cnk)$ and the final DPD distances can be calculated in $O(cn)$. Therefore, the total computational complexity for each iteration of k-means will be $O(cnd + dk(c+n)\log(n) + cnk + cn) = O((d+k)cn + dkn\log n)$. k is a constant but d can be arbitrarily large depending on the underlying data. So, we can assume $d + k \approx d$. Consequently, the total complexity of k-means clustering using DPD will be $O(idcn + idkn\log n)$. In cases that the dataset consists of large number of tiny clusters (e.g. most image recognition datasets such as fingerprint identification or face recognition), c will become larger than $k\log n$.

Therefore, $idcn > idkn \log n$ and the complexity of k-means using DPD will become $O(idcn)$ which is equal to the original k-means complexity. It means that our proposed distance measure will not change the runtime order. But usually that is not the case and we have small number of clusters with respect to sample size that means $idkn \log n > idcn$. In this scenario the computational complexity of k-means using DPD will be $O(idkn \log n)$. If we assume $k \approx c$ then DPD is slowing down the k-means clustering by a factor of $O(\log n)$. In fact, in the worst case the proposed DPD is slower than Euclidean distance by a factor of $O(\log n)$, in return it gives higher accuracy.

6   Experiment Setup

Our density-penalized distance measure can be employed in any clustering algorithm. Three different distance-based (K-means [1], K-medoids [14] and CLARANS [15]), one density-based (DBSCAN [3]) and one Spectral clustering method [16] are chosen in order to examine the effect of distance measure. For the Spectral clustering, the unnormalized version with k-nearest neighbor similarity graph is used. For the DBSCAN clustering algorithm, the same parameter settings are used for all distance functions. The $minPts$ parameter is set to 5 and the ϵ parameter is chosen adaptively in such a way that in average each point have four neighbors in its ϵ-neighborhood. Since the comparison of distance measures is the main concern and not the clustering methods, the aforementioned methods are chosen as representatives. Three well-known distance measures including Euclidean, Mahalanobis and Manhattan distances are selected for evaluation and comparison. Each of the clustering methods is performed using all four distance measures and the results are compared.

Nine popular datasets from UCI repository [22] are selected for evaluation: abalone, white wine quality, ecoli, glass identification, iris, ionosphere, libras movement, thyroid disease and handwritten digit recognition. Each dimension of the datasets are normalized independently to have zero mean and one standard deviation.

In classification datasets, the true labels are available and the performance of the clustering method is estimated based on how good the method can reproduce the labels. Well-known performance measures for clustering are Rand index [23], F-measure [24] and Normalized Mutual Information (NMI) [25]. In this work, Adjusted Mutual Information (AMI) [26] is used for evaluation. AMI is a modified version of mutual information that removes the effect of chance. The term chance here means getting a positive performance for a random labeling.

Each method is run 20 times and the average and standard deviation of the performance is used for comparing different distance functions. Student's t-test is used to determine whether the performances are significantly different.

7   Experimental Results

In all of the experiments, L_2 regularization ($p = 2$) and $k = 15$ is used in our distance function. Additionally, it is assumed that the number of clusters

Table 1. K-means performance (Adjusted MI) on UCI datasets using different distance functions. The first row on each dataset represents the average and standard deviation of the performance index and the best performing one is represented as boldface. The second row on each dataset represents the p-values of the statistical t-test against the best performing one. Whenever a method is significantly worse than the best one, it's p-value is indicated as boldface.

	Manhattan	Mahalanobis	Euclidean	DPD
Abalone	0.140 ± 0.002	0.144 ± 0.004	0.143 ± 0.002	**0.146 ± 0.002**
	$< 10^{-4}$	**0.0240**	**$< 10^{-4}$**	1.0000
Wine Q. (W)	0.057 ± 0.005	0.439 ± 0.470	0.247 ± 0.386	**1.000 ± 0.000**
	$< 10^{-4}$	**$< 10^{-4}$**	**$< 10^{-4}$**	1.0000
Ecoli	0.509 ± 0.030	0.526 ± 0.031	0.519 ± 0.033	**0.543 ± 0.022**
	0.0003	0.0641	**0.0129**	1.0000
Glass Ident.	0.276 ± 0.034	0.222 ± 0.043	0.285 ± 0.040	**0.311 ± 0.023**
	0.0005	**$< 10^{-4}$**	**0.0149**	1.0000
Iris	**0.635 ± 0.071**	0.515 ± 0.175	0.613 ± 0.058	0.618 ± 0.036
	1.0000	**0.0075**	0.2925	0.3430
Ionosphere	0.101 ± 0.002	0.067 ± 0.067	0.121 ± 0.002	**0.437 ± 0.000**
	$< 10^{-4}$	**$< 10^{-4}$**	**$< 10^{-4}$**	1.0000
Libras Movement	0.488 ± 0.026	0.137 ± 0.022	**0.511 ± 0.019**	0.501 ± 0.020
	0.0028	**$< 10^{-4}$**	1.0000	0.1290
Thyroid Disease	**0.528 ± 0.058**	0.169 ± 0.052	0.482 ± 0.006	0.520 ± 0.046
	1.0000	**$< 10^{-4}$**	**0.0011**	0.6100
Handwrtn Digit	0.512 ± 0.029	0.049 ± 0.014	0.505 ± 0.027	**0.518 ± 0.025**
	0.4492	**$< 10^{-4}$**	0.1210	1.0000

is known, so the right number is used in all the experiments. All clustering methods are executed on nine UCI datasets using all distance functions and the performances are measured using Adjusted Mutual Information.

Table 1 compares the results of different distance functions when used in k-means clustering. The performance is measured by adjusted mutual information and the average and standard deviation over 20 different runs are reported. The best performing distance function in each dataset is represented as boldface and all other methods are tested for statistical significance against the best one. The p-values of the t-test are reported in the second row of each dataset. It can be seen from the table that our method has the highest performance in six datasets and the difference is statistically significant most of the time. In four datasets, Abalone, Wine, Glass and Ionosphere, the proposed DPD is significantly better than any other distance function. Another point that can be observed from the table is that whenever DPD is not the best performing one, its performance is almost equal to the best one and their difference is not significant. In fact, DPD is never significantly worse than any other distance function.

Table 2 compares the Euclidean distance and our proposed density-penalized distance in all clustering methods evaluated by adjusted mutual information.

Table 2. Adjusted MI for different clustering algorithms using Euclidean distance and DPD on UCI datasets. The first column for each method represents the average value of the performance index and the second column is the p-value of the statistical t-test. Whenever the difference is statistically significant, the best performance and its associated p-value is indicated as boldface.

	Dist.	DBSCAN		Spectral		K-means		K-medoids		CLARANS	
		Acc	p-val	Acc	p-val	Acc	p-val	Acc	p-val	Acc	p-val
Abalone	Euc.	**.12**	$<10^{-4}$	**.14**	$<10^{-4}$	.14±.00	$<10^{-4}$	.14±.00	.6229	.14±.00	.9496
	DPD	.11		.13		**.15±.00**		.14±.00		.14±.00	
Wine (W)	Euc.	-2.92	$<10^{-4}$	1.00	1.0000	.25±.39	$<10^{-4}$	.71±.45	.7528	.62±.48	.5259
	DPD	**1.00**		1.00		**1.00±.00**		.76±.42		.71±.45	
Ecoli	Euc.	**.31**	$<10^{-4}$	.43	$<10^{-4}$	.52±.03	.0129	.47±.04	.9830	**.51±.02**	.0070
	DPD	.29		**.45**		**.54±.02**		.47±.04		.50±.02	
Glass Ident.	Euc.	.00	1.0000	.20	$<10^{-4}$	.28±.04	.0149	.24±.04	.0167	.26±.03	$<10^{-4}$
	DPD	.00		**.30**		**.31±.02**		**.27±.04**		**.31±.02**	
Iris	Euc.	.58	1.0000	.48	$<10^{-4}$	.61±.06	.7555	.64±.07	.9666	**.69±.03**	.0007
	DPD	.58		**.57**		.62±.04		.63±.04		.66±.02	
Ion.	Euc.	.26	$<10^{-4}$	.07	$<10^{-4}$	.12±.00	$<10^{-4}$	.10±.05	$<10^{-4}$	.11±.01	$<10^{-4}$
	DPD	**.27**		**.15**		**.44±.00**		**.23±.11**		**.14±.02**	
Libras Mov.	Euc.	.25	$<10^{-4}$	.50	$<10^{-4}$	.51±.02	.1290	.45±.03	.6422	.47±.02	.0002
	DPD	**.31**		**.55**		.50±.02		.46±.04		**.50±.02**	
Thyroid Disease	Euc.	.00	1.0000	.24	$<10^{-4}$	.48±.01	.0009	.31±.10	.2319	.50±.06	$<10^{-4}$
	DPD	.00		**.34**		**.52±.05**		.35±.10		**.58±.05**	
Handwrtn Digit	Euc.	**.52**	$<10^{-4}$	.54	$<10^{-4}$	.51±.03	.1210	.35±.03	.1502	.35±.03	.0413
	DPD	.46		**.57**		.52±.03		.34±.04		**.37±.02**	

Each method is run 20 times and the average performance and also the p-value of the statistical t-test are reported. Whenever the difference between the performances is significant, the best performing one and the corresponding p-value are represented as boldface. Since DBSCAN and Spectral clustering have no variation on the results, the variance is not reported and the difference is always considered significant. It can be seen that the proposed distance function outperforms Euclidean distance significantly in most cases. In DBSCAN the performance of the two distance functions are close to each other. In K-medoids, DPD is significantly better than Euclidean only in two datasets but in the rest the difference is not significant. However, in K-means, CLARANS and Spectral clustering, DPD is outperforming Euclidean distance in a significant way in six, five and seven datasets, respectively. Spectral clustering which can also be considered as a density-based approach has higher accuracy than DBSCAN in average. But DPD still has improved it a lot. It means that some clustering methods have more potential to be enhanced if they are equipped with an appropriate distance function.

8   Conclusion

In this paper, a new distance metric, called DPD is proposed. Even though this distance function is primarily designed to improve the performance of clustering methods, it is independent of the clustering algorithm and it can be used even in different contexts. A new point density estimation method, an analysis on the computational complexity of the method as well as theoretical proof for satisfying the required properties of a distance function are provided in this work. Extensive experiments in five different clustering algorithms show that our density-penalized distance measure improves the quality of clustering. Statistical tests show that our combined measure's performance is significantly better than other standard distance functions in most cases or at least it is comparable to the best performing one. Using DPD in contexts other than clustering (e.g. classification) is one of our possible future works. Another direction for the future work is to look into geometrical and statistical interpretation of DPD as well as running a sensitivity analysis on the parameter k. We will also examine the inherent characteristics of different clustering algorithms in order to identify when DPD can make the most improvement.

Acknowledgments. The authors wish to thank Natural Sciences and Engineering Research Council of Canada (NSERC) for providing support during this research.

References

1. MacQueen, J.: Some methods for classification and analysis of multivariate observations. In: Proc. of the 5th Berkeley Symp. on Math. Stat. and Prob., vol. 1, pp. 281–297 (1967)
2. Dempster, A.P., Laird, N.M., Rubin, D.B.: Maximum likelihood from incomplete data via the EM algorithm. J. of the Royal Stat. Society **39**, 1–38 (1977)
3. Ester, M., Kriegel, H.P., Sander, J., Xu, X.: A density-based algorithm for discovering clusters in large spatial databases with noise. In: Proc. of KDD, pp. 226–231 (1996)
4. Sander, J., Ester, M., Kriegel, H.P., Xu, X.: Density-based clustering in spatial databases: The algorithm gdbscan and its applications. Data Mining and Knowledge Discovery **2**, 169–194 (1998)
5. Ankerst, M., Breunig, M.M., Kriegel, H.P., Sander, J.: Optics: Ordering points to identify the clustering structure. ACM SIGMOD Record **28**, 49–60 (1999)
6. Ye, J., Zhao, Z., Liu, H.: Adaptive distance metric learning for clustering. In: IEEE Conference on Computer Vision and Pattern Recognition (CVPR), pp. 1–7 (2007)
7. Chen, J., Zhao, Z., Ye, J., Liu, H.: Nonlinear adaptive distance metric learning for clustering. In: Proc. of the 13th ACM SIGKDD Int. Conf. on Knowledge Discovery and Data Mining, pp. 123–132 (2007)
8. Charalampidis, D.: A modified k-means algorithm for circular invariant clustering. IEEE Trans. on Pattern Analysis and Machine Intelligence (PAMI) **27**, 1856–1865 (2005)
9. Bandyopadhyay, S., Saha, S.: GAPS: A clustering method using a new point symmetry-based distance measure. Pattern Recognition **40**, 3430–3451 (2007)
10. Tenenbaum, J.B., De Silva, V., Langford, J.C.: A global geometric framework for nonlinear dimensionality reduction. Science **290**, 2319–2323 (2000)
11. Roweis, S.T., Saul, L.K.: Nonlinear dimensionality reduction by locally linear embedding. Science **290**, 2323–2326 (2000)
12. Belkin, M., Niyogi, P.: Laplacian eigenmaps for dimensionality reduction and data representation. Neural Computation **15**, 1373–1396 (2003)

13. Niyogi, X.: Locality preserving projections. Neural Information Processing Systems **16**, 153 (2004)
14. Kaufman, L., Rousseeuw, P.J.: Finding groups in data: an introduction to cluster analysis. John Wiley & Sons (2009)
15. Ng, R.T., Han, J.: CLARANS: A method for clustering objects for spatial data mining. IEEE Transactions on Knowledge and Data Engineering **14**, 1003–1016 (2002)
16. Luxburg, U.V.: A tutorial on spectral clustering. Statistics and Computing **17**, 395–416 (2007)
17. Scott, D.W.: Multivariate density estimation: theory, practice, and visualization. John Wiley & Sons (2009)
18. Kim, J., Scott, C.D.: Robust kernel density estimation. The Journal of Machine Learning Research (JMLR) **13**, 2529–2565 (2012)
19. Seber, G.A.: Multivariate observations. John Wiley & Sons (1984)
20. Vaidya, P.M.: An $O(n \log n)$ algorithm for the all-nearest-neighbors problem. Discrete & Computational Geometry **4**, 101–115 (1989)
21. Clarkson, K.L.: Fast algorithms for the all nearest neighbors problem. In: IEEE 24th Annual Symposium on Foundations of Computer Science (FOCS), pp. 226–232 (1983)
22. Bache, K., Lichman, M.: UCI machine learning repository. Univ. of California, Sch. of Inf. and Computer Science, Irvine (2013)
23. Rand, W.M.: Objective criteria for the evaluation of clustering methods. J. of the American Stat. Assoc. **66**, 846–850 (1971)
24. Hripcsak, G., Rothschild, A.S.: Agreement, the f-measure, and reliability in information retrieval. J. of the American Medical Informatics Assoc. **12**, 296–298 (2005)
25. Strehl, A., Ghosh, J.: Cluster ensembles–a knowledge reuse framework for combining multiple partitions. The Journal of Machine Learning Research (JMLR) **3**, 583–617 (2003)
26. Vinh, N.X., Epps, J., Bailey, J.: Information theoretic measures for clusterings comparison: is a correction for chance necessary?. In: Proc. of the 26th Int. Conf. on Machine Learning (ICML), pp. 1073–1080 (2009)

Learning Paired-Associate Images with an Unsupervised Deep Learning Architecture

Ti Wang and Daniel L. Silver[⊠]

Jodrey School of Computer Science, Acadia University,
Wolfville, NS B4P 2R6, Canada
danny.silver@acadiau.ca

Abstract. This paper presents an unsupervised multi-modal learning system that learns associative representation from two input modalities, or channels, such that input on one channel will correctly generate the associated response at the other and *vice versa*. In this way, the system develops a kind of supervised classification model meant to simulate aspects of human associative memory. The system uses a deep learning architecture (DLA) composed of two input/output channels formed from stacked Restricted Boltzmann Machines (RBM) and an associative memory network that combines the two channels using a simple back-fitting algorithm. The DLA is trained on and pairs of MNIST handwritten digit images to develop hierarchical features and associative representations that are able to reconstruct one image given its paired-associate. Experiments show that the multi-modal learning system generates models that are as accurate as back-propagation networks but with the advantage of a bi-directional network and unsupervised learning from either paired or non-paired training examples.

1 Introduction

Humans learn knowledge from the environment by data that is provided in several forms, or *modalities*, such as audio and visual signals. Psychologists define multi-modal learning as learning new knowledge from multiple sensory modalities [15]. Researchers have shown that people's understanding of new concepts is enhanced with mixed-modality knowledge representations [14]. The human brain has adapted to fuse associated sensory signals so as to learn more effectively and efficiently. The long-term goal of this research is to develop a learning system that simulates aspects of the multi-modal learning ability of humans. In particular, we investigate unsupervised learning methods that can create generative models capable of generalization and classification from one input or output modality to another (eg. from visual to verbal). We are interested in how this can be done without resorting to multi-layer supervised learning that always requires matching modality values and must backward propagate an error signal from one layer to another. As we will discuss, back-propagation solutions do not scale well as the number of modalities increases.

© Springer International Publishing Switzerland 2015
D. Barbosa and E. Milios (Eds.): Canadian AI 2015, LNAI 9091, pp. 250–263, 2015.
DOI: 10.1007/978-3-319-18356-5_22

Deep learning is a sub-area of machine learning, which typically uses Restricted Boltzmann Machines (RBM), a type of stochastic associative artificial neural network (ANN), to develop a multi-layer generative models [8]. Deep learning architectures, or DLA, provide an exciting new substrate upon which to explore new computational and representational models of how knowledge can be acquired, consolidated and used [1]. Prior work has investigated the use of DLAs and unsupervised learning methods to develop models for a variety of purposes including auto-associative memory, pattern completion, and clustering as well as generalization and classification [11].

This paper takes a first step toward developing a multi-modal sensory/motor learning system by examining a DLA that is capable of learning images presented to either of two input modalities (channels) and associating paired images when presented at both modalities. The challenge for the DLA is to reconstruct the matching image at channel A when it observes it's paired image at channel B, and *vice versa*. By doing so the system creates a supervised classification model from one channel to another. The system uses unsupervised learning to develop generative RBM channels for non-paired images and then associates paired images across these channels by using a simple back-fitting algorithm that fine-tunes the weights leading to a large RBM layer shared by both channels. Thus, the DLA can learn not only paired-associate examples, but also non-paired independent examples at each sensory modality. Experimentation shows quantitatively and qualitatively that the system generates models that accurately generates associated images as compared to models developed using traditional supervised back-propagation networks.

2 Background

Artificial neural networks (ANN) are widely used to solve classification problems such as image and speech recognition, however many do not work in the same fashion as the human nervous system. For example, back-propagation ANNs are good for modeling complex mapping relations between input and output data, but are not as good for reconstructing, or recalling a pattern. Humans have the ability to recover complete information from partial information; this is referred to as associative memory [5]. When a child watches a tennis game, he or she learns the appearance of the tennis ball and the racket. Next time when the child sees a picture of a tennis ball, the child may recall an image of a racket and of the game. Associations are clearly a major part of learning about the world.

Associative ANNs are inspired by cognitive psychology and are designed to mimic the way that collections of biological neurons may store and recall associative memories [16]. Geoffrey Hinton, University of Toronto, advocates using Boltzmann Machine associative networks to simulating human brain structure. After a Boltzmann Machine has been trained on a set of patterns, it has the ability to reconstruct any one of those patterns from a partial or noisy pattern. However, learning is slow in large Boltzmann Machines because of the many weights in a fully connected network and the iterative sampling of node activities required for each weight update. The time complexity of a BM in order to reach its equilibrium distribution grows exponentially with the number of neurons. This limitation leads to the impracticality of using BMs to achieve learning in practical applications using current computing hardware.

2.1 Restricted Boltzmann Machine

A Restricted Boltzmann Machine (RBM) is a variant of a BM that is meant to overcome long training times by limiting the number of connections in its network and using a modified learning algorithm. RBMs have both visible and hidden layers of neurons just like BMs, however there are no intra-layer connections, so they can be characterized as a bipartite graph (see Figure 1) [11]. When settling to equilibrium, neuron h_j turns on with the probability $p_j = \frac{1}{1+exp(-b_j - \sum_i w_{ij}v_i)}$, and neuron v_i turns on with the probability $p_i = \frac{1}{1+exp(-b_i - \sum_j w_{ij}h_j)}$. The states v_i, h_j of neuron i and j keep changing with probabilities p_i and p_j. The system computes the activation energy $E = -\sum_i b_i v_i - \sum_j b_j h_j - \sum_i \sum_j v_i h_j w_{ij}$ where b_i and b_j are the bias terms for their respective nodes [12]. The global energy E will be reduced more quickly in an RBM compared to a BM because of the reduced number of connections. The goal of training is to modify the weights of the network to establish low energy states that correspond with training patterns at the visible nodes. Similar input patterns will have energy states closer to each other, whereas two orthogonal patterns (e.g. patterns that share few common pixels) will have energy states more distant from each other.

The method of weight update we use for this research is called Contrastive Divergence, or CD [11]. The weights of the network are initialized to small random values. When training data x_i is given to the visible neuron v_i, the RBM clamps the states of visible neurons and frees the states of hidden binary neuron h_j (see Figure 1). Each weight w_{ij} of the RBM is updated as per the following formula $\Delta w_{ij} = \eta(<v_i h_j>^0 - <v_i h_j>^1)$, where η is the learning rate, $<v_i h_j>$ is the expectation over all possible pairs of visible and hidden node values, and the 0 and 1 superscripts indicate the expectation based on the training example and its reconstruction, respectively. This equation approximates the gradient of the log probability of a training example with respect to a weight. Weight w_{ij} is updated until the global energy E (for all training examples) reduces below a threshold. With probability p_i, neuron i will then reconstruct the input data x_i. The derivative of the log probability of training data simplifies the training process, such that the system updates the weight set by taking the difference between the pairwise statistics $<v_i h_j>_0$ (from the original input data) and $<v_i h_j>_1$ (from the data reconstructed by doing Gibbs Sampling) [9]. The CD training procedure is an approximation of the Maximum-Likelihood learning algorithm [7,11] that is guaranteed to not get stuck in a local minimum [9].

The more Gibbs sampling steps used, the better model an RBM will develop [4,9]. CD_n denotes the training process with n full steps of Gibbs sampling. Hence the weight change of CD_n is updated by the formula $\Delta w_{ij} = \eta(<v_i h_j>^0 - <v_i h_j>^n)$. During the alternating processes of learning and reconstruction, the global energy $E<v, h>$ and the global temperature of the RBM

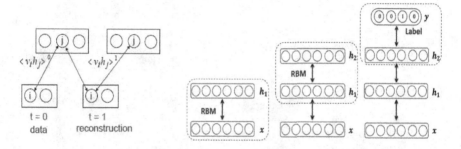

Fig. 1. RBM Training Process   **Fig. 2.** Stacking Multi-level RBMs

will drop quickly [9]. The system is trained until the hidden layer is capable of reconstructing the given input pattern from its hidden representation with the desired amount of accuracy. After training, the hidden layer weights of the RBM represent the hidden neuron feature distribution over the input space; that is $w_{ij} = p(h_j|v_i)$, gives the posterior probability of feature h_j given input v_i.

To test its ability to recall a pattern, the RBM is presented with all or some of the inputs x_i of a test example at its visible units v_i. These cause activations at each of the hidden units h_j as described above, and then the visible units are freed to generate new activations. If training has been successful, the reconstructed outputs at v_i are close to the complete pattern of the original test example.

2.2 Deep Learning Architectures

Humans tend to organize ideas and concepts hierarchically [6]. Abstract concepts are learned and recalled through the composition of simpler concepts [1]. This approach makes sense in a world where most objects are made from parts which are in turn composed of smaller features. For instance, a car is a combination of smaller parts like wheels and a frame. And a wheel is made up of smaller features like a tire and a rim. Deep learning methods aim at learning feature hierarchies with features from higher levels of the hierarchy formed by the composition of lower level features [1]. A mainstream theory is that the mammalian brain uses a deep learning architecture with multiple levels of abstraction corresponding to different areas of the neocortex [18].

Deep learning architectures, or DLA, is a sub-area of machine learning that places heavy emphasis on hierarchical composition and unsupervised learning methods. Deep Belief DLAs can be developed by stacking layers of RBMs one on top of another [11], and they can be trained sequentially from bottom to top, as shown in Figure 2. The hidden layer, h_1, of the bottom RBM learns the feature distribution of the input patterns and attempts to reconstruct them [12]. The training of this layer continues until either the system reaches a maximum number of iterations or the MSE between the reconstructed pattern and the input signal reaches a predetermined threshold. Then the next hidden layer, h_2, learns a more abstract feature distribution of layer h_1. Similarly, the posterior

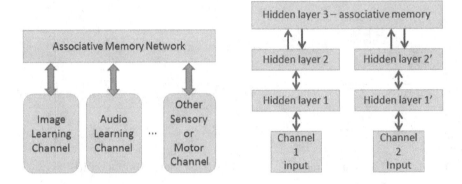

Fig. 3. Multi-modal data learning system   **Fig. 4.** Two channels DLA

probabilities of hidden layer h_2 of the second RBM will be used as the visible layer for the third RBM. The top RBM will learn an even more abstract feature distribution of h_2 in the top hidden layer h_3 [13].

RBM-based DLA systems have been successfully used to develop models for recognizing hand-writing images of digits in a manner that simulates the human visual cortex [8]. They are capable of doing unsupervised *clustering* of unlabeled data based on a hierarchy of features. As shown in Figure 2, the hidden layer of one RBM can be used as the input layer for a higher level RBM [1]. The highest level features can be used to achieve classification, if so desired. Subsequently, researchers feel that DLAs develop a hierarchy of features in a fashion similar to the mammalian brain.

Deep Belief DLAs present a new way at looking at systems that learn. RBM DLAs can be used as an auto-encoder to model high-dimensional data, such as that of image and audio files [3]. Bengio reports that deep architectures are more expressive than shallow ones by analyzing the depth-breadth trade-off of architecture representation [2]. Perhaps most importantly, Deep Belief DLAs learn representative hierarchies directly from the data [1]. This is in contrast to alternative approaches, also considered DLAs, such as convolutional networks that use receptive fields and modified back-propagation methods that rely heavily on known topological characteristics of the input space [17].

3   Multi-modal Learning Using an Unsupervised DLA

The objective of this research is to develop a learning system that can memorize and recall multi-channel data using an associative memory network. The learning system should be able to recall the pattern from the associative network on one sensory modality given data on another sensory modality. The long-term goal of our research is to create a system that can learn concepts using three or more sensory/motor modalities, such as audio, optical, and vocal (see Figure 3).

3.1   Learning Paired-Associate Images

Consider the problem of learning paired-associate images at two input modalities (channels). We propose to use a DLA network that, after training, will be able to generate a paired image on one channel when prompted with an image on another channel. The process is meant to simulate human sensory modalities and associative memory, and to provide insights into how classification can be done using an unsupervised learning approach. As shown in Figure 4, the learning system is composed of two major parts, a associative memory network and two unsupervised RBM sensory channel networks. The sensory channel networks are designed for the recognition and reconstruction of sensory data. The associative memory network ties the sensory channel networks together to form a multimodal associative memory. Both the channels and the layer 3 associative memory can be based on RBMs.

Because of its reduced representation, the recall capacity of an RBM is not as high as a fully-connected BM. We have determined for our learning problem that an RBM is unable to accurately recall patterns when only half of the visible neurons are given correct pattern values [20] unless it contains a large amount of neurons in the hidden layer. In either case when an RBM is used as the top associative memory network, we take an additional step after the CD algorithm has completed training to fine-tune the weights connecting to this network.

To achieve pattern recall in the associative memory, the weights that connect to the top layer should be able to generate the top layer features. A fine-tuning algorithm is required to adjust the learned model such that it can generate the full features with half of the input provided. However, fine-tuning the bi-directional weights of the RBM may destroy their ability to generate lower level features. To protect the accuracy of the generative model, it is necessary to untie the weights between the top layer of each channel and the associative memory network layer and create two sets of weights - recognition weights and generative weights (see Figure 5) [10,11]. The recognition weights are used in the bottom-up pass which receives an input pattern and the generative weights are used in the top-down pass to reconstruct an output pattern. The generative weights are left as trained by the CD algorithm. The recognition weights are fine-tuned using a back-fitting algorithm as per [8], such that the associative memory network can generate a relatively accurate full set of associative memory features with only input from one channel.

Following RBM training, to fine-tune channel 1, the recognition weights w_{ij}, where i is a neuron in hidden layer 2 and j is a neuron in hidden layer 3, are used as the initial weight values for a gradient descent regression over all paired patterns. For each training pattern, the posterior probabilities $\{p_i\}$ of hidden layer 2 are used as the input attribute, and the posterior probabilities $\{p_j\}$ of hidden layer 3 are used as the target output. A new set of posterior probabilities $\{p'_j\}$ for hidden layer 3 are computed using $p'_j = \frac{1}{1+exp(-\sum_i w_{ij}p_i)}$, and the weights are updated using gradient descent to minimize the error between $\{p_j\}$ and $\{p'_j\}$. The gradient descent algorithm attempts to find a local minimum of the error function. In this way the recognition weights which pass the input

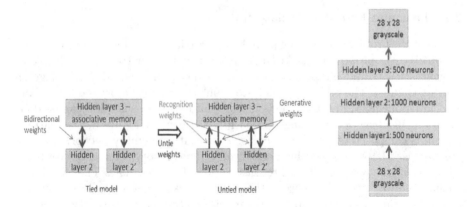

Fig. 5. Untieing the weights

Fig. 6. BP ANN in Experiment 1

signal from sensory channel 1 to the associative memory network are fine-tuned to generate a full set of associative memory features which channel 2 can use to generate the appropriate output.

We seek an unsupervised learning approach to multi-channel learning the avoids the problems that come with the use of fine-tuning weights using back-propagation supervised learning such as that used by Srivastava [19]. With local back-fitting, the multi-modal DLA should be able to achieve similar associative accuracy, Without supervised learning from h_2 to $h_{2'}$ through h_3, the performance of the DLA is unlikely to exceed that of a traditional BP ANN approach; however, we do expect it to do as well. The hierarchical feature learning of the sensory channels and the local back-fitting of the recognition weights between two layers means that our method can be used to fine-tune the associative network as the number of channels grows beyond two. This is not true of the supervised back-propagation approach because a separate set of recognition to associative network to generative weights would be needed for each combination of input-output modalities.

3.2   Impact of Learning Non-paired Patterns

Sensory data does not always come in pairs in real life. For example, one can see a cat meowing, see an image of a cat, or hear meowing without seeing a cat. In this case, the sound "meow" is the audio signal and the image of the cat is the visual signal. These two sensory channels can come together to allow paired-associate learning, but their individual channel representations can be learned and improved upon separately. We propose that learning each sensory modality with non-paired examples will help to improve the associative memories ability to generate the correct image on one channel when given its paired-associate on the other. It would be informative to have an experiment to test the impact on the multi-channel learning systems performance after separately training the sensory channels with non-paired examples.

4   Empirical Studies

Three empirical studies were carried out using two different data sets. The first and third experiments used paired images from the MNIST dataset of handwritten numeric digits. The second experiment used paired images from a synthetic dataset of numeric digits. In all experiments, five pairs of odd and even digits were associated with each: 1-2, 3-4, 5-6, 7-8, 9-0.

4.1   Experiment 1

Objective: The objective of this experiment is to compare the unsupervised DLA with a supervised BP ANN approach to learning paired-associate images. Each learning system is trained such that when a handwritten digit image is provided, the system will generate its paired digit image.

Material and Methods: This experiment uses a dataset of paired MNIST handwritten digits as the learning domain. The experiment is repeated four times with different training sets, validation sets and test sets. Each of these datasets contains 1,000 paired-associate examples that are randomly selected from the MNIST dataset.

A deep learning architecture of RBMs is used to develop an unsupervised learning model for the problem. The architecture is in accord with Figure 4. Each channel network is composed of two RBM layers, each of which contains 500 hidden neurons. Hidden layers 1 and 1' and then layers 2 and 2' will develop more abstract features of the original images [11]. The associative top layer contains 1,000 neurons. The unsupervised DLA uses back-fitting to fine-tune the weights of the associative top layer after the CD algorithm training is finished.

When training the DLAs, the training process of each sensory channel stops when the maximum iteration of 60 is reached, and the associative memory network is trained to 100 iterations. Validation sets are used to monitor the back-fitting to avoid over-fitting. The odd digit part of a test example is used to test the reconstruction of its corresponding even digit image, and *vice versa*.

Note that BP networks are feed-forward networks, hence two BP networks are required to learn the bidirectional mappings. One network is trained to map odd digit images to even digits, the other *vice versa*. Both BP networks use the architecture shown in Figure 6. The BP networks use the same training set, validation set and testing set as the DLA. The validation set is used to prevent the BP algorithm from over-fitting to the training set.

The accuracy of reconstruction is measured by testing the output images using Hinton's DLA handwritten digits classification software as an Oracle. This software is known to classify MNIST dataset of handwritten digits with only 1.15% errors [11]. One can pass the input images and the reconstructed images through *Hinton's classifier* to determine their digit category. The accuracy of the models is then based on the number of correctly paired images.

Table 1. Accuracy of test set reconstruction (%)

	1→2	2→1	3→4	4→3	5→6	6→5	7→8	8→7	9→0	0→9	Avg
DLA	95.25	95.88	82.63	94.63	92.38	88.75	90.5	79.75	91.63	93	90.74
BP ANNs	98.0	72.5	83.75	95.13	90.38	82.88	91.13	82.88	89.0	92.5	88.82

Fig. 7. Examples of reconstruction results with the DLA and BP ANNs

Results and Discussion: Using Hinton's software, the reconstruction accuracy was checked on the testing set. The average results of four replications of the experiments are shown in Table 1. On average, the unsupervised DLA (model 1) generated images that were 90.74% accurate, and the BP ANNs (model 2) generated images that were 88.82% accurate. One can see that the two models did equally well. This suggests that the unsupervised DLA models are able to achieve the same level of accuracy as the supervised BP approach.

Figure 7 shows examples of reconstructed images produced by the DLAs and the BP ANNs. One can see that the images generated by the DLAs are clearer than those generated by the BP ANNs. We suspect this because the DLA models are able to better differentiate features from noise. This will be investigated further in the next experiment.

4.2 Experiment 2

Objective: The objective of this experiment is to develop auto-associative models that can overcome noise injected into synthetic training examples. An unsupervised DLA with back-fitting and supervised BP ANNs will be developed from a noisy dataset, and the quality of their regenerated images will be compared.

Material and Methods: This experiment uses a synthetic dataset that contains five different sets of 10 x 5 paired images from Figure 8. 10% random noise was added to each template image to produce 60 instances of each category, or 300 in total. The first 100 of these images are used as a training set, the next 100 are used as a validation set, while the remaining 100 are used as a test set.

A DLA architecture, in accord with the previous experiment, is used to develop an unsupervised learning model. Each of the sensory channel layers contains 50 hidden neurons, and the associative top layer contains 100 neurons. The training process of the sensory channel networks stops when the maximum iteration of 60 is reached; the associative memory network trains for 100 iterations.

Templates **1234567890**

Fig. 8. Templates of the synthetic dataset

Table 2. RMSE of test set reconstruction (out of 1)

	1→2	2→1	3→4	4→3	5→6	6→5	7→8	8→7	9→0	0→9	Avg
DLA	0.012	0.071	0.046	0.029	0.004	0.01	0.008	0.0	0.04	0.015	0.032
BP ANNs	0.162	0.216	0.209	0.081	0.06	0.115	0.135	0.165	0.11	0.106	0.144

As in Experiment 1, two BP networks were developed to learn the same paired-associate mapping. Both BP networks used an architecture similar to that shown in Figure 6 with 50 neurons in layers 1 and 3 and 100 neurons in layer 2. The BP networks uses the same training set, validation set and test set as the DLA.

The accuracy of reconstruction was measured by comparing the similarity between the generated images and their corresponding template images for a set of test examples. The template images represent the common features across all input images. The DLA learns features from the training set, that is, it learns the template images during training. Hence the more similarities a generated image and its corresponding template image share, the better the DLA learned to regenerate this image. The RMSE between the pixels of each reconstructed image and its corresponding template (without noise) was computed to give an average error over all examples (image pixels are normalized to the range [0,1]).

Results and Discussion: The RMSE of the reconstructed images for the test set is shown in Table 2. The DLA with back-fitting out-performs the BP networks in generating the images in the presence of noise. Figure 9 shows examples of reconstructed images from the DLA and the BP ANNs. The generated images from the DLA are quite similar to the template images of Figure 8, while there is significant noise on the generated images from the BP network. DLAs attempt to probabilistically differentiate features from noises, whereas BP ANNs attempt to map input pixels to output pixels. Features are formed in BP networks, but they are for the purpose of mapping and not reconstruction of the original images. Hence a DLA is a better choice if the objective is to construct a noiseless category example as a form of classification.

4.3   Experiment 3

Objective: The preceeding experiments used paired-associate examples to develop neural network models, however, sensory data does not always come in pairs in real life. The objective of this experiment, in accord with Section 3.2, is to develop an associative learning system with both paired associative examples

Input:

Unsupervised
DLA output:

BP ANN
output:

Fig. 9. Examples of reconstruction results with DLA and BP ANNs

and independent non-paired examples. The experiment is designed to test if the performance of an associative learning system can be improved by separately training the sensory channels with non-paired examples.

Material and Methods: This experiment uses the database of MNIST examples as in Experiment 1. The experiment is repeated four times with different training sets, validation sets and test sets. For each repetition, four models are built using the same architecture but with different amounts of training examples. The first model is built with 100 paired-associate examples. The second model is built with 100 paired-associate examples, and 100 non-paired examples of even digit images. The third model is built with 100 paired-associate examples, 100 non-paired examples of even digit images, and 100 non-paired examples of odd digit images. The last model is built with 200 paired-associate examples. Figure 10 shows the number of paired and non-paired examples in each training set. All the odd digits images are used to train the odd channel and all the even digits are used to train the even channel, but only the paired-associate examples are used to develop the associative memory.

The four models use the same 3-layered architecture, parameters, validation sets and test sets as in Experiment 1. While doing back-fitting, validation sets are used to monitor overfitting. Test sets are used to examine the associative learning performance of the learning system. The odd digits are used to test the recall of even digits, and *vice versa*. The recalled images are classified by *Hinton's classifier* to examine the accuracy of the models.

Results and Discussion: The performance (averaged over four repetitions) of the four models at recalling even digits from odd digits, odd digits from even digits, and the average of them are shown in Figure 11; the error bars represent the 95% confidence over the repeated studies. The mean accuracy increases marginally (the error bars show that the improvements are not significant) from model 1 to model 3, which means that using non-paired examples to better develop one of the channels representation may improve the overall performance of an associative learning system. We conjecture that this is because both the

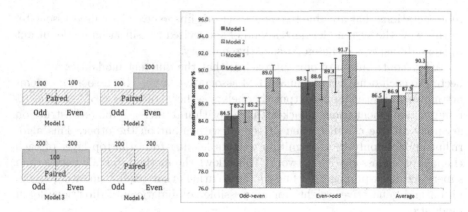

Fig. 10. The number of exam- **Fig. 11.** Accuracy comparison between four models
ples used to train four models

recognition weights and the generative weights of this channel are optimized. Improving the recognition weight performance of the odd digits channel will provide better features to the associative memory network to generate the corresponding even digits. Better generative weights for the odd digits channel will generate more accurate odd digits when even digits are provided. In general, this result suggests that improving one of the sensory channel networks of a multi-channel learning system which contains more than two channels will improve any recall that involves that channel.

It is also important to note that the reconstruction accuracy clearly increases from model 3 to model 4. This demonstrates that using more paired-associate examples to develop the associative memory network can improve the performance of the system over the equivalent number of non-paired examples. In a system with three or more channels we conjecture that paired-associate examples for any two channels will be of benefit to the entire associative memory network.

5   Conclusion

This paper presents recent work on an unsupervised multi-modal learning system that can develop an associative memory structure that connects two input/output modalities or channels. Our long-term goal is to develop learning systems that are able to develop knowledge representations from multiple sensory input and/or motor output modalities in a manner similar to humans.

The system uses a deep learning architecture (DLA) composed of two input/output channels formed from stacked Restricted Boltzmann Machines (RBM) and an associative memory network that connects the two channels. The DLA is trained using matched pairs of MNIST handwritten digit images at either channel as well as non-paired images using the unsupervised Contrastive Divergence (CD) algorithm. The DLA develops hierarchical features and associative representations that are able to reconstruct one image given its paired-associate. In this way, the system develops a kind of supervised classification model meant to simulate aspects of human associative memory.

A key contribution of our approach is that the different modalities are connected to each other via one final RBM associative layer. This associative memory network ties the input/output channels together. Following RBM training, it requires refinement using a back-fitting algorithm to increase the reconstruction accuracy on one channel when presented with input on the other. This algorithm updates only the recognition weights leading from the top layer of each channel to the associative network. This allows the method to scale better than supervised back-propagation fine-tuning solutions that require updating recognition and generative weights for all possible combinations of directions for n modalities.

Experimentation shows quantitatively (using an independent classification method) and qualitatively (by viewing the generated images) that the multimodal learning system generates models that are able to reconstruct accurate paired images as compared to supervised back-propagation network models. They also have the advantage of using bi-directional generative components and unsupervised learning from either paired or non-paired training examples.

In future work, we will move to train our system to associate three or more RBM channels with mixed modalities such as sound, image and motor input and outputs.

Acknowledgments. This research has been funded in part by the Government of Canada through NSERC.

References

1. Bengio, Y.: Learning deep architectures for ai. Found. Trends Mach. Learn. **2**(1), 1–127 (2009)
2. Bengio, Y., Lecun, Y.: Scaling learning algorithms towards AI. MIT Press (2007)
3. Deng, L., Seltzer, M.L., Yu, D., Acero, A., Mohamed, A.R., Hinton, G.E.: Binary coding of speech spectrograms using a deep auto-encoder. In: Kobayashi, T., Hirose, K., Nakamura, S. (eds.) Interspeech, pp. 1692–1695. ISCA (2010)
4. Desjardins, G., Courville, A., Bengio, Y.: Tempered markov chain monte carlo for training of restricted boltzmann machines. Technical Report 1345, Département d'Informatique et de Recherche Opérationnelle, Université de Montréal, October 2009
5. Gerrig, R.J., Zimbardo, P.G.: Psychology and Life. MyPsychLab Series. Pearson/ Allen and Bacon (2007)
6. Gouws, S.: Deep unsupervised feature learning for natural language processing. In: Proceedings of the 2012 Conference of the North American Chapter of the Association for Computational Linguistics: Human Language Technologies: Student Research Workshop, NAACL HLT 2012, Stroudsburg, PA, USA, pp. 48–53. Association for Computational Linguistics (2012)
7. Hinton, G.E.: Training products of experts by minimizing contrastive divergence. Technical report, Gatsby Computational Neuroscience Unit, University College London (2002)

8. Hinton, G.E.: Learning multiple layers of representation. Trends in Cognitive Sciences **11**, 428–434 (2007)
9. Hinton, G.E.: A practical guide to training restricted boltzmann machines, Technical report (2010)
10. Hinton, G.E., Dayan, P., Frey, B.J., Neal, R.M.: The wake-sleep algorithm for unsupervised neural networks. Science **268**(5214), 1158–1161 (1995)
11. Hinton, G.E., Osindero, S., Teh, Y.-W.: A fast learning algorithm for deep belief nets. Neural Computation **18**, 1527–1554 (2006)
12. Hinton, G.E., Sejnowski, T.J.: Learning and relearning in boltzmann machines. In: Parallel Distributed Processing: Explorations in the Microstructure of Cognition, vol. 1, pp. 282–317. MIT Press, Cambridge (1986)
13. Le, Q.V., Monga, R., Devin, M., Corrado, G., Chen, K., Ranzato, M., Dean, J., Ng, A.Y.: Building high-level features using large scale unsupervised learning. CoRR, abs/1112.6209 (2012)
14. Mayer, R.E: Multimedia Learning. Cambridge University Press (2009)
15. Paivio, A.: Mental representations. Oxford University Press, Incorporated (1990)
16. Nther Palm, G.: Neural associative memories and sparse coding. Neural Netw. **37**, 165–171 (2013)
17. Ranzato, M., Boureau, Y.I., Lecun, Y.: Sparse feature learning for deep belief networks. In: NIPS-2007 (2007)
18. Serre, T., Kreiman, G., Kouh, M., Cadieu, C., Knoblich, U., Poggio, T.: A quantitative theory of immediate visual recognition. Prog Brain Res., 33–56 (2007)
19. Srivastava, N., Salakhutdinov, R.: Multimodal learning with deep boltzmann machines. In: Bartlett, P., Pereira, F.C.N., Burges, C.J.C., Bottou, L., Weinberger, K.Q. (eds.) Advances in Neural Information Processing Systems 25, pp. 2231–2239 (2012)
20. Wang, T.: Classification Via Reconstruction Using A Multi-Channel Deep Learning Architecture. Masters Thesis, Acadia University, Wolfvillle, NS, Canada (2013)

Incremental Cluster Updating
Using Gaussian Mixture Model

Elnaz Bigdeli[1(✉)], Mahdi Mohammadi[2], Bijan Raahemi[2],
and Stan Matwin[3,4]

[1] School of Electrical Engineering and Computer Science,
University of Ottawa, Ottawa, ON, Canada
ebigd008@uottawa.ca
[2] Knowledge Discovery and Data Mining Lab,
Telfer School of Management, University of Ottawa, Ottawa, ON, Canada
{mmohamm6,braahemi}@uottawa.ca
[3] Department of Computing, Dalhousie University, Halifax, NS, Canada
[4] Institute of Computer Science, Polish Academy of Sciences, Warsaw, Poland
stan@cs.dal.ca

Abstract. In this paper, we present a new approach for updating clusters incrementally. The proposed incremental approach preserves comprehensive statistical information of the clusters in form of Gaussian Mixture Models (GMM). As each GMM needs the number of Gaussian (component) as an input parameter, we proposed a method to determine the number of components automatically with introducing the concept of core points. In the updating phase, instead of processing each new sample individually, we collect the new incoming samples and cluster them. By employing the concepts of core points and GMMs, we build a number of GMMs for the new samples and we label the new GMMs based on their similarity to the already existing GMMs. To find the similarity among GMMs, we introduce a new modified version of Kullback-Leibler as a distance function. For merging the current GMMs and the new GMMs, we proposed a new merging mechanism in which the closest components in both GMMs are merged to create a new GMM. Since GMM structure is a compact representation of clusters, there is no increase in the time neither in clustering side nor in updating phase. We measured the accuracy of clusters based on different clustering validity metrics (DB, Dunn, SD and purity) and the results show that our algorithm outperforms other incremental clustering algorithms in terms of quality of the final clusters.

Keywords: Incremental clustering · Gaussian Mixture Model · Stream data clustering

1   Introduction

Nowadays, numerous applications ranging from business and market analysis to computer networks deal with overwhelming quantities of data. The demand to work with large and fast incoming data creates a new trend to tailor the traditional algorithms to the online environment. Additionally, the main part of the data is unlabelled this is why the clustering algorithms receive attention to deal with the unlabeled data [1] [2].

© Springer International Publishing Switzerland 2015
D. Barbosa and E. Milios (Eds.): Canadian AI 2015, LNAI 9091, pp. 264–272, 2015.
DOI: 10.1007/978-3-319-18356-5_23

Since in online environment the behavior of data changes by passing time, there is a great need to propose algorithms which are able to update themselves with new trends in data behavior. The primitive way to update clusters based on the new incoming data is to add the new data to the existence data and then apply the clustering algorithm on the entire data. Obviously, this approach is not practical in many online applications. To solve this problem incremental clustering algorithms are suggested in which the clustering algorithms keep a history of data and then add the new data to the current data in an incremental way. Incremental clustering algorithms need to have specific characteristics. First, the incoming sample has to be processed quickly in order to find the closest cluster. Second, the clusters have to be updated very fast to be adapted to the new changes in data. Third, the clusters have to be compact and well separated over the time, and they should not be grown fast by every incoming sample. Another important feature, which needs to be considered, is to make the incrementally generated clusters as close as possible to the clusters generated using the whole data.

Incremental clustering algorithms such as STREAM [3] and Clustream [4] update clusters efficiently in terms of memory and time complexity. In these algorithms, finding the closest cluster to the new incoming sample is based on finding the distance of the new sample to the clusters centers. There is another trend in incremental clustering that estimates the entire data with one GMM in which each component is a representative of a cluster. Considering a single GMM for the whole data has its own problems. The most important problem is that the model is complex and updating step involves the whole samples in the dataset. To conquer this problem, in this paper, we employ the advantages of both trends in incremental clustering algorithms. In our proposed approach, instead of finding a single GMM for the whole data, first, the data is clustered and then each cluster is represented using a GMM. It has two advantages: first, the GMM is able to generate the original data with a good accuracy; second, the GMM formula finds the closest cluster to the new sample with a simple calculation. From updating perspective, in our approach instead of updating clusters based on each new incoming sample, the new coming samples are collected, and a group of samples are used to update GMMs. In this way, updating phase is less sensitive to noise. Moreover, since the updating is happening offline, it is able to deal with a huge amount of data in the online applications.

The rest of the paper is organized as follows: In Section 2 a review of incremental clustering algorithms is presented. The general architecture of our GMM-based incremental clustering algorithm is presented in Section 3. The clustering approach used in the paper is discussed in Section 4. The new algorithm for creating GMM is presented in Section 5. GMM-based updating approach is introduced in Section 6. The experimental results and the all the comparisons with all other algorithms are presented in Section 7. The conclusion and future work of the paper are presented in Section 8.

2 Related Work

There are different approaches for clustering data in online applications. The first well-known algorithm in online clustering is BIRCH [5] that was introduced by Zhang et al. The BIRCH algorithm creates a hierarchical structure of clusters in the form of a tree that makes this algorithm fast for searching purposes. In spite of being fast, this algorithm has its own disadvantages. First, creating a hierarchy of data is usually a difficult task which involves many parameter setting, plus the fact that updating a tree structure is not a straightforward task. Moreover, BIRCH employs the idea of the centers and a radius to create a cluster which makes it inappropriate for non-convex clusters. In addition, the center-based approaches are sensitive to noise and also they have low accuracy in clustering new samples. STREAM [3] is an incremental algorithm that uses the idea of preserving weighted medians over the time. In this algorithm in each step the LSEARCH algorithm finds the clusters medians. Every new incoming data is added to the current medians and LSEARCH is applied to find the new medians. To have an accurate clustering, a large number of medians need to be preserved which is not appealing in the online environments. Clustream [4] is an incremental algorithm that combines the idea of BIRCH and STREAM in a framework. In pyramidal time steps, the clusters are updated, and the current clusters are replaced by new ones. Clustream still carries the problems related to BIRCH and STREAM.

In the area of density-based clustering algorithms, DenStream [6] is used to make DBSCAN an incremental clustering algorithm. The main algorithm for initialization and updating stages is DBSCAN. Each new coming sample is assigned to a cluster with distance less than a radius to a core micro cluster. Dealing with the whole dataset makes the algorithm slow and unpleasant for online applications. Grid-based algorithms are also changed to be used in incremental and online environments [7]. The idea is pretty much similar to DenStream while finding dense regions and connecting them is applied to the grid that has its own problems. Incremental Gaussian Mixture Model algorithms are another group of incremental clustering algorithms that consider all data is generated based on a Gaussian Mixture Model [8]. In this group, a single GMM is found for the entire dataset. After generation of the first GMM, each new sample is fed to the current GMM and the GMM is updated based on the new sample [9] [10] [11]. Our approach is a combination of both GMM-based incremental clustering and traditional incremental clustering algorithms, and it also introduces a new structure which is fast and accurate for online data clustering and is explained in next section.

3   The Proposed Incremental Cluster Updating

In this section, we present the proposed method CUGMM (Cluster Updating based on GMM) in more details. Our approach has three main phases which are shown in Figure 1. As shown in this figure, the input dataset is available and a clustering algorithm is applied to partition the dataset into some clusters (cluster creation phase is explained in Section 4).

In the second phase, for each cluster, a GMM is employed to estimate the distribution and shape of the cluster. In this paper, we modify the standard GMM estimation phase in order to improve the time and memory complexity of the previous methods. The proposed approach is described in Section 5. In updating phase, instead of updating the current clusters using each new sample, we collect the incoming samples and after collecting a specific number of samples, we cluster the collected samples into some clusters. Then we apply the same procedure on the new clusters and represent each cluster by a GMM as shown in Figure 1. In this step, we compare the newly generated clusters with the exciting ones. If the new GMM is close enough to one of the existing cluster GMMs, they are merged and they create an updated cluster; otherwise, it is either added as a new cluster or deleted. In the following, we discussed each step in more detail.

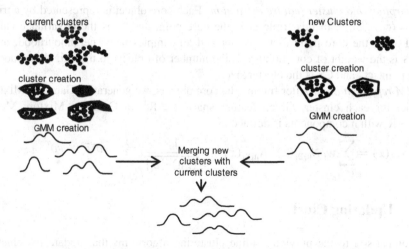

Fig. 1. The General Structure of Incremental cluster updating

4   Creating Clusters

In each dataset, based on the requirements of an application and nature of data, different kinds of clustering algorithms are applied. In the proposed model, any kind of clustering algorithms can be employed. This means that the proposed algorithm is able to work with the clusters produced by any clustering algorithm. To preserve the shape of a cluster either the majority of samples in each cluster should be stored or the boundary of each cluster must be detected. Either way, the memory complexity for these methods is not negligible. Moreover, we need to use a technique that not only preserves the shape of the cluster, but also retains all valuable information for that cluster to use for further investigation of clusters. Therefore, we proposed our GMM-based approach called SGMM [2] to summarize the clusters as a GMM. In the next section, we explain the proposed SGMM algorithm in details.

5   Creating Gaussian Mixture Model

Summarization based on Gaussian Mixture Model (SGMM), has three main steps to represent each cluster as a GMM. First, a set of objects called core objects are detected. These objects are representative of a cluster, and they are able to generate the original cluster as needed. After detecting the backbone objects, there comes the absorption step, where the attached objects to the core point are absorbed and represented by the core object. Then by introducing a new feature set for each core point, the cluster is summarized while its original distribution is preserved using a GMM.

-Finding core objects: In this phase, a radius is considered as an input parameter for the algorithm. The radius is employed to find dense regions and the core points which are the center of these dense regions. Every point with a greater number of neighbors in comparison with other objects is a good candidate to be a final core point.

-*Absorption and cluster feature extraction:* Each core object is represented by a triple $CF_i = \langle c_i, \Sigma_i, \omega_i \rangle$. In this triple c_i is the core point and Σ_i is the covariance calculated using the core point as the center and all samples in its neighbourhood. $\omega_i = n/CS$ is the weight of core point, n is the number of objects in the neighbourhood of core point c_i , and CS is the cluster size.

-*GMM representation:* After finding the core objects, we generate a Gaussian Mixture Model for each cluster. Given feature space, $f \subset R^d$ a Gaussian Mixture Model $g: f \to R$ with n components is defined as:

$$g(x) = \sum_{i=1}^{n} w_i\, N_{\mu_i \Sigma_i}(x); \; N_{\mu_i \Sigma_i}(x) = \frac{1}{\sqrt{(2\pi)^d |\Sigma_i|}} e^{-\frac{1}{2}(x-\mu_i)\Sigma_i^{-1}(x-\mu_i)^T} \tag{1}$$

6   Updating Cluster

In comparison to the previous online clustering algorithms that update the clusters based on each new sample, our algorithm updates clusters in a batch way. To update the clusters, first, the new incoming samples are collected, and some new clusters are generated and by GMMs. The updating procedure consists of two main steps; finding two closest clusters and then merging the close clusters.

6.1   Finding Two Closest GMMs

To update clusters, first we need to find the closest clusters. Based on the definition of GMM in Equation (1), the Kullback-Leibler [12] distance for two normal distributions is:

$$KL(N_0 | N_1) = \frac{1}{2} (tr(\Sigma_1^{-1}\Sigma_0) + (\mu_1 - \mu_0)^T \Sigma_1^{-1}(\mu_1 - \mu_0) - k$$
$$- \ln\left(\frac{\det(\Sigma_0)}{\det(\Sigma_0)}\right) \tag{2}$$

This distance is used to measure the distance of two normal distributions not two mixtures of the normal distributions. To propose a symmetric distance, we need to find both $kl(N^b | N^a)$ and $kl(N^a | N^b)$, and then take the average. In our proposed distance measure, we first find the distance of first GMM components to the second GMM components using weights of the first GMM. Then, we find the distance of second GMM's components to the first GMM's components using weights of the second GMM.

$$D(G_b | G_a) = \sum_{i=1}^{n} \sum_{j=1}^{m} w_i^b\, kl(N_i^b | N_j^a) \qquad D(G_a | G_b) = \sum_{i=1}^{m} \sum_{j=1}^{n} w_i^a\, kl(N_i^a | N_j^b) \qquad D_{cluster} = \frac{(D(G_a | G_b) + D(G_b | G_a))}{2} \tag{3}$$

6.2   Merging Two GMMs

After finding the two closest GMMs, the next step is to merge them. According to our proposed approach, the new data samples $\{x_1,...,x_m\}$ are clustered to clusters $\{c_1,...,c_k\}$. SGMM algorithm finds all equivalent GMMs; $\{G_1,..,G_k\}$, for clusters $\{c_1,...,c_k\}$. Each GMM consists of a set of components $G_i = \{g_i^1, \cdots, g_i^s\}$ where s is the number of GMM components. Each GMM component is represented by $g_i^j = \{\mu_i^j, \Sigma_i^j, \omega_i^j\}$. Considering that G_a and G_b are two close GMMs. In G_a the i_{th} component that is g_a^i is close to j_{th} component of G_b that is g_b^j. Therefore, the new GMM component is created based on merging $g_a^i = \{\mu_a^i, \Sigma_a^i, N_a^i\}$ and $g_b^j = \{\mu_b^j, \Sigma_b^j, N_b^j\}$.

$$\mu_{new} = \frac{N_a^i\mu_a^i + N_b^j\mu_b^j}{N_a^i + N_b^j} \quad (4)$$

$$\Pi = \frac{N_a^i + N_b^j}{\sum_{i=1}^s N_a^i + \sum_{j=1}^l N_b^j} \quad (5)$$

$$\Sigma_{new} = \frac{N_a^i\Sigma_a^i + N_b^j\Sigma_b^j}{N_a^i + N_b^j} + \frac{N_a^i\mu_a^i\mu_a^{i^T} + N_b^j\mu_b^j\mu_b^{j^T}}{N_a^i + N_b^j} - \mu_{new}\mu_{new}^T \quad (6)$$

6.3   Discussion on the Efficiency of the Proposed Method

Summarization is the main feature of the proposed method. Instead of keeping the entire samples in each cluster for preserving the arbitrary shape of the cluster, the proposed method summarizes each cluster into a set of Gaussian models. Each Gaussian model contains a mean, a variance, and a weight. In terms of memory complexity, summarizing the clusters with Gaussian mixtures are more beneficial than holding the entire samples in each cluster. It can estimate the shape of the clusters as well as distribution of samples in each cluster by estimating each cluster with a set of Gaussian models. In arbitrary shape clustering algorithms which keep the entire samples, to label the new incoming sample, the distances between the new test samples and entire samples in the cluster (or at least a part of the samples) have to be calculated. In the proposed method, a few number of core points are kept based on which a GMM is estimated on each cluster. Summarizing the clusters with a set of core points and estimating the characteristics of the clusters by the GMMs benefits the proposed method in a noisy environment. If the updating process is triggered whenever a new sample is introduced to the model, the noisy samples may bias or change the attributes of the clusters based on their noisy behavior. It may end to inaccurate final clusters which do not follow the real behavior of the clean data. But the proposed method collects the incoming samples, and then group them into some clusters and consider the extracted clusters to represent the new samples. After clustering the new samples, the outliers and the clusters which have the number of the samples less than a given threshold are removed from the data. It prevents the final clusters to be persuaded by the input noisy data.

In some cases, the number of noisy samples, which are placed in the same neighborhood, is enough to shape a component of a GMM. In this case, the noisy component should be close to any other clean components in other clusters. If the algorithm decides to combine two GMMs (one of the GMMs consists of the noisy component), there will be two approaches for the algorithm. First, while the noisy component is not close enough to any components in the other GMM and the number of samples in the component is less that a specific value, the proposed method eliminates the component from its GMM. In the second approach, if the noisy component is not close to any component in the other GMM but the number of samples is not small enough to be eliminated, the noisy component is departed from the GMM and forms a new GMM (cluster) which has one component representing the whole samples in the new cluster. If the newly formed cluster is not really a noisy cluster, by adding new samples, it may have the chance to find the clusters close enough to be merged and shape a new cluster, otherwise it may   not have a chance to be merged with other clusters.

7   Experimental Results

In this part, we evaluate the proposed method based on different criteria consisting of five different metrics as Dunn [13], DB [14], SD, SSQ [4], and Purity [4]. In this paper, we first divide the input dataset into 6 parts and pick the first part to generate the initial clusters and the rest is added to the existing clusters incrementally in different rounds (iterations). That is to say, in each round we introduce a new part of data to the previously seen data to examine the learning of the proposed method. Some of the UCI repository datasets are used in our experiments, as well as a 2D synthetic dataset to visualize the final clustering results. The datasets, which are chosen from UCI repository, are KDDCup99, MagicGamma. Figure (2) shows the result of SSQ for KDDCup99 dataset for CUGMM and Clustream [3] and Clustering the Whole data in each Round (CWR).

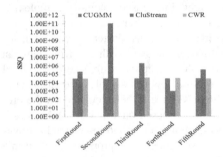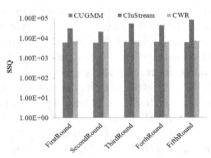

Fig. 2. SSQ for KDD dataset   **Fig. 3.** SSQ for MagicGamma dataset

In CWR, in the first step, we cluster the first part of data using K-means algorithm. In second round, we add the second part of the data into the existing one and apply the K-means algorithm on the whole data. That is to say, in each round we forget about the previous clusters and build the new clusters based on the whole data received by the current round. This means that there is no incremental learning process in CWR that is why we expect the CWR outperforms other algorithms in most cases. But the main point is that we can measure how far we are from the CWR by introducing the CUGMM, which works in an incremental way. These results indicate the accuracy of clusters by considering each cluster as a Gaussian Mixture Model. In the Clustream, the clustering is based on considering core clusters and the distance of the new sample to the center of the cluster. As shown in the Figure 3, CUGMM shows better performance in clustering the new incoming samples incrementally since the results is very close that of CWR while CWR considers the whole samples in each round in clustering and CUGMM updates the clusters incrementally. Figure (3) shows the same result on MagicGamma dataset which confirms that the CUGMM works better than Clustream algorithm, and it almost ties with the CWR method. For example, in the first round the CUGMM SSQ is 6221 in comparison to SSQ 32210 for Clustream and SSQ 7100 for the basic model. The lower error rate shows that the approach does not lose any accuracy in terms of clustering over the time. In next experiment, we evaluate the proposed method using Dunn, DB, SD, and Purity. For DB and SD the lowest value shows the best performance while for Dunn and Purity the highest is the best (. 1). For KDD dataset in each round the CUGMM algorithm has a better result than Clustream, and it is close to the results of CWR. For DB index, the CUGMM shows a stable performance during the whole rounds while Clustream and CWR fluctuate between [0.4-0.8] and [0.2-0.7] respectively. In terms of SD, most of the time the CUGMM shows better results than that of Clustream and also CUGMM is very close to CWR. It shows the proposed method (CUGMM) outperforms the Clustream, and it is able to generate the same cluster as CWR. The result for DB index shows that the generated clusters are good in all rounds. The result for SD confirms the previous results. Result shows that with CUGMM algorithm, the purity inside clusters based on label of the data is higher than that of Clustream, and it is close to the CWR.

Table 1. Performance of clustering based on Dunn, DB, SD, Purity indexes

Clustering index		Dunn			DB			SD			Purity		
DataSet/ Algorithm		CUGMM	Clu-stream	CWR	CUGMM	Clu-stream	CWR	CUGMM	Clu-stream	CWR	CUGMM	Clu-stream	CWR
KDD	R1	1e-6	2e-8	0.0003	0.613	0.81	0.35	2.01	3.2	2.22	0.76	0.63	0.79
	R4	0.00025	1e-6	0.0006	0.634	0.71	0.69	2.3	2.27	2.27	0.74	0.66	0.73
Shuttle	R1	7e-6	2e-8	3e-05	0.332	0.49	0.2	6.35	7.15	6.42	0.96	0.81	0.95
	R4	9 e-7	2e-8	2e-05	0.363	0.59	0.29	3.88	4.98	3.98	0.96	0.85	0.95
Magicc Gamma	R1	0.00489	0.0001	0.0222	0.777	0.87	0.66	1.89	2.89	1.9	0.76	0.64	0.73
	R4	0.01960	0.0039	0.0339	0.727	0.94	0.7	1.89	2.71	1.9	0.75	0.64	0.75
Synthetic	R1	2e-09	1e-011	1e-05	0.40	0.45	0.35	2.25	3.25	2.08	0.89	0.79	0.91
	R4	2e-07	4e-09	6. e-06	0.38	0.68	0.28	2.33	4.47	1.77	0.9	0.81	0.9

8   Conclusions

This paper presents a new approach to update clusters based on the new incoming samples. In our approach, all clusters are represented using a GMM that keeps a summary of each cluster and is able to be updated very fast and more accurate. Instead of processing each new incoming sample individually, we first collect the samples and then group them into some clusters. For each cluster a GMM is defined to estimate the samples scattering and distribution. In updating phase, we proposed a modified Kullback-Leibler distance measure to find the closest cluster to the new clusters. After finding a pair of close clusters from current and new clusters, their corresponding GMMs are merged. We introduce a merging strategy which enables the algorithm to detect noisy samples and remove them from the rest of the process. We evaluate the proposed method in comparison with two other clustering algorithms, CWR and Clustream based on different cluster validity indices. The result shows that our updating approach tends to follow the original shape of clusters during time. The result on different dataset based on different factors shows that updating clusters using GMMs is more accurate than Clustream.

References

1. Mohammadi, M., Akbari, A., Raahemi, B., Nasersharif, B., Asgharian, H.: A fast anomaly detection system using probabilistic artificial immune algorithm capable of learning new attacks. Evolutionary Intelligence 6(3), 135–156 (2014)
2. Bigdeli, E., Mohammadi, M., Raahemi, B., Matwin, S.: Arbitrary shape cluster summarization with Gaussian Mixture Model. In: KDIR, Roma (2014)
3. O'Callaghan, L., Mishra, N., Meyerson, A., Guha, S., Motwani, R.: Streaming-Data Algorithms for High-Quality Clustering. IEEE Conference on Data Engineering (2001)
4. Aggarwal, C.C., Han, J., Wang, J., Yu, P.S.: A framework for clustering evolving data streams. In: VLDB 2003 (2003)
5. Zhang, T., Ramakrishnan, R., Livny, M.: BIRCH: A New Data Clustering Algorithm and Its Applications. Data Mining and Knowledge Discovery 1, 141–182 (1997)
6. Cao, F., Ester, M., Qian, W., Zhou, A.: Density-based clustering over an evolving data stream with noise. In: SIAM Conference on Data Mining (2006)
7. Chen, Y., Tu, L.: Density-based clustering for real time stream data. In: ACM SIGKDD (2007)
8. Hajji, H.: Statistical Analysis of Network Traffic forAdaptive Faults Detection. Transactions on Neural Networks, 16(5) (2005)
9. Song, M., Wang, H.: Highly efficient incremental estimation of Gaussian Mixture Models for online data stream clustering. In: SPIE Conference on Intelligent Computing: Theory And Applications III, Orlando (2005)
10. Declercq, A., Piater, J.H.: Online learning of Gaussian Mixture Models: a two-level approach. In: Third International Conference on Computer Vision Theory and Applications (2008)
11. Hennig, C.: Methods for merging Gaussian mixture components. Advances in Data Analysis and Classification 4(1), 3–34 (2010)
12. Kullback, S., Leibler, R.A.: On Information and Sufficiency. Annals of Mathematical Statistics, 79–86 (1951)
13. Dunn, K., Dunn, J.: Well separated clusters and optimal fuzzy partitions. Cybernetics 4, 95–104 (1997)
14. Davies, L.D., Bouldin, W.D.: A cluster separation measure. IEEE Trans. Pattern Anal. Machine Intell 1(4), 224–227 (1979)

Emotional Affect Estimation Using Video and EEG Data in Deep Neural Networks

Arvid Frydenlund[1]([⊠]) and Frank Rudzicz[1,2]

[1] Department of Computer Science, University of Toronto, Toronto, ON, Canada
arvie@cs.toronto.edu, frank@spoclab.com
[2] Toronto Rehabilitation Institute-UHN, Toronto, ON, Canada

Abstract. We present a multimodel system for independent affect recognition using deep neural networks. Using the DEAP data set, features are extracted from EEG and other physiological signals, as well as videos of participant faces. We introduce both a novel way of extracting video features using sum-product networks, and a unique method of creating extra training examples from data that would have otherwise been lost in downsampling. Deep neural networks are used for estimating the emotional dimensions of arousal, valence, and dominance, along with favourability and familiarity. This work lays the foundation for future work in estimating emotional responses from physiological measurements.

Keywords: Affect recognition · Deep neural networks · Emotional analysis · EEG

1 Introduction

Emotion- or affect-recognition is the task of estimating emotional responses to some stimulus. Humans typically recognize emotion by integrating multiple modalities, whether from vocal intonation, perspiration, or facial expressions. Different modalities may also offer different benefits and challenges than one another for affect recognition, both in terms of their descriptive power, but also in terms of how practical or economical they are to capture.

Electroencephalographs (EEGs) measure activity in the brain by amplifying electrical brain waves created in psycho-physiological processes such as experiencing an emotion. One motivation for using EEG for affect recognition is its use for augmented human-computer interfaces, especially in instances where physical impairments prevent the use of traditional interfaces.

In this work we present a multimodal system for subject-independent affect recognition using deep neural networks and the DEAP data set [9]. Our novel approach is significantly more accurate than two baselines on this task.

1.1 The DEAP Dataset

This work uses the DEAP data set introduced by Koelstra *et al.* [9]. DEAP is a multi-modal dataset created for affect recognition based on emotional responses

© Springer International Publishing Switzerland 2015
D. Barbosa and E. Milios (Eds.): Canadian AI 2015, LNAI 9091, pp. 273–280, 2015.
DOI: 10.1007/978-3-319-18356-5_24

by participants to music videos. The modalities recorded include 32-channel EEG and various other physiological signals including galvanic skin response, electromyogram, and electrooculography. Some of the data set also includes video of participant faces during stimuli, along with links to the music video stimuli.

Emotional response to the music videos is quantified by the participants themselves after watching each video. This quantification is described using a three-dimensional affect space described by Russell [13]. Each dimension ranges from negative to positive which, in DEAP, is represented on a continuous scale of 1 to 9. The first dimension is **valence**, which is a measure of intrinsic qualities of attractiveness/repulsion or pleasantness/unpleasantness, and ranges from 'sad' (1) to 'happy' (9). The second dimension is **arousal**, which is measure of activity or extent ranging from 'bored' (1) to 'excited/alert' (9). Finally, the third dimension is **dominance**, which is a measure of the feeling of control one has and ranges from 'helpless' (1) to 'empowered' (9). Specific emotions are then areas or points in this affect space; for example, *sadness* would have low valence, low arousal, and low dominance while *fear* would have low valence, high arousal, and low dominance. Participants were also asked to rate their **favourability** or how much they 'liked' the music video on a 1 to 9 scale and their **familiarity** with the video on a 1 to 5 scale.

1.2   Related Work

Much of the previous work using DEAP involved classification. For this, classes were formed by partitioning the affect space into categorized regions. Works that only use EEG and the other non-video signals include [4,7,8,19,20]. Wichakam and Vateekul used separate binary classes for valance and arousal [20] where each dimension was bisected and classified separately. Chung and Yoon used both binary and trinary classes and reported 66.6% and 66.4% accuracy for valance and arousal, respectively, for the binary class problem and 53.4% and 51.0% for the three-class problem [4]. Jie *et al.* partitioned the valance and arousal space into quadrants [7]. They reported a 95% accuracy for subject-dependent classification and 70% for subject-independent classification. They used sequential backwards-selection for feature reduction and used pseudo-inverse linear discriminant analyses for classification. Jirayucharoensak *et al.* used separate trinary classes and a deep-learning model based on stacked autoencoders [8], and Wang and Shang used separate binary classes and deep-belief networks to obtain values of 60.9% and 51.2% for arousal and valance, respectively [19].

Torres-Valencia *et al.* used videos of the participants' faces [17], where they extracted the mean shape of the face using fiducial points. They found that no features derived from the videos survived feature selection given their methods.

Both Koelstra *et al.* [9] and Acar [1] included acoustic features of the music videos themselves, and Acar (*et al.*) would later also add video features from those data [2]. In both of Acar's experiments, using features extracted from the music videos improved classifier performance. *Favourability* was classified by Wichakam and Vateekul [20], and by Wang and Shang [19] with accuracies of 66.8%. and 68.4%, respectively.

Relatively little work has been done using DEAP for regression tasks. Garcia *et al.* measured autoregression of signals, and passed those as observations into hidden Markov models which were mapped by Fisher kernels into a new space where support vector regression was used to perform inference [6]. They did subject-independent cross-validation and regress on six output variables of negative and positive valance/arousal, pleasant and unpleasant valence, and active and passive arousal. They reported values of approximately 0.68 root mean squared error (RMSE). Similarly, Torres *et al.* used structured support vector regression to jointly model valence and arousal [18]. They used recursive feature elimination and reported RMSE of around 0.24 and 0.25 for arousal and valence.

MAHNOB-HCI [16] is another multi-modal affect recognition data set used by [10] in a regression task using both EEG and video features. They converted arousal and variance to the range [0-1] and reported mean square error (*not* RMSE) as low as 0.063 for arousal and 0.069 for valence.

2   Preprocessing and Feature Extraction

In our experiments, we preprocess features extracted from EEG and other physiological signals in addition to the associated facial videos. The DEAP dataset consists three parts, only two of which are used here: 1) the EEG and other physiological signals; 2) the videos of the participants' faces, and; 3) the music videos the participants viewed, which are not used in this work. While DEAP consists of 32 subjects, there is only facial video data from 22 of these, which constitutes the subset used in this work. Details can be found in [9].

2.1   Physiological Signal Preprocessing

The first 25 seconds and the last second of data were removed. For each subject, each trial was z-score normalized and then test and training partitions were created. A leave-two-out partitioning was employed with one of the partitions used as a validation set.

It is common to down-sample the data down to around 1/4 of the sampling rate [4,6], which unfortunately throws out a lot of data. Instead, we create new trials from the samples which would have otherwise been discarded. Specifically, let $1/n$ be the fraction of data one wishes to retain per trial. For simplicity, assume each trial consists of only a single signal, s, which is a vector of N samples, and that n evenly divides N. The trial for s is therefore replaced by n new trials, $s_1...s_n$ by including every n^{th} sample offset by $0...(n-1)$.

For example, down-sampling s by $1/2$ gives two sequences using every second sample, the first starting at $s[0]$ and the other at $s[1]$. This process increases the number of trials to train a prediction model where each resulting trial appears as a down-sampled but unique version of some true trial. This may potentially prevent or decrease over-fitting by decreasing data sparseness. This is similar to a method used in machine vision where multiple instances of an image that have been rotated and/or rescaled are included in training in order to make the model scale- or rotation-invariant. Each image in that case acts as a noisy version of some true image in the correct scale and position. Our method can be seen as adding slight temporal invariance, which might be important if subjects have similar reactions to the music videos but at slightly different times.

Using this scheme, the data were down-sampled to 1/5 of the original rate. For each partition, independent component analysis (ICA) was applied to the set of EEG channels. Ocular artifacts were removed using EEGLAB [5]. A bandpass filter was applied to the EEG channels to get the theta band (4-8 Hz), slow alpha (8-10 Hz), alpha (8-12 Hz), beta (12-30 Hz), and gamma (30-100 Hz) [9].

Galvanic skin response (GSR) was smoothed using a 256-point moving average, as in [9]. The non-EEG physiological channels were bandpass filtered on [0.1-0.2] Hz for skin temperature, for the multiple ranges of [0.01-0.08] Hz, [0.08-0.15] Hz, and [0.15-0.5] Hz for blood pressure, and [0.05-0.25] Hz and [0.25-5] Hz for the respiration channel.

All the original channels, along with the bandpass-filtered channels, were used for feature extraction. For each channel, each trial was split into non-overlapping semi-equal time segments and the following features were extracted: average, standard deviation, range for the segment, for the given signal and the first two derivatives of the signal. For GSR, the average of negative values of the first derivative, and proportion of negative derivatives to positive ones were also found, as in [9]. The non-overlapping time segments were set empirically to the whole trial, 1/4, 1/10, and 1/20 of the trial.

2.2   Video Preprocessing

Videos were preprocessed by finding the bounding square that contains the face for each frame, converting to grey-scale, and then resizing the image to 64×64 pixels. Images were then z-scored over each trial independently.

For each partition described above, the training set was divided into classes based on the quadrants formed by dividing valence into *high* and *low* and arousal into *high* and *low*. Figures 1 and 2 show these class boundaries as red lines. Arousal was split into at 4.5, and valence was split at 5.5. These class boundaries where chosen so that classes represented semi-equal areas of affect space while minimizing the distance of sample points to the class boundaries. Dominance was not considered, as it would have created more classes thereby decreasing the amount of training data per class. These classes are used to produce four video-based features for each image frame, i.e., the log-likelihood of an image belonging to one of those classes, as determined by trained sum-product networks (SPNs).

SPNs, introduced by Poon and Domongos [12], are graphical models in which exact and tractable inference is achieved by placing limitations on the model's structure. In this paper, each SPN represents a distribution over faces in different quadrants of the two-dimensional affect space. This is done by training four SPN models, one for each of the valence-arousal classes in figure 1, using the same graph structure as in previous work [12]. The average and standard deviations of the log-likelihood values for each of the four SPNs, over the previously described time segments, are then used as features in a larger neural-network based regression engine.

Scatter and density plots of arousal and valence across trials

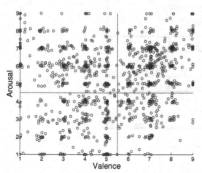

Fig. 1. Class boundaries of the four SPN models, shown as red lines

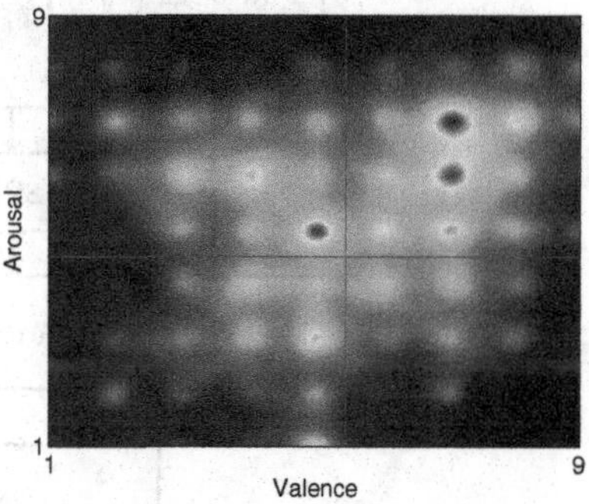

Fig. 2. The density of arousal and valence, indicating that much of the data is on integer values in the range of [1-9] for both

Other features are extracted from the facial videos using principal component analysis. To extract features, we calculate the std of images in the time segments defined above, mapped to the 1st through 20th principal components.

Finally, feature selection is run over each channel by correlating the channel with the variable we wish to estimate, e.g., valence etc. and taking the top number of correlating features. Figure 3 shows an overview of the data processing pipeline for inference.

3   Experiments

The affect and favourability dimensions are rounded to whole numbers since much of the data already falls on integer values, as shown in figure 2. The RMSE introduced between the rounded values and the original values over all the data (on a 1 to 9 scale) is: 0.1849 for valence, 0.1799 arousal, 0.1763 dominance, and 0.1798 favourability. Familiarity is already on an discrete scale of 1 to 5.

Linear regression and deep neural networks (DNNs) are used for the regression tasks. For every test partition, a linear regression and a neural network model are created for each affect dimension, favourability, and familiarity. The neural network consists of two hidden layers, a sigmoidal activation function for interior layers, and softmax for the output layer. The number of features ranges empirically from 200 to 700, depending on the regressed variable. A combination of Spearmint [15] and hand-tuning is used to set the neural network's hyper-parameters. The same data are used to train both the neural networks that produce our output regressions, and the SPNs.

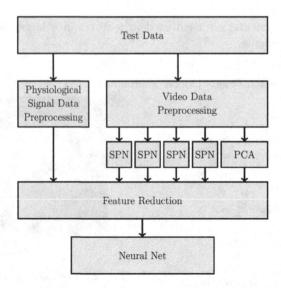

Fig. 3. Flowchart of the inference pipeline

Baselines for the models are formed by plurality classification. Table 1 presents RMSE means ± std. dev. across the test partitions in both the [1-9] and the [0-1] ranges for each affect dimension. This is done as RMSE is dependent on the scale of the values and allows for potential comparison with other work, such as that of Koelstra and Patras [10]. Table 2 shows the RMSE for the favourability and familiarity regressed variables. For an $\alpha = 0.05$, a one-sided t-test was conducted between the neural net results and the baseline, as shown in table 3.

4   Discussion

Our proposed model significantly out-performs baselines across affect, dominance, and favourability, and is merely more accurate for valence. Our model is not significantly different than the baseline for familiarity. Linear regression failed to get better results than the other baseline in all cases.

This work acts as a feasibility study for our 'downsampling extension' scheme, and further work needs to be done to examine its affect. Feature selection is another possible line of inquiry, since we have selected features based on linear relations, where non-linear relations may be more appropriate for DNNs.

Emotion is an essential part of communication and, as computer systems become increasingly sophisticated, emotion recognition could be incorporated as part of natural language applications. For example, effective speech recognition for emergency response and medical consultation may be improved by accommodating emotional speech [14]. Affect recognition from physiological signals will also be important for applications such as recommendation systems [3], which will complement ongoing applications of sentiment analysis. This work provides a multimodal step in that direction.

Table 1. RMSE ± standard deviation n both a [1-9] range of possible values and a [0-1] range across the test partitions for each of the regressed variables in the affect space

Model	Scale	Arousal	Valence	Dominance
Baseline	[1-9]	1.9062 ± 0.4051	1.7795 ± 0.2462	2.6591 ± 0.2557
	[0-1]	0.2383 ± 0.0506	0.2224 ± 0.0308	0.3324 ± 0.0320
Logistic Regression	[1-9]	3.8275 ± 0.6686	3.5465 ± 0.8793	3.9718 ± 0.6367
	[0-1]	0.4784 ± 0.0836	0.4433 ± 0.1099	0.4964 ± 0.0796
Deep Neural Net	[1-9]	1.5778 ± 0.5301	1.5396 ± 0.8797	2.0181 ± 0.4489
	[0-1]	0.1972 ± 0.0663	0.1925 ± 0.1100	0.2523 ± 0.0561

Table 2. RMSE ± standard deviation in both a [1-9] range and for [0-1] range across the test partitions for favourability and familiarity

Model	Scale	Favourability	Familiarity
Baseline	[1-9]	2.7618 ± 0.3995	2.0072 ± 0.4065
	[0-1]	0.3452 ± 0.0499	0.5018 ± 0.1016
Logistic Regression	[1-9]	4.0557 ± 0.5482	2.3883 ± 0.31994
	[0-1]	0.5070 ± 0.0685	0.5971 ± 0.0800
Deep Neural Net	[1-9]	2.2366 ± 0.5975	2.0432 ± 0.4010
	[0-1]	0.2796 ± 0.0747	0.5108 ± 0.1002

Table 3. Results of one-sided t-tests conducted between the neural network results and the baseline, specifically the t-statistic, p-value, and confidence interval

	Arousal	Valence	Dominance	Favourability	Familiarity
t(40)	-2.2557	-1.2034	-5.6866	-3.3486	0.2887
P	0.014814	0.11795	0.000007	0.0008896	0.61285
CI	-0.08326	0.09578	-0.45122	-0.26111	0.24577

Acknowledgments. This paper makes use of the following software: Deep Learning Toolbox [11], Spearmint [15], EEGLAB [5], and modified SPN code from [12].

References

1. Acar, E.: Learning representations for affective video understanding. In: Proceedings of the 21st ACM International Conference on Multimedia, pp. 1055–1058. ACM (2013)
2. Acar, E., Hopfgartner, F., Albayrak, S.: Understanding affective content of music videos through learned representations. In: Gurrin, C., Hopfgartner, F., Hurst, W., Johansen, H., Lee, H., O'Connor, N. (eds.) MMM 2014, Part I. LNCS, vol. 8325, pp. 303–314. Springer, Heidelberg (2014)
3. Chen, Y.A., Wang, J.C., Yang, Y.H., Chen, H.: Linear regression-based adaptation of music emotion recognition models for personalization. In: Acoustics, Speech and Signal Processing (ICASSP) (2014)

4. Chung, S.Y., Yoon, H.J.: Affective classification using bayesian classifier and supervised learning. In: 2012 12th International Conference on Control, Automation and Systems (ICCAS), pp. 1768–1771. IEEE (2012)
5. Delorme, A., Makeig, S.: EEGlab: an open source toolbox for analysis of single-trial EEG dynamics including independent component analysis. Journal of neuroscience methods **134**(1), 9–21 (2004)
6. Garcia, H.F., Orozco, A.A., Alvarez, M.A.: Dynamic physiological signal analysis based on fisher kernels for emotion recognition. In: 2013 35th Annual International Conference of the IEEE Engineering in Medicine and Biology Society (EMBC), pp. 4322–4325. IEEE (2013)
7. Jie, X., Cao, R., Li, L.: Emotion recognition based on the sample entropy of EEG. Bio-medical materials and engineering **24**(1), 1185–1192 (2014)
8. Jirayucharoensak, S., Pan-Ngum, S., Israsena, P.: EEG-based emotion recognition using deep learning network with principal component based covariate shift adaptation. The Scientific World Journal 2014 (2014)
9. Koelstra, S., Muhl, C., Soleymani, M., Lee, J.S., Yazdani, A., Ebrahimi, T., Pun, T., Nijholt, A., Patras, I.: Deap: A database for emotion analysis; using physiological signals. IEEE Transactions on Affective Computing **3**(1), 18–31 (2012)
10. Koelstra, S., Patras, I.: Fusion of facial expressions and EEG for implicit affective tagging. Image and Vision Computing **31**(2), 164–174 (2013)
11. Palm, R.B.: Prediction as a candidate for learning deep hierarchical models of data. Technical University of Denmark (2012)
12. Poon, H., Domingos, P.: Sum-product networks: a new deep architecture. In: 2011 IEEE International Conference on Computer Vision Workshops (ICCV Workshops), pp. 689–690. IEEE (2011)
13. Russell, J.A.: A circumplex model of affect. Journal of personality and social psychology **39**(6), 1161 (1980)
14. Shneiderman, B.: The limits of speech recognition. Communications of the ACM **43**(9), 63–65 (2000)
15. Snoek, J., Larochelle, H., Adams, R.P.: Practical bayesian optimization of machine learning algorithms. In: Advances in Neural Information Processing Systems, pp. 2951–2959 (2012)
16. Soleymani, M., Lichtenauer, J., Pun, T., Pantic, M.: A multimodal database for affect recognition and implicit tagging. IEEE Transactions on Affective Computing **3**(1), 42–55 (2012)
17. Torres, C.A., Orozco, A.A., Alvarez, M.A.: Feature selection for multimodal emotion recognition in the arousal-valence space. In: 2013 35th Annual International Conference of the IEEE Engineering in Medicine and Biology Society (EMBC), pp. 4330–4333. IEEE (2013)
18. Torres-Valencia, C.A., Alvarez, M.A., Orozco-Gutierrez, A.A.: Multiple-output support vector machine regression with feature selection for arousal/valence space emotion assessment. In: 2014 36th Annual International Conference of the IEEE Engineering in Medicine and Biology Society (EMBC), pp. 970–973. IEEE (2014)
19. Wang, D., Shang, Y.: Modeling physiological data with deep belief networks. International Journal of Information and Education Technology **3** (2013)
20. Wichakam, I., Vateekul, P.: An evaluation of feature extraction in EEG-based emotion prediction with support vector machines. In: 2014 11th International Joint Conference on Computer Science and Software Engineering (JCSSE), pp. 106–110. IEEE (2014)

Active Learning with Clustering
and Unsupervised Feature Learning

Saul Berardo[✉], Eloi Favero, and Nelson Neto

Instituto de Ciências Exatas e Naturais, Universidade Federal do Pará,
Belém, PA 66075-900, Brazil
{saulberardo,favero,nelsonneto}@ufpa.br

Abstract. Active learning is a type of semi-supervised learning in which the training algorithm is able to obtain the labels of a small portion of the unlabeled dataset by interacting with an external source (e.g. a human annotator). One strategy employed in active learning is based on the exploration of the cluster structure in the data, by using the labels of a few representative samples in the classification of the remaining points. In this paper we show that unsupervised feature learning can improve the "purity" of clusters found, and how this can be combined with a simple but effective active learning strategy. The proposed method shows state-of-the art performance in MNIST digit recognition in the semi-supervised setting.

Keywords: Active learning · Clustering · Unsupervised feature learning

1 Introduction

Active learning algorithms can exploit labeled and unlabeled data, as in the general semi-supervised setting, but instead of using a predefined set of labeled examples, the training algorithms is allowed to query an oracle during training (e.g. a human annotator) to obtain the labels of the training examples considered the most informative. This approach is particularly useful in problems in which unlabeled data is abundant, but obtaining labels is expensive, such as in protein classification [18].

There are two main strategies employed in active learning. The most commonly explored in previous works is based on the use of labeled examples near the decision boundary of a discriminative classifier to direct the search in the hypothesis space in a efficient way. This strategy assumes that points near the decision boundary are more informative and gives little importance to points farther away [5]. The second strategy is based on the manifold hypothesis, according to which points of real data concentrate in the neighborhood of a low-dimensional manifold embedded in a high-dimensional space [4,12]. In an ideal scenario, a clustering algorithm would isolate points of different classes in their own clusters and one labeled sample for each would be enough to correctly classify the remaining examples [6]. In practice finding good clusters in complex and high-dimensional data such as images is not easy.

© Springer International Publishing Switzerland 2015
D. Barbosa and E. Milios (Eds.): Canadian AI 2015, LNAI 9091, pp. 281–290, 2015.
DOI: 10.1007/978-3-319-18356-5_25

It has been long suggested that images could be recognized by multiple layers of feature detectors [15], but it has not been until recently that effective methods for training models with multiple layers have been devised [3,8]. The first successful techniques relied on a greedy layer-wise unsupervised procedure (pre-training) for initializing the network before applying backpropagation. One perspective of pre-training procedure is that each layer learns to extract good features from the layer below. In image recognition, for example, a linear classifier in the output layer can more easily separate different classes in terms of higher level features, such as object parts, than in terms of the original input, such as pixels.

In this work we explore the intuition that the same layer-wise unsupervised training procedure used in deep learning can also be used to transform the input from the original space to a feature space in which clusters are more easily identifiable. Our method is composed by the following steps:

1. We start by applying an unsupervised feature learning technique to extract features from unlabeled data.
2. We then convert the raw input into features and use a clustering algorithm to find clusters in the feature space.
3. Then we query the labels of cluster representatives and classify all points in the same clusters with the same labels.
4. We optionally use a probabilistic model to determine the quality of the classification and remove the "excess" of points classified with low probabilities.
5. Finally, we use the training examples with the labels found as the training set of a neural network.

Our experiments demonstrate that unsupervised feature learning can improve clustering purity and we show how the learned features can be combined in a simple but effective active learning strategy.

2   Unsupervised Feature Learning

Every clustering algorithm needs a measure of dissimilarity (or similarity), such as the euclidean distance used in k-means. In certain datasets it is easy to notice that the euclidean distance does not represent well our intuitive notion of what similar points are. In digit images for example a shifted version of a digit will stop sharing most pixels in common with the original version and thus their distance will be large, although they still belong to the same class. Unsupervised feature learning techniques can be used to convert the raw input to a feature space where the distance metric can more easily represent the actual notion of dissimilarity between data points. If we have features which represent objects parts, for example, it is easier to tell whether two images are similar or not by checking if they share the same object parts, rather than by comparing their raw pixels.

2.1   Denoising Auto-Encoder

Unsupervised feature learning has been one of the main factors behind the recent breakthroughs in machine learning [2]. Denoising auto-encoder (DAE) is a technique that has been proven to yield useful representations despite its simplicity. A DAE is a neural network trained to reconstruct a clean input from a noisy version [16]. More formally, let $\tilde{\mathbf{x}} \in [0,1]^d$ be a corrupted version of the input $\mathbf{x} \in [0,1]^d$, where d is the size of the input, and $\theta = \{\mathbf{W}, \mathbf{b}, \mathbf{W}', \mathbf{b}'\}$ the encoder and decoder parameters, respectively. The network computes the reconstruction $\mathbf{z}$, given by

$$\mathbf{z} = g(\mathbf{W}'f(\mathbf{W}\tilde{\mathbf{x}} + \mathbf{b}) + \mathbf{b}') \tag{1}$$

where f and g are the activation functions. Usual choices are the logistic function $s(x) = (1 + exp(-x))^{-1}$ for both encoder and decoder, and the linear activation function for the decoder when the data is real valued. The network parameters are optimized such that

$$\theta^* = arg\min_{\theta} \frac{1}{n} \sum_{i=1}^{n} L\left(\mathbf{x}^{(i)}, \mathbf{z}^{(i)}\right) \tag{2}$$

where L is a loss function, usually the squared error $L(\mathbf{x}, \mathbf{z}) = \|\mathbf{x} - \mathbf{z}\|^2$ for real data, and the reconstruction cross-entropy

$$L_H(\mathbf{x}, \mathbf{z}) = -\sum_{i=1}^{d} \mathbf{x_i} \log \mathbf{z_i} + (1 - \mathbf{x_i}) \log (1 - \mathbf{z_i}) \tag{3}$$

for binary inputs. To generate $\tilde{\mathbf{x}}$, one common option is to add masking noise to the input (i. e. to set each element of $\mathbf{x}$ to zero with probability p).

2.2   Stacking Denoising Auto-Encoders

A DAE can be seen as a mapping from a lower to a higher level representation. Using the output from a previously trained DAE to train a new DAE, we can build a Stacked Denoising Auto-encoder (SDAE), which is capable of learning still higher levels representations [17]. Although this greedy unsupervised layer-wise procedure has been shown to be useful for initializing deep networks for subsequent fine-tuning, it has fallen out of favor for this purpose since the publication of better optimization techniques, such as Hessian-Free Optimization, which can achieve comparable (or even better) results [11]. However, as shown in this work, it is still useful as a feature extraction technique.

3   Active Learning by Clustering

Finding one single big cluster per class in the data in a purely unsupervised way is intrinsically hard or even impossible. Different criteria used to judge similarity

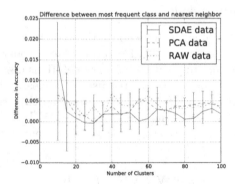

Fig. 1. Difference in accuracy by classifying points in each cluster with the most frequent class within it and with the label of the points closest to cluster centroids. For a number of clusters slightly higher than the number of classes (10) the difference is close to zero. The difference tend to be slightly lower when using SDAE to extract features.

(e.g. weights given to different features) can lead to different clusters, which do not necessarily correspond to the classes we want to find. We can though assume that in general nearby examples belong to the same classes independently of the distance metric used, forming locally small homogeneous clusters. We can explore this local structure for active learning by querying the labels of representative samples from each cluster and assuming their neighbors share the same classes.

In principle any clustering algorithm could be used, however we concentrated our experiments in the use of k-means as clustering algorithm for two reasons: first, k-means is an efficient algorithm, suitable for large datasets; and second, points near cluster centroids are intuitively good representative samples. When using other clustering strategies, such as hierarchical clustering (which create cluster with irregular shapes), the notion of what constitutes good "representative samples" is not immediately clear.

3.1   Finding Representative Samples

By using k-means as the clustering algorithm, we can use as the representative sample from each cluster the data points closest to their centroids. We then assign to all points in the same cluster the same label. This is equivalent to classifying the dataset with a k-nearest neighbors algorithm using the cluster centroids as the training examples (with k equals to one). This procedure leads to results almost identical to assigning to each cluster the most frequent label within it, as shown in Fig. 1. On MNIST dataset using just 10 clusters the difference between the two procedures is in average less than 1%. Using more than 15 clusters, this difference falls quickly to less than 0.005%. This shows that we can with high precision identify the most prevalent class in each cluster by querying the label of a single sample.

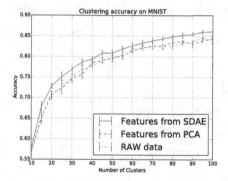

Fig. 2. Classification accuracy with 10000 training examples from MNIST. Each image is labeled with the most frequent class in the cluster. Applying PCA is not better than using the original input, while features extracted by a SDAE lead consistently to higher accuracies. Each point is the average of 10 runs of k-means with random initialization. The error bar shows the 95% confidence interval.

3.2 Clustering Evaluation on MNIST

The clustering evaluation measure which best reflects our goal of obtaining clusters containing elements from a single class is the *Purity*. It is a external criterion (i. e. compares cluster assignments against a gold standard) which is applicable even when the number of clusters is different from the number of known classes. It is given by

$$Purity = \frac{1}{N} \sum_k \max_j n_j^k \tag{4}$$

where N is the total number of data points and n_j^k is the number of points in cluster k from class j. If we treat the most frequent class of each cluster as the label given to every element of the cluster by a predictive model, this measure is equivalent to the common notion of classification accuracy. Whereas we are most interested in the final prediction, we will therefore use the term accuracy throughout the paper as a synonym of purity.

To assess the improvements of clustering after feature extraction, we compared the accuracy found by using different features as input. Fig. 2 shows the increase in classification accuracy in function of the number of clusters used (i.e. the number of labeled training examples) on a portion of MNIST training set. While Principal Component Analysis (PCA) did not improve the accuracy, SDAE consistently led to lower error rates. We tested PCA with number of components ranging from 50 to 500 and did not find any significant difference in the results. A major concern in semi-supervised learning is the lack of validation data for hyperparameter optimization. We assume the premise that labeled data is scarce, thus any additional labeled data that could be used for validation would be in most real scenarios better spent as part of the training set. We have

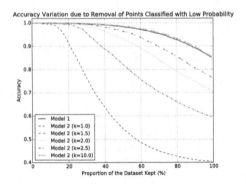

Fig. 3. After classification, points were sorted accordingly to the difference between the two highest probabilities. With both models we can obtain near 100% accuracy by keeping about 40% of points classified with highest margin between first and second guesses. Model 2 is worse than Model 1 for low values of K. For K = 10 both model are almost equivalent.

therefore tested a reduced set of hyperparameters to prevent overfitting (one and two layers, corruption level 0.1 and 0.2, and layer size 300 and 500) and used the best performing model in all our experiments (SDAE with 2 layers, 0.2 corruption level, and 500 units per layer).

3.3   Determining Classification Probability

Since the most central training example in a cluster obtained by k-means is a good representative of the most frequent class in the cluster, we can presume that the most central points have higher probabilities of being correctly classified, while points farther from cluster centroids have lower probabilities. However points which lay in-between clusters of the same class are likely to be correctly classified, despite of being peripheral, because two neighbor clusters of the same class suggest the existence of a bigger cluster in which both are contained.

We tested two models to evaluate the conditional probability of a point belonging to a class. Let $d(\mathbf{x}, \mathbf{p})$ be the euclidean distance from $\mathbf{x}$ to $\mathbf{p}$ and C_k the set of cluster centroids from class k, we define the probability of point $\mathbf{x}$ belonging to class j as the normalized inverse distance from $\mathbf{x}$ to the nearest neighbor from class k, given by

$$p(j \mid \mathbf{x}) = \frac{\left(\min_{\mathbf{p} \in C_j} d(\mathbf{x}, \mathbf{p})\right)^{-1}}{\sum_i \left(\min_{\mathbf{p} \in C_i} d(\mathbf{x}, \mathbf{p})\right)^{-1}} \tag{5}$$

which accounts as evidence for a class just the distance of the single closest point from the class. Our second model takes into consideration the sum of the inverse distances to every point in the class. The probability is given by

$$p(j \mid \mathbf{x}) = \frac{\sum_{\mathbf{p} \in C_j} d(\mathbf{x}, \mathbf{p})^{-K}}{\sum_i \sum_{\mathbf{p} \in C_i} d(\mathbf{x}, \mathbf{p})^{-K}} \tag{6}$$

where K is a constant which regulates how quickly the influence of a point decays with distance. For large values of K the influence of the nearest points tend to dominate over the influence of more distant points, and this model when used for ranking the best classifications gives similar results as the first, but with less meaningful probabilities.

We can use these models to identify points classified with low probabilities. This can be useful in the labeling process of a large dataset. It makes more sense to spend the most resources labeling the "hard" cases instead of the easier ones. Removing the excess of wrongly labeled points can also be useful in the next step, when training a neural network. Wrong labels can degrade the performance of a classifier. Fig. 3 shows how the accuracy varies when points classified with low probabilities are removed. With high values of K the results of both models are almost identical, for low values of k the first model is better. This difference occurs because the cluster structures are not confined in convex regions in space. The information of far points is not as meaningful as the information of closer points, thus in general considering just the nearest neighbors is better.

3.4 K-means Initialization

The qualities of clusters found by k-means can vary significantly with the initialization scheme used. An independent-samples t-test was conducted to compare k-means++ against random initialization. k-means++, which chooses as initial centroids data points more evenly spread than by random initialization [1], results in average in higher accuracies ($t(58) = 3.66$, $p <= 0.001$). There is a modest negative correlation ($r = -0.26$, $p < 0.001$) between clustering accuracy and the k-means distortion function (sum of squared distances from points to cluster centroids to which they are assigned), given by

$$\sum_{i=0}^{N} \|\mathbf{x_i} - \mu_{\mathbf{x_i}}\|^2 \tag{7}$$

where N is the number of points and $\mu_{\mathbf{x_i}}$ is the centroid of the cluster to which $\mathbf{x_i}$ belongs. Repeating the clustering step many times with k-means++ initialization and selecting the run with the lowest distortion results in a significant improvement in final accuracy.

4 Neural Network Training

We can improve further our results by using the labeled examples obtained so far as train data of a supervised model. We trained two architectures of neural networks: Multilayer Perceptron (MLP) and Convolutional Neural Network

Table 1. Error rate comparison (on MNIST test set)

Method	100	600	1000
Supervised Neural Network	25.13	11.21	9.84
Manifold Tangent Classifier [14]	12.6	5.13	3.64
Semi-Supervised Embedding [19]	**7.75**	3.82	**2.73**
Pseudo-label [9]	10.49	4.01	3.46
AL + MLP	8.17	4.57	3.86
AL + ConvNet	7.98	**3.73**	3.32

(ConvNet). In Table 1 we compare the error rate of our active learning strategy (AL) with other semi-supervised models on MNIST test set when using 100, 600 and 1000 labeled examples. To represent the expected performance in a more realistic setting, we ran the clustering step with 30 random initializations and selected the model with the median accuracy to provide the labeled examples used in the neural network training. We chose not to use an additional validation data, as before, for hyperparameters optimization and used instead 10000 samples from the training examples with the labels found in the previous step. As a negative effect of adopting this validation set "contaminated" with wrong labels, early stopping had its efficacy reduced. Even though, we achieved the lowest error rates reported for 600 labeled examples. For the other values, our results were just slightly worse than the results obtained by Semi-Supervised Embedding.

We have observed that removing the excess of incorrectly labeled data, excluding the training examples classified with low probabilities has somewhat unpredictable effects. Although the training data created this way has fewer wrong labels, it also has less diversity. In general this procedure does not seem to improve the final accuracy and in some cases the results are even worse, showing that the diversity of the dataset can be as important as its size.

5   Related Work

One of the first attempts to explore the clustering structure in data for active learning was made by Xu et al. [20], who employed k-means clustering to draw representative samples from unlabeled examples which were used to speed up the convergence of Support Vector Machines. Zhu et al. [21] explored the manifold structure in the data by modeling the probability distribution of points belonging to classes as a Gaussian Random Field built on a graph whose nodes are the training examples and the edges encode the distances between them. Nguyen et al. [12] addressed some shortcomings of the two previous works by taking measures to avoid repeatedly labeling samples in the same clusters. They used the representative samples from clusters to train a discriminative classifier to obtain the labels from the remaining unlabeled points (differently from our work which uses a distance based classifier to propagate labels). The idea of using stacked auto-encoders before clustering was explored by Lefakis et al. [10], but this work focused in the use of under-complete representations and "regular" auto-encoders (without further regularization, such as noise), which usually learns poorer features.

6   Conclusion and Future Work

In this work, we explored the use of unsupervised feature learning together with clustering to build a simple and effective active learning strategy which shows state-of-the-art performance on MNIST in the semi-supervised setting. Our results raises a series of questions to orient future investigations. Unsupervised feature learning using only MNIST data is too limited. Humans learn from a lot of unlabeled data before being able to recognize handwritten digits and more complex objects. This idea of learning features from one kind o data and using them in a different context has already been successful explored in the "self-taught learning" setting [13] and we expect it to be useful in active learning as well. It is also worth to explore other metrics and clustering algorithms. K-means is intimately related to the euclidean distance, which could be possibly replaced by better metrics for comparing objects in a feature space. We briefly experimented some k-means variations, such as spherical k-means, [7] which is based on the cosine similarity, but it seems to suffer from the same k-means and euclidean space problems and did not lead to improvements.

References

1. Arthur, D., Vassilvitskii, S.: k-means ++ : the advantages of careful seeding. In: Proceedings of the Eighteenth Annual ACM-SIAM Symposium on Discrete Algorithms, vol. 8, pp. 1027–1035 (2007)
2. Bengio, Y., Courville, A., Vincent, P.: Representation learning: A review and new perspectives. IEEE Transactions on Pattern Analysis and Machine Intelligence **35**(8), 1798–1828 (2013)
3. Bengio, Y., Lamblin, P., Popovici, D., Larochelle, H.: Greedy layer-wise training of deep networks. Advances in Neural Information Processing Systems **19**(1), 153–160 (2007)
4. Cayton, L.: Algorithms for manifold learning. Univ. of California at San Diego Tech. Rep pp. 1–17 (2005). http://www.vis.lbl.gov/romano/mlgroup/papers/manifold-learning.pdf
5. Dasgupta, S.: Two faces of active learning. Theoretical Computer Science **412**(19), 1767–1781 (2011)
6. Dasgupta, S., Hsu, D.: Hierarchical sampling for active learning. In: Proceedings of the 25th International Conference on Machine Learning, pp. 208–215 (2008)
7. Dhillon, I.S.: Concept Decompositions for Large Sparse Text Data using Clustering. Machine Learning **42**(1–2), 143–175 (2004)
8. Hinton, G.E., Osindero, S., Teh, Y.W.: A fast learning algorithm for deep belief nets. Neural Computation **18**(7), 1527–1554 (2006)

9. Lee, D.H.: Pseudo-label: the simple and efficient semi-supervised learning method for deep neural networks. In: Workshop on Challenges in Representation Learning, ICML (2013)

10. Lefakis, L., Wiering, M.: Semi-supervised methods for handwritten character recognition using active learning. In: Proceedings of the Belgium Netherlands Conference on Artificial Intelligence, pp. 205–212 (2007)

11. Martens, J.: Deep learning via hessian-free optimization. In: Proceedings of the 27th International Conference on Machine Learning (ICML-2010), pp. 735–742 (2010)

12. Nguyen, H.T., Smeulders, A.: Active learning using pre-clustering. In: Proceedings of the Twenty-First International Conference on Machine Learning, p. 79. ACM (2004)

13. Raina, R., Battle, A., Lee, H., Packer, B., Ng, A.Y.: Self-taught learning: transfer learning from unlabeled data. In: Proceedings of the 24th International Conference on Machine Learning, pp. 759–766. ACM (2007)

14. Rifai, S., Dauphin, Y.N., Vincent, P., Bengio, Y., Muller, X.: The manifold tangent classifier. In: Advances in Neural Information Processing Systems, pp. 2294–2302 (2011)

15. Selfridge, O.G.: Pandemonium: a paradigm for learning. In: Proceedings of the Symposium on Mechanisation of Thought Processes, vol. 1, pp. 511–529. HMSO (1959)

16. Vincent, P., Larochelle, H., Bengio, Y., Manzagol, P.A.: Extracting and composing robust features with denoising autoencoders. In: Proceedings of the 25th International Conference on Machine Learning, pp. 1096–1103. ACM (2008)

17. Vincent, P., Larochelle, H., Lajoie, I., Bengio, Y., Manzagol, P.A.: Stacked denoising autoencoders: Learning useful representations in a deep network with a local denoising criterion. The Journal of Machine Learning Research 11, 3371–3408 (2010)

18. Weston, J., Leslie, C., Ie, E., Zhou, D., Elisseeff, A., Noble, W.S.: Semi-supervised protein classification using cluster kernels. Bioinformatics 21(15), 3241–3247 (2005)

19. Weston, J., Ratle, F., Mobahi, H., Collobert, R.: Deep learning via semi-supervised embedding. In: Montavon, G., Orr, G.B., Müller, K.-R. (eds.) NN: Tricks of the Trade, 2nd edn. LNCS, vol. 7700, pp. 639–655. Springer, Heidelberg (2012)

20. Xu, Z., Yu, K., Tresp, V., Xu, X.W., Wang, J.: Representative Sampling for Text Classification Using Support Vector Machines. Springer, Heidelberg (2003)

21. Zhu, X., Ghahramani, Z., Lafferty, J., et al.: Semi-supervised learning using gaussian fields and harmonic functions. In: ICML, vol. 3, pp. 912–919 (2003)

STOCS: An Efficient Self-Tuning Multiclass Classification Approach

Yiming Qian[1], Minglun Gong[1]($\boxtimes$), and Li Cheng[2,3]

[1] Deparment of Computer Science, Memorial University, St. John's, NL, Canada
yq4048@mun.ca
[2] Bioinformatics Institute, A*STAR, Singapore, Singapore
gong@cs.mun.ca
[3] School of Computing, National University of Singapore, Singapore, Singapore
chengli@bii.a-star.edu.sg

Abstract. A simple, efficient, and parameter-free approach is proposed for the problem of multiclass classification, and is especially useful when dealing with large-scale datasets in the presence of label noise. Grown out of one-class SVM, our approach enjoys several distinct features: First, its decision boundary is learned based on both positive and negative examples; Second, the internal parameters and especially the kernel bandwidth are self-tuned. Our approach is compared side-by-side with LIB-SVM, arguably the most widely-used multiclass classification system, in a sequence of empirical evaluations, where our approach is shown to perform almost as well as their optimal parameter settings tuned for individual datasets, while consuming only a fraction of the processing time.

1   Introduction

Multiclass classification is a fundamental problem in machine learning and has been extensively studied (e.g. [8,11,14,30,35]). The existing methods often include tuning parameters, such as the cost matrix and regularization parameters [14], that are application-dependent and are often non-intuitive to machine learning practitioners. This issue is particularly pronounced with large-scale applications [15], where even moderate amount of tuning parameters might be computationally too expensive. Besides, human annotations tend to be error-prone especially when working with problems of large-scale and many classes [13]. This motivates us to consider a novel multiclass classification approach that is parameter-free, efficient, and capable of dealing with label noise. To achieve this, an online Self-Tuning One-Class SVM algorithm (or STOCS) is developed. The key insight of our approach is to train our model using both positive and negative examples. This allows us to adaptively chose the optimal parameters for each support vector so that it does not provide strong support to negative examples. In what follows, we will review the related research efforts and relate them with our approach.

© Springer International Publishing Switzerland 2015
D. Barbosa and E. Milios (Eds.): Canadian AI 2015, LNAI 9091, pp. 291–306, 2015.
DOI: 10.1007/978-3-319-18356-5_26

Parameter Tuning (aka Estimation of Internal Parameters or Adaptive Bandwidth). Cross-validation [9,19,32] is probably the most widely used method for estimating the internal parameters. Kohavi [24] empirically compares cross-validation with bootstrap [19] and finds that the latter one tends to introduce extremely large bias sometimes, while the former often performs significantly better. A typical cross-validation strategy is to perform coarse grid search over the space of internal parameters, which is however often time-consuming. Poggio and Cauwenberghs [29] accelerate cross-validation by reduce the cost of a single pass of the dataset. Izbicki [21] proposes efficient algebraic methods aiming at its speedup. In [26], the initial cross-validated parameter values are further refined by coordinate descent on induced objective functions. Despite of its popularity and usefulness, cross-validation also possesses several major issues and limitations, as e.g. discussed in [34]. Meanwhile other methods have also been studied: From the view of Bayesian evidence maximization, Gold *et al.* [16] instead consider a hybrid Monte Carlo approach based on the evidence gradients. Zou [37] presents a adaptive Lasso method to adjust coefficient shrinkage for individual variables for regression related problems. The method in [33] is based on variable selection stability and is dedicated to problems involving penalized regression models.

The idea of adaptive or variable bandwidth has also been studied for density estimation [23], regression [20], and classification [10] problems. Existing works usually focus on being adaptive in term of *only* locality and is agnostic to different class labels, while the variable bandwidth considered in our approach is sensitive to its location, as well as adversary classes from its spatial vicinities. Furthermore, unlike existing approaches that try to pick the same set of parameters for all support vectors, we allow different support vectors having different parameters. We argue that: 1) allowing support vectors that are far away from the decision boundaries having larger influence area can effectively reduce the number of support vectors needed; and 2) assigning small influence areas to support vectors that are close to the decision boundaries help to reduce the level of confusion.

Prediction with Noisy Labels. In real-life applications it is of great importance to make reliable predictions even in the present of noisy labels. The problem has been studied by numerous research efforts, starting from the early works [1,36] with theoretical analysis [3]. A kernel LDA method is considered in [25], while the method in [12] focuses on exploiting the positive instances. Following that of [3], various noise-resistant variants of the perceptron method are also proposed [4,22]. Very recently, there are several attempts (e.g. [27,31]) to provide a more rigorous account of the theoretical understanding and analysis of this problem. Interested readers may refer to review articles [13,28] for further details. Our approach tries to address the label noise problem through adaptively selecting cut-off parameters for individual support vectors. We notice that outliers are often surrounded by correctly labeled examples. By assigning low cut-off values to the corresponding support vectors, we can effectively limit their impact to the final decision.

Our approach builds upon online one-class SVM [6] and inherits its efficiency of handling large-scale data. It differs from it in several ways, though. First, conventional one-class SVM works only with positive examples, while our approach learns the decision boundary based on both positive and negative examples. Second, the challenging issue of manually tuning kernel weights as in [6] is self-adjusted in our approach, which makes it handy to use even for first-time users.

2   The STOCS Algorithm

2.1   Online Learning Model

Training a SVM using a large set of examples is a classical batch learning problem, the solution of which can be found through minimizing a quadratic objective function. Previous studies [2] have shown that a similar or even better generalization performance can be achieved using online learning with a *much less* computational cost, by showing all examples repetitively to an online learner, when comparing to that of batch learning.

The online learner we use follows the one proposed by [6,7]. Let $f_t(\cdot)$ be a score function of examples at time t, $k(\cdot,\cdot)$ be a kernel function, and α_t be a non-negative weight of example of time t. When a new example x_t with label y_t arrives, the score function becomes:

$$f_t(x_t) = \sum_{i=1}^{m} \alpha_i k(\hat{x}_i, x_t), \tag{1}$$

and the update rule for weights is:

$$\alpha_t = \text{clamp}\left(\frac{\gamma - (1-\tau)f_t(x_t)}{k(x_t, x_t)}, 0, (1-\tau)\chi\right),$$
$$\alpha_i \quad \leftarrow \quad (1-\tau)\alpha_i \quad \forall i = 1, \ldots, m, \tag{2}$$

where $\gamma := 1$ is the margin, $\tau \in (0,1)$ the decay parameter, and $\chi > 0$ the cut-off value. $\hat{x}_i$ is the i_{th} support vector of the class y_t, and m is the total number of support vectors in the class. clamp$(\cdot, A, B)$ is an identical function of the first argument bounded by A and B. x_t is selected as a new support vector if and only if $\alpha_t > 0$.

The above online learner is used to train a dedicated one-class SVM classifier for each input class. To label a test example afterward, we simply compute its scores under different classes and label it with the one that gives the highest score.

When using the above online learning model for batch learning, the parameter τ needs to be carefully adjusted throughout the iterative process. This is because at the beginning of the training, the active set is empty and the score function of any input data will be zero. Consequently, the first group of support vectors added into the active set tend to have large α values, which needs to be lowered to their proper values in the later iteration. As shown in Eq.(2), once a support vector (x_t, α_t) is added to the active set, over the time its weight α_t is only affected by the decay parameter τ. Hence, the initial value for τ needs to be large to effectively reduce the alpha values of existing support vectors. On the other hand, the iterative training process cannot converge unless $\tau \approx 0$. As a result, τ needs to be gradually reduced throughout the iterative process. Conventionally, the decay parameter is set based on an exponential function: $\tau = \exp(-\frac{t}{\xi})$, where parameter ξ controls how fast the decay parameter τ decreases.

2.2 Adjustable Kernel Functions

The kernel function $k(u, v)$ used in Eq.(1) computes the dot product of two high-dimensional vectors to which examples u and v are mapped. It determines how much one example, if chosen as support vector, will provide support to the other examples. A good kernel function should output high values when applied to two similar examples, and low values for dissimilar ones. Here we call a kernel function a *normalized* kernel if it satisfies the following property:

$$\begin{cases} k(x_t, x_t) = 1, & \forall x_t \\ 0 \le k(x_t, x_s) \le 1, & \forall x_t, x_s \end{cases}$$

By definition, the Gaussian kernel, $k(u, v) = \exp(-\|u - v\|^2/\sigma^2)$, is a normalized kernel. When applied to histogram vectors, the histogram intersection kernel, $k(u, v) = \sum_{i=1}^{n} \min(u_i, v_i)$, is also normalized. The linear kernel, $k(u, v) = u \cdot v$, is not normalized in general, but it becomes normalized if the input vectors are both nonnegative and normalized, i.e., $u_i \ge 0, \forall u_i$ and $\|u\| = 1, \forall u$.

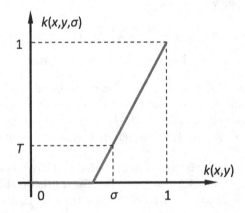

Fig. 1. Define an adjustable kernel $k(u, v, \sigma)$ based on a normalized kernel $k(u, v)$

We further call a *normalized* kernel $k(u, v, \sigma)$ with parameter σ *adjustable*, if and only if it possesses the following two properties: i) through setting σ, we can always satisfy $k(x_t, x_s, \sigma) \le T, T \in (0, 1)$ for all $x_t, x_s, x_t \ne x_s$; and ii) if $k(x_t, x_s) \ge k(x_t, x_l)$, then $k(x_t, x_s, \sigma) \ge k(x_t, x_l, \sigma)$ holds regardless the setting of σ. Gaussian kernel is adjustable since setting $\sigma = \sqrt{-\|x_t - x_s\|^2/\log T}$ would satisfy the first requirement and adjusting σ does not alter the overall shape of the Gaussian either. For any normalized kernel function $k(u, v)$ that is not inherently adjustable, we can define an adjustable version as:

$$k(u, v, \sigma) = \max\left(1 - (1 - T)\frac{1 - k(u, v)}{1 - \sigma}, 0\right) \tag{3}$$

As shown in Figure 1, for a given σ, $k(u, v, \sigma)$ is a monotonically-increasing piecewise-linear function with respect to $k(u, v)$. By definition, setting $\sigma = k(x_t, x_s)$ allows $k(x_t, x_s, \sigma) \leq T$, regardless how the original kernel function $k(u, v)$ is defined. Furthermore, since $k(u, v, \sigma)$ is a monotonic function, it does not change the score ordering among examples defined by $k(u, v)$.

2.3 Self-Tuning of the Internal Parameters

Unlike the conventional one-class SVM, which uses only positive examples from a given class to train a model, STOCS makes use of both positive and negative examples. It also requires kernel functions *adjustable* and adaptively

---

Algorithm 1 STOCS Training

Input: training instances x and labels y, kernel function $k(\cdot, \cdot)$
Output: support vector set Φ
1: **Initialize** each Φ_j as an empty set for each j_{th} class
2: **for** each training example x_t **do**
3:     find the negative example x_n that yields the largest $k(x_t, x_n)$
4:     compute σ_t based on the kernel function
5: **end for**
6: **for** each training example x_t **do**
7:     compute the positive and negative example number $h^+(x_t)$, $h^-(x_t)$
8:     compute χ_t based on Eq.(5)
9: **end for**
10: **repeat**
11:     **for** each training example x_t with label y_t **do**
12:         compute score $f_t(x_t)$ and weight α_t based on Eq.(4)
13:         **if** x_t is identical to an existing SV $\hat{x}_t$ in Φ_{y_t} **then**
14:             **if** $\alpha_t \leq 0$ **then**
15:                 erase $\hat{x}_t$ from Φ_{y_t}
16:             **else**
17:                 update $\hat{x}_t$ with the new weight α_t
18:             **end if**
19:         **else if** $\alpha_t > 0$ **then**
20:             push $(x_t, \alpha_t, \sigma_t, \chi_t)$ into Φ_{y_t}
21:         **end if**
22:     **end for**
23: **until** there are no more changes in Φ

---

sets kernel bandwidth σ and cut-off value χ for each support vector individually. Hence, at each support vector, we not only store its weight α, but also the corresponding parameters, forming a quadruple $(\hat{x}, \alpha, \sigma, \chi)$. The details for STOCS training are shown in Algorithm 1. It is worth noting that we use x_i to represent an example, which becomes $\hat{x}_i$ if it is selected as a support vector. A newly observed example is denoted as x_t.

Removal of Decay Parameter τ. Following the idea discussed in our previous work [17,18], we first eliminate the decay parameter τ through an explicitly *reweighting* scheme. That is, if a training example x_t arrives and it turns out identical to the example in an existing support vector $(\hat{x}_t, \alpha_i)$, this support vector is taken out before computing the score function and then replaced with (x_t, α_t) that carries the newly obtained weight. Further, considering that we always have $k(x_t, x_t) = 1$ for normalized kernels, the new score function and the update rule become:

$$f_t(x_t) = \sum_{i=1}^{m} \alpha_i \delta(\hat{x}_i \neq x_t) k(\hat{x}_i, x_t, \sigma_i),$$
$$\alpha_t = \text{clamp}\left(\gamma - f_t(x_t), 0, \chi_t\right), \tag{4}$$

where $\delta(\cdot)$ is an indicator function with $\delta(true) = 1$ and $\delta(false) = 0$. Intuitively, this modified online learning method resets the weight component of a particular support vector (x_t, α_t), based on how well the separating hyperplane defined by the remaining support vectors is able to classify example x_t. This reweighting process can either increase or decrease α_t and hence decay is not necessary any more.

Adaptive Kernel Bandwidth σ. As mentioned above, the adjustable kernel function used in STOCS contains a parameter σ. We now discuss how to tune σ_i automatically and individually for each support vector $(\hat{x}_i, \alpha_i, \sigma_i, \chi_i)$. The core idea here is that a support vector should not provide strong supports to negative examples, i.e., we require $k(\hat{x}_i, x_n, \sigma_i) \leq T$ for any negative example x_n. Considering that the kernel function used is normalized, we further argue that a common and constant T value exists for different datasets. Figure 5 confirms the assumption, where a constant $T = 0.25$ is used among all experiments conducted in this paper. Using this constraint, we can easily compute σ_i based on the definition of the adjustable kernel function. That is, as soon as we meet a negative example x_n that yields $k(\hat{x}_i, x_n, \sigma_i) > T$ during the online learning, we update σ_i to ensure $k(\hat{x}_i, x_n, \sigma_i) \leq T$. Since adjusting σ does not affect the score ordering among examples, the supports to previously seen negative examples can only decrease, which ensures the training process converges. It is also worth noting that for static datasets, σ_i can be precomputed for each example x_i using the negative example x_n that yields the largest $k(x_i, x_n)$ value as shown in Algorithm 1.

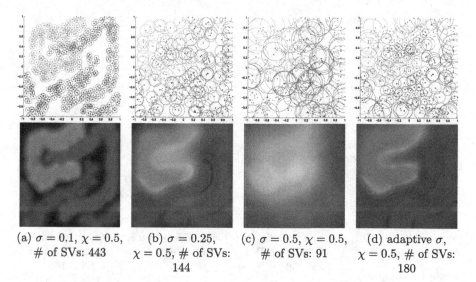

(a) $\sigma = 0.1$, $\chi = 0.5$, # of SVs: 443

(b) $\sigma = 0.25$, $\chi = 0.5$, # of SVs: 144

(c) $\sigma = 0.5$, $\chi = 0.5$, # of SVs: 91

(d) adaptive σ, $\chi = 0.5$, # of SVs: 180

Fig. 2. Visualization of the classification results for the "threeclass" dataset. Top row shows the selected support vectors with the corresponding σ values illustrated using circles. Bottom row shows the decision maps, which encode the output scores from the three one-class SVMs using red, green, and blue channels respectively. Both black color in (a-b) and orange/cyan/magenta colors in (b-c) indicate regions with ambiguities.

Without losing generality, here we use the Gaussian kernel to illustrate the idea. For Gaussian kernel, the kernel bandwidth σ is the standard deviation of the Gaussian function and controls the radius of the influence area of each support vector. As shown in Figure 2(a), when σ is too small, the obtained classifier can only recognize data that are close to one of the provided training examples, resulting over-fitting. On the other hand, large σ value causes fuzzy decision boundaries among different classes; see Figure 2(c). Hence, to ensure a proper σ value is used, previous approaches often rely on cross validation.

As shown in Figure 2(d), through automatically selecting different Gaussian support for different support vectors, the classifier obtained provides a sharp decision boundaries between the three training classes.

Adaptive Cut-off Value χ. The last standing parameter is the cut-off value χ, which is used to limit the effects of outliers. Comparison between Figure 2(b) and Figure 3(c), which are generated using the same set of parameters, suggests that the presents of label noises can severely distort the decision boundaries. Using a smaller χ values helps to reduce the impact of outliers, but at the expense of using more support vectors; see Figure 3(a-b). The use of adaptive σ also helps to limit the impact of outliers to their own neighborhoods; see Figure 3(e-g). Nevertheless, it does not fully address the problem.

Following the idea of tuning parameters for individual support vectors, here we also adaptively determine a proper χ value for each support vector. The key

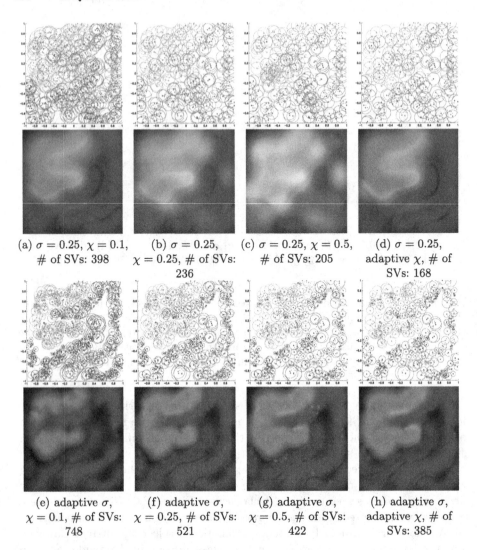

Fig. 3. Visualization of the classification results when the dataset is corrupted with label noise, i.e., 5% of the random data have their labels changed. Under both fixed (a-d) and adaptive (e-h) Gaussian support situations, using adaptive cut-off values helps to limit the impacts of outliers.

idea is that a support vector should have a higher χ if there are many positive examples within its support region and a smaller χ if there are many negative examples surrounding it. Hence we set:

$$\chi_t = 0.5 + \frac{h^+(x_t) - h^-(x_t)}{2(h^+(x_t) + h^-(x_t))} \tag{5}$$

where $h^+(x_t)$ and $h^-(x_t)$ are the numbers of positive and negative examples that satisfy $k(x_t, x_i, \sigma_t) \geq 0.6T$, respectively. Note that $h^-(x_t) \geq 1$ due to the way σ_t is calculated, ensuring the denominator being a non-zero value.

As shown in Figure 3(d & h), using adaptive χ values can effectively limit the impacts of outliers. Note that, similar to σ, the χ values for different examples can also be precomputed when the dataset is static.

3   Experiments

Here we first evaluate the effectiveness of reweighting and parameter self-tuning by comparing STOCS with the conventional online one-class SVM training technique [6]. Comparisons with benchmark approach [5] in terms of classification accuracy, support vector number, and training time, are then reported. STOCS was implemented in C++ and we ran experiments on a PC with 3.2GHz Intel Core i5 CPU and 4GB RAM. Our implementation can be downloaded from http://www.cs.mun.ca/~gong/research/SelfTuningClassifier.html.

3.1   The Effectiveness of Reweighting

Figure 4 compares the convergence speed between the reweighting scheme and the conventional training approach under different decay rate settings using a

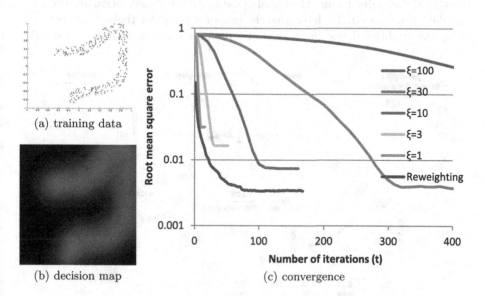

Fig. 4. Convergence of the conventional approach and the reweighting scheme. For a given set of examples (a), the decision map (b) obtained using the conventional approach under a very slow decay change rate ($\xi = 100$) is used as the ground truth. Comparing the decision maps generated after different iterations with the ground truth shows the convergence speeds of different approaches (c).

Gaussian kernel. For the input dataset shown in Figure 4(a), we first generate an one-class SVM classifier using Eq.(1) and (2) under fixed parameters ($\sigma = 0.25$; $\chi = 0.5$) and a very slow decay change rate ($\xi = 100$). This classifier is then used to generate a decision map (see Figure 4(b)), which is used as the ground truth. The decision maps obtained after different iterations and under different settings are compared with the ground truth. The results shown in Figure 4(c) suggest that the conventional approach generally converges after 10ξ iterations, i.e., $\tau < \exp(-10)$. Furthermore, when a faster decay change rate is used, the final decision map defers from the ground truth, indicating a premature convergence. The classifier obtained using reweighting with the same σ and χ setting yields almost identical decision map as Figure 4(b), but it converges much faster than the conventional approach under $\xi = 100$.

3.2   The Effectiveness of Parameter Self-Tuning

We now evaluate the presented self-tuning scheme by comparing it with the conventional one-class SVMs trained using fixed parameters. Seven multiclass classification datasets downloaded from the LIBSVM website are used. Due to space limits, we refer readers to the LIBSVM website for detailed descriptions on these datasets.

First we test the effectiveness of self-tuning σ using a Gaussian kernel and under a fixed cut-off value ($\chi = 0.5$). Both STOCS and the conventional one-class SVM are trained using the training data and the classification accuracy on test data is measured. To determine the proper σ range for these datasets when using the traditional one-class SVM, we preform a coarse-to-fine grid search.

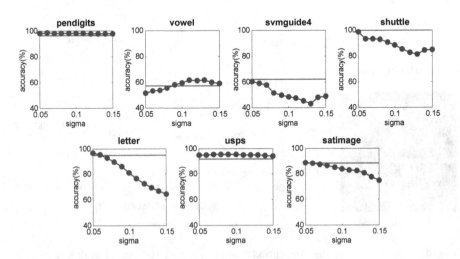

Fig. 5. The effectiveness of adaptive selecting kernel bandwidth σ for seven datasets. Red line shows the classification accuracy of using self-tuned σ based on a constant $T = 0.25$, whereas the blue curve plots the accuracy by fixing the σ to different values.

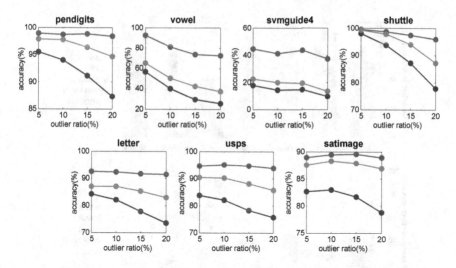

Fig. 6. The effectiveness of adaptive selecting cut-off value. The classification accuracy of self-tuned (red), $\chi = 0.1$ (green), and $\chi = 0.5$ (blue) under increasing label noise are plotted for seven datasets.

Figure 5 plots the classification accuracy of the traditional one-class SVMs when σ varies within the proper range found. As expected, the performance highly depends on the σ value and there is no single σ value that works well for all datasets. STOCS adaptively selects σ for each support vector based on a fixed threshold value $T = 0.25$, which confirms our assumption that a common T exists for different datasets when normalized kernel is used. Its performance is close to the traditional one-class SVM under optimal settings in all cases, and even outperforms the latter in "svmguide4" and "shuttle" datasets. We attribute such performance gain to adjusting kernel bandwidth for support vectors individually, rather than using the same tuned parameters for all.

Next, we evaluate the effectiveness of self-tuning χ on datasets corrupted with label noises. For each dataset, we randomly and incrementally select training examples as outliers and alter their labels. These outlier examples with their original labels are then used as testing data to evaluate whether the classifiers can correct labeling errors. The test is repeated 10 times for each dataset to compute the average accuracy. The results of self-tuning χ and fixed χ are plotted in Figure 6, where σ is self-tuned in all tests. The comparison clearly shows that the performance of conventional one-class SVM drops as the more label noise is introduced. Using self-tuned χ values not only improves the accuracy, but also makes the performance of classifier more robust against the increases of labeling noise.

3.3 Comparison with LIBSVM [5]

Finally we compare the performances of STOCS with a benchmark approach, the LIBSVM [5] using the one-vs-one training approach, under both Gaussian and

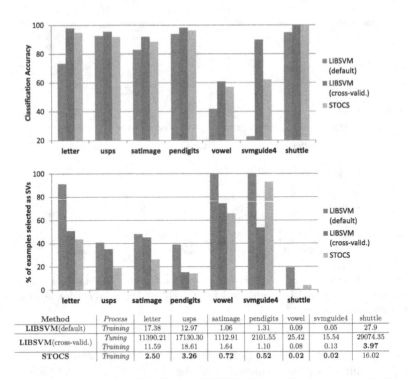

Method	Process	letter	usps	satimage	pendigits	vowel	svmguide4	shuttle
LIBSVM(default)	*Training*	17.38	12.97	1.06	1.31	0.09	0.05	27.9
LIBSVM(cross-valid.)	*Tuning*	11390.21	17130.30	1112.91	2101.55	25.42	15.54	29074.35
	Training	11.59	18.61	1.64	1.10	0.08	0.13	**3.97**
STOCS	*Training*	**2.50**	**3.26**	**0.72**	**0.52**	**0.02**	**0.02**	16.02

Fig. 7. Comparison between STOCS and LIBSVM under Gaussian kernel. Top row shows the classification accuracy, middle row shows the number of support vectors used, and bottom row gives the processing time needed.

Linear kernels. Notice that LIBSVM is arguably the most widely used multiclass classification method in practice.

Gaussian Kernel. We start with comparing the performances of STOCS and LIBSVM on the aforementioned seven datasets. For LIBSVM, both default and cross-validated parameter settings are tested. Note that here LIBSVM is tuned for each dataset individually, resulting different parameter settings for different datasets. STOCS, on the other hand, uses the same self-tuning procedure with the threshold $T = 0.25$ for all datasets.

Figure 7 shows that STOCS is more accurate than the LIBSVM under default settings in six out of the seven datasets. While LIBSVM with parameters tuned through cross-validation outperforms STOCS, the performance difference is less than 4% in six datasets. To achieve this performance gain, LIBSVM requires much longer processing time for parameter tuning. It is worth noting that cross-validation can be accelerated using the methods in [21,29], whereas they both cannot handle parameter overfitting problem as shown in Figure 9. Furthermore, for most datasets, STOCS uses fewer support vectors than LIBSVM with both default and tuned parameters, allowing faster labeling of incoming examples.

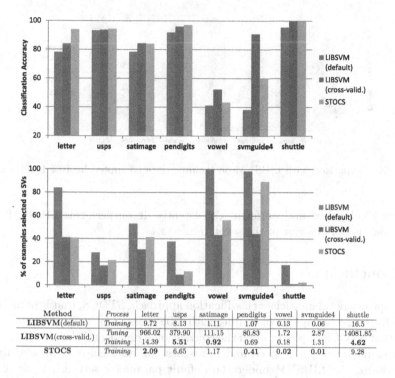

Method	Process	letter	usps	satimage	pendigits	vowel	svmguide4	shuttle
LIBSVM(default)	Training	9.72	8.13	1.11	1.07	0.13	0.06	16.5
LIBSVM(cross-valid.)	Tuning	966.02	379.90	111.15	80.83	1.72	2.87	14081.85
	Training	14.39	5.51	0.92	0.69	0.18	1.31	4.62
STOCS	Training	2.09	6.65	1.17	0.41	0.02	0.01	9.28

Fig. 8. Comparison between STOCS and LIBSVM under linear kernel

Linear Kernel. To evaluate whether the presented approach works well for kernel functions that are not inherently adjustable, here we also compare STOCS with LIBSVM under linear kernel. To satisfy the nonnegative and normalized conditions that makes linear kernel normalized, we first map all examples feature-wise to range [0,1] and then normalize individual example vector to unit length. The same normalized datasets are used to train both LIBSVM and STOCS. However, LIBSVM uses the original linear kernel, whereas STOCS uses its adjustable version; see Eq(3). The same threshold value $T = 0.25$ is used here for STOCS.

From Figure 8 we observe that, not only STOCS outperforms LIBSVM under the default parameter setting in all cases, it also outperforms LIBSVM with tuned parameters for two datasets and performs on par for two other datasets. This is remarkable since LIBSVM uses the best parameters specifically tuned for the given datasets.

Label Noise. When the training data contains label noises, parameter tuning through cross-validation may cause the parameters being overfit to the corrupted data, leading to poor classification accuracy. This is evidenced by Figure 9, where the LIBSVM with tuned parameters is outperformed by the default parameters for "usps" and "satimage" datasets. Our approach, on the other hand, is robust

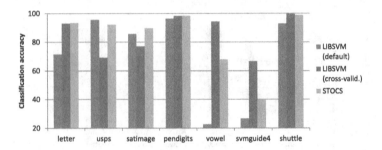

Fig. 9. Comparison with LIBSVM on training data corrupted by 10% label noises

against label noises and performs consistently. It outperforms LIBSVM with tuned parameters in four of the seven datasets.

4   Conclusions

We propose a novel multiclass classification approach, STOCS, that is parameter-free, efficient, and capable of dealing with label noise. Our empirical study suggests that STOCS improves upon the conventional online one-class SVM training in terms of convergence speed and robustness to label noise. It almost always outperforms the LIBSVM under the default parameter settings, while at the same time uses much fewer support vectors. Its performance is also very close to LIBSVM's optimal results obtained by cross-validation for individual datasets, but uses only a fraction of the processing time. Future research includes extending the technique to broader range of problems, including semi-supervised and multiple-label problems.

Acknowledgments. The authors gratefully acknowledge the constructive comments that they received from Dr. Zhi-Hua Zhou and anonymous reviewers.

References

1. Angluin, D., Laird, P.: Learning from noisy examples. Mach. Learn. **2**(4), 343–370 (1988)
2. Bottou, L., Bousquet, O.: The tradeoffs of large scale learning. In: NIPS (2008)
3. Bylander, T.: Learning linear threshold functions in the presence of classification noise. In: COLT (1994)
4. Cesa-Bianchi, N., Shalev-Shwartz, S., Shamir, O.: Online learning of noisy data. IEEE Transactions on Information Theory **57**(12), 7907–7931 (2011)
5. Chang, C., Lin, C.: LIBSVM: A library for svms. ACM Trans. on Intelligent Systems and Technology **2**, 27:1–27:27 (2011)
6. Cheng, L., Gong, M., Schuurmans, D., Caelli, T.: Real-time discriminative background subtraction. In: TIP (2011)

7. Cheng, L., Vishwanathan, S., Schuurmans, D., Wang, S., Caelli, T.: Implicit online learning with kernels. In: NIPS (2007)
8. Crammer, K., Singer, Y., Cristianini, N., Shawe-Taylor, J., Williamson, B.: On the algorithmic implementation of multiclass kernel-based vector machines. JMLR **2**, 2001 (2001)
9. Craven, P., Wahba, G.: Smoothing noisy data with spline functions: Estimating the correct degree of smoothing by the method of generalized cross-validation. Numer. Math. **31**, 377–403 (1979)
10. Dai, J., Yan, S., Tang, X., Kwok, J.: Locally adaptive classification piloted by uncertainty. In: ICML (2006)
11. Dietterich, T., Bakiri, G.: Solving multiclass learning problems via error-correcting output codes. JAIR **2**, 263–286 (1995)
12. Elkan, C., Noto, K.: Learning classifiers from only positive and unlabeled data. In: KDD (2008)
13. Frenay, B., Verleysen, M.: Classification in the presence of label noise: a survey. IEEE Trans. NNLS (2014)
14. Gao, T., Koller, D.: Multiclass boosting with hinge loss based on output coding. In: ICML (2011)
15. Garcia-Pedrajas, N., Haro-Garcia, A.: Scaling up data mining algorithms: review and taxonomy. Progress in Artificial Intelligence, 71–87 (2012)
16. Gold, C., Holub, A., Sollich, P.: Bayesian approach to feature selection and parameter tuning for support vector machine classifiers. Neural Networks **18**(5–6), 693–701 (2005)
17. Gong, M., Qian, Y., Cheng, L.: Integrated foreground segmentation and boundary matting for live videos. IEEE Trans. Image Processing (TIP) (2015)
18. Gong, M., Cheng, L.: Foreground segmentation of live videos using locally competing 1svms. In: 2011 IEEE Conference on Computer Vision and Pattern Recognition (CVPR), pp. 2105–2112. IEEE (2011)
19. Hastie, T., Tibshirani, R., Friedman, J.: The Elements of Statistical Learning. Springer (2009)
20. Herrmann, E.: Local bandwidth choice in kernel regression estimation. J. of Computational and Graphical Statistics **6**(1), 35–54 (1997)
21. Izbicki, M.: Algebraic classifiers: a generic approach to fast cross-validation, online training, and parallel training. In: ICML (2013)
22. Khardon, R., Wachman, G.: Noise tolerant variants of the perceptron algorithm. JMLR **8**, 227–248 (2007)
23. Kim, J., Scott, C.: Variable kernel density estimation. Ann. of Statistics **20**, 1236–1265 (1992)
24. Kohavi, R.: A study of cross-validation and bootstrap for accuracy estimation and model selection. In: IJCAI (1995)
25. Lawrence, N., Scholkopf, B.: Estimating a kernel fisher discriminant in the presence of label noise. In: ICML (2001)
26. Lorbert, A., Ramadge, P.: Descent methods for tuning parameter refinement. In: AISTATS (2010)
27. Natarajan, N., Dhillon, I., Ravikumar, P., Tewari, A.: Learning with noisy labels. In: NIPS (2013)
28. Nettleton, D., Orriols-Puig, A., Fornells, A.: A study of the effect of different types of noise on the precision of supervised learning techniques. Artificial Intelligence Review **33**, 275–306 (2010)

29. Poggio, T., Cauwenberghs, G.: Incremental and decremental support vector machine learning. In: Advances in Neural Information Processing Systems 13: Proceedings of the 2000 Conference, vol. 13, p. 409. MIT Press (2001)
30. Rifkin, R., Klautau, A.: In defense of one-vs-all classification. JMLR **5**, 101–141 (2004)
31. Scott, C., Blanchard, G., Handy, G.: Classification with asymmetric label noise: Consistency and maximal denoising. In: COLT (2013)
32. Stone, M.: Cross-validatory choice and assessment of statistical predictions. J. of Roy. Stat. Soc. (JRSS) Series B **36**, 111–147 (1974)
33. Sun, W., Wang, J., Fang, Y.: Consistent selection of tuning parameters via variable selection stability. JMLR **14**, 3419–3440 (2013)
34. Wang, W., Gelman, A.: A problem with the use of cross-validation for selecting among multilevel models. Tech. rep., Columbia Univ., New York (2013)
35. Zhang, T.: Statistical analysis of some multi-category large margin classification methods. The Journal of Machine Learning Research **5**, 1225–1251 (2004)
36. Zhu, X., Wu, X.: Class noise vs. attribute noise: A quantitative study of their impacts. Artif. Intell. Rev. **22**(3), 177–210 (2004)
37. Zou, H.: The adaptive lasso and its oracle properties. J. of the American Statistical Association **101**, 476 (2006)

Consolidation Using Sweep Task Rehearsal: Overcoming the Stability-Plasticity Problem

Daniel L. Silver[✉], Geoffrey Mason, and Lubna Eljabu

Jodrey School of Computer Science, Acadia University, Wolfville,
NS B4P 2R6, Canada
danny.silver@acadiau.ca

Abstract. This paper extends prior work on knowledge consolidation and the stability-plasticity problem within the context of a Lifelong Machine Learning (LML) system. A *context-sensitive* multiple task learning (*cs*MTL) neural network is used as a consolidated domain knowledge store. Prior work has demonstrated that a *cs*MTL network, in combination with *task rehearsal*, can retain previous task knowledge when consolidating a sequence of up to ten tasks from a domain. However subsequent experimentation has shown that the method suffers from scaling problems as the learning sequence increases resulting in the loss of prior task accuracy and a growing computational cost for rehearsing prior tasks using larger training sets. A solution to these two problems is presented that uses a *sweep* method of rehearsal that requires only a small number of rehearsal examples (as few as one) for each prior task per training iteration in order to maintain prior task accuracy.

1 Introduction

Lifelong machine learning, or LML, an area of machine learning research, is concerned with the persistent and cumulative nature of learning [15]. LML considers situations in which a learner faces a series of different tasks and develops methods of retaining and using prior knowledge to improve the effectiveness (more accurate hypotheses) and efficiency (shorter training times) of learning. We focus on the learning of concept tasks, where the target value for each example is either zero or one. Knowledge retention is necessary but is not sufficient for LML. Consolidation is the act of retaining knowledge, in an integrated representational form. This is necessary for the efficient and effective retention of knowledge and the efficient and effective indexing of that knowledge for later use [13]. The challenge for an LML system is consolidating the knowledge of a new task while retaining and possibly improving knowledge of prior tasks. This challenge is generally known in machine learning, neural science and psychology as the stability-plasticity problem[6].

This paper extends our work on knowledge consolidation and the stability-plasticity problem within the context of a LML system. It uses a modified multiple-task learning (MTL) neural network called *context-sensitive* multiple task learning (*cs*MTL) as a consolidated domain knowledge store. *cs*MTL networks have been

© Springer International Publishing Switzerland 2015

D. Barbosa and E. Milios (Eds.): Canadian AI 2015, LNAI 9091, pp. 307–322, 2015.
DOI: 10.1007/978-3-319-18356-5_27

demonstrated as an effective method of using prior knowledge to do transfer learning [14]. Prior work has also demonstrated that a *cs*MTL network, under the right conditions, can retain previous task knowledge when consolidating a sequence of up to ten tasks from a domain [5]. However subsequent experimentation has shown that the method suffers from scaling problems as the learning sequence increases: (1) there is loss of prior task accuracy within the consolidated model, and (2) the computational cost of consolidating each additional task grows quadratically in the total number of tasks (prior and new).

In this paper we present a solution to these two problems that combines prior findings with recent discoveries [4]. Most importantly we show that overcoming the stability-plasticity problem depends heavily upon how prior tasks are "rehearsed" within a network while consolidating the examples of a new task. We present empirical results that demonstrate that only a small number of properly chosen training examples (as few as one) from each prior task are required, per training iteration, to maintain long-term knowledge.

The paper has four remaining sections. Section 2 provides the necessary background on LML and consolidation using MTL and *cs*MTL networks. Section 3 presents new theory on the best approach to task rehearsal when using *cs*MTL networks. Section 4 provides the results of empirical studies that test the proposed approach on one synthetic and two real-world domains. Finally, Section 5 summarizes the findings and presents future work.

2   Background

2.1   Lifelong Machine Learning

A general framework for LML is shown in Figure 1. The framework is similar to that of standard inductive learning, however an LML system can save the knowledge of each accurate hypothesis it has learned in a long-term memory store called domain knowledge. When a new task is being learned, knowledge can be drawn from domain knowledge and used as inductive bias for the learning system. Ideally, the result is a more accurate hypothesis developed in a shorter period of time than would be created from just the training examples. If this new model is sufficiently accurate, it is retained in the domain knowledge store. Thus, an LML system requires a method of using prior knowledge to learn models for new tasks as efficiently and effectively as possible, and a method of retaining task knowledge after it has been learned.

Thrun writes "The acquisition, representation and transfer of domain knowledge are the key scientific concerns that arise in lifelong learning" [16]. Knowledge retention and representation will play an important a role in the development of LML systems. More specifically, the interaction between knowledge retention and knowledge transfer will be key to the design of LML agents. Lifelong learning research has the potential to make serious advances on a significant AI problem - the learning of *common background knowledge* that can be used for future learning, reasoning and planning. The work at Carnegie Mellon University on NELL is an early example of such research [2]. NELL stands for the

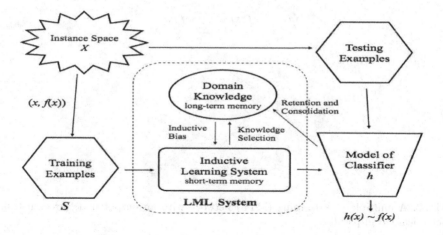

Fig. 1. A framework for Lifelong Machine Learning

never-ending language learner. Each day NELL must (1) extract, or read, information from the web to populate a growing structured knowledge base, and (2) learn to perform this task better than on the previous day. The system uses a semi-supervised multiple task learning approach in which hundreds of different semantic functions are trained together in order to improve learning accuracy.

Knowledge retention is necessary for a lifelong learning system, however, it is not sufficient. In [13] we propose that domain knowledge must be integrated for the purposes of efficient and effective retention and for more efficient and effective transfer during future learning. The process of integration we define as *consolidation*. The challenge for a lifelong learning system is consolidating the knowledge of a new task while retaining (and possibly improving) knowledge of prior tasks. This is known as the *stability-plasticity* problem and it refers to the challenge of adding new information to a system without the loss of prior information [6].

2.2   Consolidation with Multiple Task Learning

Multiple task learning (MTL) neural networks are one of the better documented methods of transfer learning [3,11]. An MTL network is a feed-forward multi-layer network with an output for each task that is to be learned. The standard back-propagation of error learning algorithm is used to train all tasks in parallel. Consequently, MTL training examples are composed of a set of input attributes and a target output for each task. Figure 2 shows an MTL network containing a hidden layer of nodes that are common to all tasks. The sharing of internal representation is the method by which inductive bias occurs within an MTL network [1]. MTL is a powerful method of knowledge transfer because it MTL allows two or more tasks to share the common feature layer to the extent to which it is mutually beneficial. The more that tasks are related, the more they will share representation and create positive inductive bias.

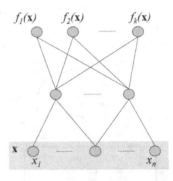

Fig. 2. A multiple task learning (MTL) network with an output node for each task being learned in parallel

Formally, let X be a set on $\Re^n$ (the reals), Y the set of $\{0,1\}$ and *error* a function that measures the difference between the expected target output and the actual output of the network for an example. MTL can be defined as learning a set of target concepts $\mathbf{f} = \{f_1, f_2, \dots f_k\}$ such that each $f_i : X \to Y$ with a probability distribution P_i over $X \times Y$. We assume that the environment delivers each f_i based on a probability distribution Q over all P_i. Q is meant to capture regularity in the environment that constrains the number of tasks that the learning algorithm will encounter. Q therefore characterizes the domain of tasks to be learned. An example for MTL is of the form $(\mathbf{x}, \mathbf{f}(\mathbf{x}))$, where $\mathbf{x}$ is a vector containing the input values $x_1, x_2, \dots, x_n$, and $\mathbf{f}(\mathbf{x}) = \{f_i(\mathbf{x})\}$, a set of target outputs. A training set S_{MTL} consists of all available examples, $S_{MTL} = \{(\mathbf{x}, \mathbf{f}(\mathbf{x}))\}$. The objective of the MTL algorithm is to find a set of hypotheses $\mathbf{h} = \{h_1, h_2, \dots, h_k\}$ within its hypothesis space H_{MTL} that minimizes the objective function $\sum_{x \in S_{MTL}} \sum_{i=1}^{k} error\,[f_i(\mathbf{x}), h_i(\mathbf{x})]$. The assumption is that H_{MTL} contains sufficiently accurate h_i for each f_i being learned.

In 2005, we developed a theory of task knowledge consolidation using MTL networks [13] based on the following ideas: A plastic network is one that can accommodate new knowledge. A stable network is one that can accurately retain previously learned knowledge. The secret to having a stable and yet plastic network is to relearn or *rehearse* examples of prior tasks to maintain stable function while allowing the underlying representation to slowly change to accommodate the learning of the new task. We demonstrated that knowledge consolidation for short sequeces of tasks can be accomplished in MTL networks if the following four conditions are met [10,13]. First, that *task rehearsal* be used to maintain prior model accuracy. Task rehearsal prevents the loss of previously learned task knowledge stored in a MTL network by relearning, or *rehearsing* synthesized *virtual examples* of prior tasks while simultaneously learning examples of the new task. Virtual examples are synthesized using the learned models in the MTL network at the beginning of each consolidation. Second, the network must have sufficient internal representation (number of hidden nodes) for learning all tasks of the domain. Third, a small learning rate must be is used to ensure slow integration of the new task into existing representation while maintaining the functional accuracy of prior tasks. Lastly, a method is needed to prevent the network from over-fitting to the training data and therefore creating a local minima from which escape is difficult. A simple regularization method such as monitoring a validation set of data for minimum error works nicely.

Despite these important findings, the MTL-based approach to consolidation is marred with several limitations related to there being a separate output for each task learned by the network [12,13]. Multiple outputs breed redundant representation for models that are closely related. A separate output per task makes the accumulation of model accuracy through practise particularly challenging because the learning system has no way to determine with which prior task a new set of examples should be associated. The learning environment should provide the contextual queues that suggest the relatedness of examples to a prior task, however this suggests additional inputs, not outputs. In general, the multiple outputs make it difficult to maintain and possibly improve related model performance.

In response to these problems, we developed *context-sensitive* MTL, or *cs*MTL, which uses only one output for all tasks and additional *context* inputs to associate an example with its task [14].

2.3   Consolidation with *cs*MTL Networks

Figure 3 presents a *cs*MTL network. It is a feed-forward network architecture of input, hidden and output nodes that uses the back-propagation of error training algorithm. The *cs*MTL network requires only one output node for learning multiple concept tasks. Similar to standard MTL neural networks, there is one or more layers of hidden nodes that act as feature detectors. The input layer can be divided into two parts: a set of *primary* input variables $(x_1, x_2, ..., x_n)$ for the tasks and a set of inputs $(c_1, c_2, ..., c_k)$ that provide the network with the *context* of each training example. The context inputs can simply be a set of task identifiers that associate each training example to a particular task.

Formally, let X be a set on $\Re^n$ (the reals), Y the set $\{0, 1\}$, and let C be a set on $\Re^n$ representing the context of the primary inputs from X. Let $\mathbf{c}$ be a particular example of this set where $\mathbf{c}$ is a vector containing the values $c_1, c_2, \ldots, c_k$; where $c_i = 1$ indicates that the example is associated with function f_i; where $f_i : X \to Y$ with a probability distribution P_i over $X \times Y$. *cs*MTL can be defined as learning a family of target concepts $f' : C \times X \to Y$; where $f' = \{f_1, \ldots, f_i, \ldots f_k\}$. We assume that the environment delivers each f_i based on a probability distribution Q over all P_i. Q is meant to capture regularity in the environment that constrains the number of tasks that the learning algorithm will encounter. Q therefore characterizes the domain of tasks to be learned. An example for *cs*MTL takes the form $(\mathbf{c}, \mathbf{x}, f'(\mathbf{c}, \mathbf{x}))$, where $f'(\mathbf{c}, \mathbf{x}) = f_i(\mathbf{x})$ when $c_i = 1$ and $f_i(\mathbf{x})$ is the target output for task f_i. A training set S_{csMTL} consists of all available examples for all tasks, $S_{csMTL} = \{(\mathbf{c}, \mathbf{x}, f'(\mathbf{c}, \mathbf{x}))\}$. The objective of the *cs*MTL algorithm is

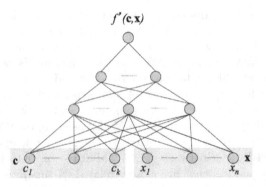

Fig. 3. csMTL neural network with a single output and additional context inputs

to find a hypothesis h' within its hypothesis space H_{csMTL} that minimizes the objective function, $\sum_{x \in S_{csMTL}} error\,[f'(\mathbf{c}, \mathbf{x}), h'(\mathbf{c}, \mathbf{x})]$. The assumption is that $H_{csMTL} \subset \{f | f : C \times X \to Y\}$ contains a sufficiently accurate h'.

With csMTL, the entire representation of the network is used to develop hypotheses for all tasks of the domain. The focus shifts from learning a subset of shared representation for multiple tasks to learning a completely shared representation for the same tasks. This presents a more continuous sense of domain knowledge and the objective becomes that of learning internal representations that are helpful to predicting the output of similar combinations of the primary and context input values. Once f' is learned, if $\mathbf{x}$ is held constant, $\mathbf{c}$ indexes over the hypothesis base H_{csMTL}. In fact, if $\mathbf{c}$ is a vector of real-valued inputs and from the environment, it provides a grounded sense of task relatedness.

Fowler addressed knowledge consolidation and the stability-plasticity problem using csMTL network[5,14]. He used the same approach as that used for consolidation with MTL networks including task rehearsal to maintain functional stability of prior tasks as the examples of the new task are being integrated. He showed that csMTL overcomes several key limitations of standard MTL methods when used for consolidation. Specifically, within a long-term csMTL network there can be an effective and efficient sharing of internal representation between related tasks, without the MTL disadvantage of having redundant outputs for near identical tasks. With just a single output, over time, practice sessions for the same task will contribute to the development of a more accurate long-term hypothesis. In fact, the long-term csMTL network can represent a fluid domain of tasks where subtle differences between tasks can be represented by small changes in the context inputs, in the case where they are real-valued. The results showed the csMTL-based consolidation can increase the accuracy of related hypotheses existing in the csMTL network as a new task is integrated. The approach can also easily accommodate tasks that have multiple outputs such as image transformation tasks.

The above results have been promising however all attempts at consolidation using MTL and task rehearsal methods have suffered from scaling limitations.

As the number of tasks increased past 10 to 20, we observed (1) a slow but steady loss of prior task accuracy, (2) problems integrating new tasks into the consolidated network, and (3) an increase in the learning time as a quadratic function of the number of training examples required to rehearse the growing number of prior tasks. The rehearsal of each of the existing domain knowledge tasks requires the creation and training of $m \cdot k$ virtual examples, where m is the number of virtual training examples per task and k is the number of tasks. The focus of the theory in the following chapter and the experiments in Section 4 are on solutions to these problems.

3   Theory and Approach

The objective is to refine our MTL and rehearsal based method of consolidation so that it can maintain the knowledge of previously learning tasks while accurately integrating in knowledge of a new task over a lengthy sequence of tasks. Our system uses a large csMTL network as the domain knowledge structure that meets the conditions for consolidation stated in Section 2.2. It uses task rehearsal as a method of maintaining the knowledge of previously learned tasks, but does so in several important new ways, by using: (1) two layers of hidden nodes to allow the network to generate a hierarchy of features that can be used by multiple tasks; (2) a training and a validation set equally weighted for all tasks so that the best overall models are retained and integrated (3) knowledge of the probability distributions over the inputs of the original training examples to generate better virtual training examples; and (4) a sweep method of task rehearsal (inspired by [8]) to reduce the training time. The following discusses each of these advances.

Work by the authors has shown that csMTL networks with two layers of hidden nodes perform better than networks with just one hidden layer [4]. Two layers of hidden nodes allow the back propagation algorithm to create a hierarchy of features that take into consideration both the primary inputs to the network as well as the context inputs. One way to characterize this is to consider the first hidden layer constructing low-level features of the primary inputs, and the second hidden layer creating more abstract features from the low-level features based and the context inputs.

When doing transfer or consolidation there is always a question as to the relative weighting of training and validation examples. We have tried various approaches in prior studies of consolidation using csMTL networks. Recent work by the authors demonstrates that it is best to ensure that the training sets for all tasks are equally weighted, as should be the validation sets for all tasks [4]. This promotes the development of good models for all tasks without bias in favour of any one to task.

An interesting suggestion made by Fowler was that virtual examples should not be developed from uniformly random input values, rather they should be based on knowledge of the probability distribution over the input space [5]. In this research we take this approach and select virtual examples for task rehearsal that are in agreement in this way with the original training examples. Our system learns and saves the probability distributions of the input values $(x_1, x_2, ...x_n)$ from the original training examples for each task learned. When generating the virtual examples for task rehearsal, the input value x_i, will be selected according to the saved probability distribution for x_i. The complete input $(x_1, x_2, \ldots, x_n)$ can then be fed through the csMTL model for each task to generate the matching output label y. The set $(x_1, x_2, \ldots, x_n, y)$ form a virtual example. These virtual examples are expected to accurately reflect the nature of the original training examples and the learned task.

In the 1990s [8,9], Anthony Robins studied the stability-plasiticity problem (also known as *catastrophic forgetting* [6,7]) in the context of pair-associate learning as would normally be performed by psychologists with human subjects. He introduces the concept of *rehearsal* with virtual examples as a solution to maintaining the knowledge of earned pairs of associated values. Robins proposed several methods of rehearsal, one of which he calls *sweep rehearsal*. Sweep rehearsal is based on the use of a small but "dynamic" set of secondary task examples where these examples are chosen at random for each iteration of training from a large pool of possible examples. In this way, during consolidation the network is trained on a relatively large fixed set of examples for the new task and a small "dynamic" set of virtual examples for the prior task. We apply this sweep approach to task rehearsal to our method of consolidation and compare it to the alternative that we have been using that employs a relative large but fixed set of examples for the prior tasks. Sweep rehearsal introduces two positive effects: (1) a small but changing set of virtual examples provides training on large set of virtual examples over many iterations without the computational cost that comes with a large set of virtual examples every iteration, and (2) a small but changing set of virtual examples provides stochastic noise to the system which can help escape local minima in the error surface.

4   Experiments

We empirically investigate the conditions to fulfil the long-term domain knowledge requirements of an LML system using a csMTL network. Our analysis will focus on the effectiveness of prior task retention and new task consolidation into a csMTL network. More specifically, the experiments examine (1) the retention of prior task knowledge as the number of virtual examples for each task varies, and (2) the scalability of the method as up to 20 tasks are sequentially consolidated within a csMTL network.

4.1   Task Domains

The Logic task domain is synthetic, consisting of 20 tasks. It has 23 real-valued primary inputs in the range $[0, 1]$ and 20 binary context inputs labelled a through w. An example for a task is calculated by a logical conjunction of two disjunctions, involving four of the primary inputs. For example, the first task, for which context

Table 1. Task definitions of the Logic Domain

Task name	Logical Expression for task
T_1	$(a > 0.5 \lor b > 0.5) \land (c > 0.5 \lor d > 0.5)$
T_2	$(b > 0.5 \lor c > 0.5) \land (d > 0.5 \lor e > 0.5)$
$\vdots$	$\vdots$
T_{20}	$(t > 0.5 \lor u > 0.5) \land (v > 0.5 \lor w > 0.5)$

Table 2. Metadata on the CoverType and Dermatology Domains

CoverType		Dermatology	
Task name	Examples	Task name	Examples
Spruce/Fir	1834	psoriasis	222
Lodgepole Pine	1880	semboreic dermatitis	120
Ponderosa Pine	1726	lich planus	142
Aspen	2608	pityriasis rosea	96
Douglas-fir	1924	cronic dermatitis	96
Krummholz	2726		

input $c_1 = 1$, is defined as $(a > 0.5 \lor b > 0.5) \land (c > 0.5 \lor d > 0.5)$. For each new task, the set of inputs shifts one letter to the right in the alphabet (see Table 1). We generate 500 examples for training, 500 for validation, and 1000 for a test set. All sets have equal numbers of positive and negative examples.

The CoverType and Dermatology data are from real world domains of 6 and 5 tasks, respectfully. The CoverType domain involves determining the forest cover type from 51 cartographic inputs plus 6 context inputs and the Dermatology domain involves determining the type of erythemato-squamous skin disease from 34 clinical inputs plus 5 context inputs. Each data set is balanced with randomly chosen negative examples from the other tasks of their domain so that each task has an equal number of positive and negative examples. The training, validation, and test sets were obtained by selecting 60%, 20% and 20% of the examples, respectfully. See Table 2 for details.

4.2   Common Methods and Approach

All experiments were performed using a neural network-based LML software system, called RAML3, developed for Matlab and Octave at Acadia University. RAML3 allows a sequence of tasks to be learned using either MTL or csMTL neural network methods. The system is capable of short-term transfer learning and long-term knowledge consolidation using task rehearsal.

Preliminary experiments investigated the network structure and learning parameters required to develop accurate multiple task network models of all tasks of each domain using actual examples. The Logic, CoverType, and Dermatology networks each contained two layers of hidden and had input-hidden-hidden-output architectures of 43-30-30-1, 57-10-10-1, and 39-20-20-1, respectively.

For all three networks the learning rate was set to 0.0001 and the momentum term set to 0.9.

In all experiments, the primary performance statistic is the accuracy on the independent test set. Repeat trials were conducted of each experiment so as to determine confidence in the performance of each method. Each sequential learning run (training and consolidating up to 20 tasks) required a lot of computing time, restricting us to five repetitions for each sequence. For brevity, only results for certain tasks and the mean of all tasks (including the new task being consolidated) are provided in this paper.

It is important to note that examples for the new task being consolidated have their back-propagated errors weighted to match the number of virtual examples for each prior task for a run. This is to ensure that all tasks, prior and new, get an equal opportunity to effect the weight updates in the csMTL network.

4.3 Experiment 1: Single Parallel csMTL Model

Objective: This experiment examines the change in generalization accuracy of previously learned tasks as new tasks are consolidated within a csMTL network. This experiment provides a baseline accuracy for how well we could consolidate multiple tasks together into a single model.

Methodology: As a baseline, for each domain a model was developed for all tasks using a csMTL network and the actual examples. The training, validation, and test set splits were the same as described in Section 4.2.

Results and Discussion: The Logic domain models had at least 80% accuracy for all 20 tasks; all tasks except 5 had more than 90% accuracy. The Covertype domain models had at least 75% accuracy for all 6 tasks; all tasks except 2 had more than 85% accuracy. The Dermatology domain models had at least 70% accuracy for all 5 tasks; all tasks except 3 had more than 85% accuracy.

4.4 Experiment 2: Consolidation of the Logic Domain Using Standard and Sweep Rehearsal

Objective: This experiment examines the change in generalization accuracy (stability) of previously learned tasks as new tasks are consolidated within a csMTL network and the ease with which a new task can be integrated (plasticity). We compared three different techniques when building these models: (1) using all actual examples for all tasks both prior and new, (2) using up to 500 virtual examples for the secondary tasks (standard rehearsal) and 500 actual examples for the new task, and (2) generating 500 virtual examples for the secondary tasks but only using a small random number of them for each training iteration (sweep rehearsal) and 500 actual examples for the new task.

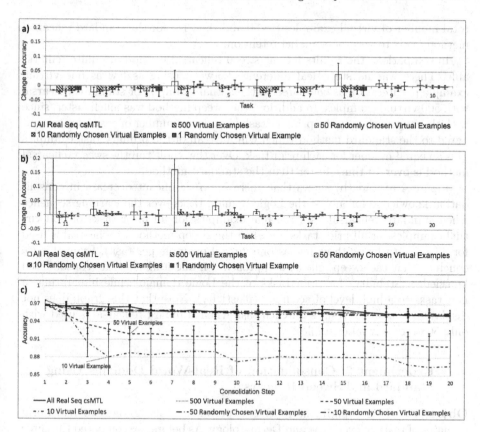

Fig. 4. Results for the Logic Domain

Methodology: Repeated runs of learning all 20 tasks sequential were conducted. For all runs the same numbers of actual examples were used for each new task and the same number of virtual examples were used for each prior task. Sequential consolidation with standard rehearsal method generated 500 virtual training and validation examples for each of the secondary tasks, while the new primary task used real examples for training and validation. All other parameters were the same as described in section 4.2. Sequential consolidation with sweep rehearsal generated 500 virtual training and validation examples for each of the secondary tasks. However, we randomly chose a small number (1, 10 or 50) of these virtual training examples for sweep rehearsal. The validation set had 500 examples selected for sweep rehearsal. All other parameters were the same as described in section 4.2.

Results and Discussion: Figure 4(c) shows that with 500 virtual examples per prior task that the standard rehearsal method is able to maintain the level of task 1's accuracy while consolidating the 19 other tasks. This is as good a result as continually rehearsing the actual examples. However, decreasing the total number of virtual examples (to 50 and then 10) for the standard rehearsal method leads to very poor model retention.

In comparison the results for the sweep rehearsal method are quite remarkable. There was no statistical difference between using as few as one virtual example per iteration as compared to the actual examples or standard rehearsal with 500 virtual examples. In addition, the sweep method was much faster, since with this approach only the primary task has a large number of examples as compared to the standard method which will have all of the primary task examples plus an equal number for each prior task. Over the whole run the sweep rehearsal method is over 10 times faster than the standard method.

Figure 4(a) and (b) shows that there is no significant decrease in consolidation knowledge for any of the Logic domain tasks over the learning sequence when standard rehearsal uses the real examples for all prior tasks or 500 virtual examples for all prior tasks, or when the sweep method of rehearsal is used for all prior tasks (with as few as one virtual example per task). From this we conclude that the sweep rehearsal approach can maintain accurate consolidated knowledge (stability) for prior tasks of the Logic domain while integrating in new tasks to a high level of accuracy (plasticity). Note that this graph does not display the far worse results for each task when standard rehearsal is used with only 50 or 10 virtual examples.

4.5   Experiment 3: Consolidation of Real-World Domains Using Standard and Sweep Rehearsal

Objective: This experiment repeats experiment 2 using the two real-world domains of tasks; CoverType and Dermatology. As before, we compared the three different techniques when building these models: (1) using all actual examples for all tasks both prior and new, (2) using large amounts of virtual examples for the secondary tasks (standard rehearsal) and actual examples for the new tasks, and (2) selecting a small random number of examples for each training iteration from a large pool virtual examples for the secondary tasks (sweep rehearsal) and actual examples for the new tasks.

Methodology: As with the Logic domain, repeated runs of learning all tasks sequential were conducted. For all runs the actual examples were used as show in Table 2 for each new task and a consistent number of virtual examples were used for each prior task. All parameters were the same as described in section 4.2 for the two domains unless otherwise specified below.

All actual examples: We used real examples for training and validation for all tasks. All other parameters were the same as described in section 4.2 for both the CoverType and the Dermatology domains.

Sequential Consolidation with Standard Rehearsal: For the Covertype domain, we generated 1000 virtual examples for each of the training and validation sets for all the secondary tasks, while the new primary task used actual examples for training and validation. For the Dermatology domain, we generated 140, 500 virtual examples for each of the training and validation sets for all the secondary tasks, while the new primary task used real examples for training and validation.

Sequential Consolidation with Sweep Rehearsal: For the Covertype domain, we generated 10000 virtual examples for each of the training and validation sets for all the secondary tasks. From this set we randomly chose 100 virtual training and validation examples per iteration, while the new primary task used actual examples for training and validation. For the Dermatology domain, we generated 10000 virtual examples for each of the training and validation sets for all the secondary tasks. From this set we randomly chose 100 virtual training and validation examples per iteration, while the new primary task used actual examples for training and validation.

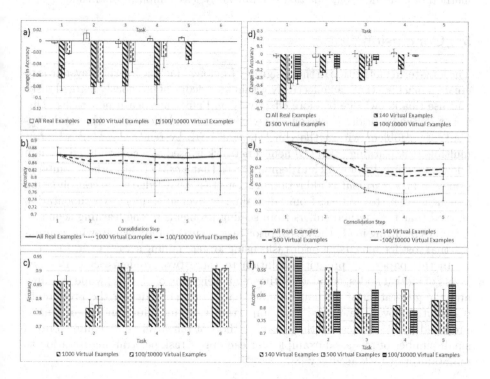

Fig. 5. Results for the CoverType (a,b,c) and Dermatology (d,e,f) domain

Results and Discussion: Figure 5 (b) shows that the sweep rehearsal method produces the best results for task 1 of the CoverType domain over the learning sequence, as compared to the standard rehearsal method. But it does not do quite as well as continued rehearsal of the actual training examples, which is an ideal baseline. Over all, as shown in Figure 5 (a) the sweep rehearsal method does at least as well as the standard method for all tasks except task 2. Figure 5 (c) shows that both methods are able to integrate new task knowledge equally well.

Similarly, Figure 5 (e) shows that the sweep rehearsal method produces the best results for task 1 of the Dermatology domain over the learning sequence, as compared to the standard rehearsal method when either 140 or 5000 fixed virtual examples are used. But neither method does as well as continued rehearsal of the actual training examples. Over all, as shown in Figure 5 (d) the sweep rehearsal method does at least as well as the standard method for all tasks with only 100 examples being rehearsed each iteration. Figure 5 (f) shows that both methods are able to integrate new task knowledge equally well.

Most importantly, as found in with the logic domain, these results confirm that the sweep rehearsal method scales much better than the standard rehearsal method. The sweep method works at least as well as the standard rehearsal method of consolidation, but takes significantly less computational time.

5 Conclusion

The long-term motivation in the design of a Lifelong Machine Learning system is to create a machine that can effectively retain knowledge of previously learned tasks and use this knowledge to more effectively and efficiently learn new tasks. This paper extends prior work on knowledge consolidation and the stability-plasticity problem within the context of a LML system. A *context-sensitive* multiple task learning (*cs*MTL) neural network is used as a consolidated domain knowledge store. Prior work had demonstrated that a *cs*MTL network, in combination with *task rehearsal*, could retain previous task knowledge when consolidating a sequence of up to ten tasks from a domain. However subsequent experimentation showed that the method suffers from scaling problems as the learning sequence increases resulting in (1) the loss of prior task accuracy and (2) a growing computational cost for rehearsing prior tasks using larger training sets.

In this paper, we present a solution to these two problems that uses two layers of hidden nodes, equal weighting of all weighting of all tasks while training, and the generation of virtual examples based on knowledge of the probability distributions over the inputs. Most importantly, we introduce an idea originally proposed by Robins called the *sweep* method of rehearsal that requires only a small number of rehearsal examples for each prior task per training iteration to maintain prior task accuracy [8].

5.1 Findings

The experiments have demonstrated a mix of results in terms of maintaining the accuracy of prior task knowledge when using both the standard and sweep rehearsal methods. However the sweep method has been demonstrated to be at least as good as the standard method at a considerable savings in terms of computation. This can be seen particularly on the synthetic Logic and the real-world Cover-Type domain where the sweep rehearsal method comes close to the ideal case of continually rehearsing prior tasks using the original training examples.

The 20 task sequence of the Logic domain clearly indicates the power of the sweep rehearsal method to overcome the computational scaling problem. The 20th task takes 19 times longer to training using 500 virtual examples for all prior tasks than it does under the sweep rehearsal method which requires only the 500 actual examples for the new task and as few as one virtual example per iteration for each of the prior tasks. Over the entire 20 task sequence the computation savings averages just over 10 times in favour of the sweep rehearsal method.

We conclude that we have made a significant step toward reducing the computational complexity of task rehearsal when using a *cs*MTL-based solution. However we have fallen short of resolving the loss of prior task knowledge, as evidenced by the results on the Dermatology domain, and this is at the heart of the stability-plasticity problem. We remain confident that this line of research will discover a method by which consolidated network models can maintain functionality accuracy while changing representation to accommodate new knowledge.

5.2 Future Directions

The findings of this research have confirmed that the choice of virtual examples is critical to the success of rehearsal. The method that we employed in this research to capture the probability distribution of the input variables ignored the conditional probability between those variables. This is likely the reason why the knowledge of the real-world domains was not retained as well as the synthetic domain. We feel this can be corrected by modifying our LML system to use deep learning methods in combination with our *cs*MTL and task rehearsal approach. We are currently in the planning stages for the development of a new *cs*MTL-based LML system capable of sequential transfer learning and knowledge consolidation and will incorporate those findings into its design.

Acknowledgment: This research has been funded in part by the Government of Canada through NSERC.

References

1. Baxter, J.: Learning model bias. In: Touretzky, D.S., Mozer, M.C., Hasselmo, M.E. (eds.) Advances in Neural Information Processing Systems, vol. 8, pp. 169–175 (1996)
2. Carlson, A., Betteridge, J., Kisiel, B., Settles, B., Hruschka, Jr., E.R., Mitchell, T.M.: Toward an architecture for never-ending language learning. In: Fox, M., Poole, D. (eds.) AAAI. AAAI Press (2010)
3. Caruana, R.A.: Multitask learning. Machine Learning **28**, 41–75 (1997)
4. Eljabu, L.: Knowledge Consolidation Using Multiple Task Learning: Overcoming The Stability-Plasticity Problem. Masters Thesis, Jodrey School of Computer Science, Acadia University, Wolfville, NS (2014)
5. Fowler, B., Silver, D.L.: Consolidation using context-sensitive multiple task learning. In: Butz, C., Lingras, P. (eds.) Canadian AI 2011. LNCS, vol. 6657, pp. 128–139. Springer, Heidelberg (2011)

6. Grossberg, S.: Competitive learning: From interactive activation to adaptive resonance. Cognitive Science **11**, 23–64 (1987)
7. McCloskey, M., Cohen, N.J.: Catastrophic interference in connectionist networks: the sequential learning problem. The Psychology of Learning and Motivation **24**, 109–165 (1989)
8. Robins, A.V.: Catastrophic forgetting, rehearsal, and pseudorehearsal. Connection Science **7**, 123–146 (1995)
9. Robins, A.V.: Consolidation in neural networks and in the sleeping brain. Connection Science Special Issue: Transfer in Inductive Systems **8**(2), 259–275 (1996)
10. Silver, D.L.: The consolidation of task knowledge for lifelong machine learning. In: AAAI Spring Symposium 2012 (2012)
11. Silver, D.L. Mercer, R.E.: The parallel transfer of task knowledge using dynamic learning rates based on a measure of relatedness. Learning to Learn, 213–233 (1997)
12. Silver, D.L., Mercer, R.E.: The task rehearsal method of life-long learning: overcoming impoverished data. In: Cohen, R., Spencer, B. (eds.) AI 2002. LNCS, vol. 2338, pp. 90–101. Springer, Heidelberg (2002)
13. Silver, D.L., Poirier, R.: Sequential consolidation of learned task knowledge. In: Tawfik, A.Y., Goodwin, S.D. (eds.) Canadian AI 2004. LNCS (LNAI), vol. 3060, pp. 217–232. Springer, Heidelberg (2004)
14. Silver, D.L., Poirier, R., Currie, D.: Inductive tranfser with context-sensitive neural networks. Machine Learning **73**(3), 313–336 (2008)
15. Thrun, S.: Is learning the n-th thing any easier than learning the first? In: Advances in Neural Information Processing Systems, pp. 640–646. The MIT Press (1996)
16. Thrun, S.: Lifelong learning algorithms. Learning to Learn, ch. 1, pp. 181–209. Kluwer Academic Publisher (1997)

Graduate Student Symposium

Resolving Elections with Partial Preferences Using Imputation

John A. Doucette$^{(\boxtimes)}$

Cheriton School of Computer Science, University of Waterloo,
Waterloo, ON, Canada
j3doucet@uwaterloo.ca

Abstract. We propose a novel approach to deciding the outcome of elections when voters are unable or unwilling to state their complete preferences. By viewing the problem as an exercise in imputation, rather than direct aggregation, elections can be decided with an empirically supported guess of how voters would have voted, if they had complete information about the alternatives. We show that when certain classification algorithms are used to generate imputations, the process can be viewed as a form of voting rule in its own right, allowing application of existing results from the field of computational social choice. We also provide an analytical relationship linking the error rate of the classifier used with the election's margin of victory, and extensive empirical support for the model using real-world electoral data. The described techniques both make extensive use of, and have applications throughout Multiagent Systems and Machine Learning.

1 Topic Area

Within the field of Mutliagent Systems, the study of Computational Social Choice is concerned with the analysis and improvement of general techniques for allowing societies (of humans or of computational agents) to reach collective decisions [1]. One such technique is the process of voting, where a set of agents select from among a set of actions that must be taken *collectively*. A familiar example comes in the form of a Canadian federal election. The society of agents (Canadian voters) must decide which group will be assigned to run the country until the next election. Although each voter has an individual preference, all voters will be bound by the collective decision reached. However, voting systems also have more direct applications to problems in artificial intelligence research. Coordinating dispersed groups of computational agents (like a robot swarm), or learning techniques like bagging could also be viewed as voting.

2 Topic and Research to Date

In the context of this work, we are concerned primarily with voting systems that make use of a *ranked ballot* [1]. The electoral system used in Canadian

© Springer International Publishing Switzerland 2015
D. Barbosa and E. Milios (Eds.): Canadian AI 2015, LNAI 9091, pp. 325–329, 2015.
DOI: 10.1007/978-3-319-18356-5_28

elections asks voters to name their favorite candidate, and selects the action that was named by the largest number of voters. This system is called "Plurality". Although Plurality is a simple system (which is a virtue when dealing with human participants), it collects very little information from the voters, and so can make very poor decisions. An important property Plurality fails to satisfy is that it can sometimes select a "Condorcet Loser" as the most preferred alternative. A Condorcet Loser is an alternative that a majority of the electorate dislike so much that they would prefer any other winner. For instance, consider the preferences represented by Figure 1. Alternative X is the most preferred choice of 8 voters, but the least preferred of the other 13. If we apply Plurality to this set of preferences, X is selected as the winner. However, if we asked the electorate "Do you prefer Y to X?" or "Do you prefer Z to X?" a majority will answer yes.

Using a voting system that elicits and uses more of the voters' preferences (i.e. a full ranking over all the candidates) can avoid situations like those just described, and a plethora of such systems exist. For instance, in an election with k alternatives, the Borda Count system would give $k-1$ points to each voter's first choice, $k-2$ to the second, and so on, selecting the alternative with the most points in total as the winner. This often results in the selection of a compromise that everyone can live with, rather than a polarizing alternative.

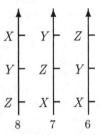

Fig. 1

There has been considerable recent interest in new techniques for deciding elections with partial preference information. Notable approaches include Lu and Boutilier's Minimax Regret (MMR) approach, which uses techniques from robust optimization to select the winner which would dissatisfy voters least if their true (unstated) preferences were adversarially completed [3]; and a variety of techniques that model ballots as incomplete or noisy impressions of an assumed ground truth (e.g. [5], [6]). While MMR is applicable to many problem domains, it

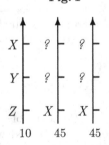

Fig. 2

operates in a conservative way, picking the candidate with the least potential be offensive, rather than the one most likely to win. Consider the vote profile presented in Figure 2. Under Plurality, MMR will pick X as the winner. This is because if we picked Y as the winner, an adversary could complete all 90 ballots so that Z was at the top, and Y got no votes at all. A similar argument could be made against the selection of Z as the winner, so X must win. However, X cannot possibly receive more than 10% of the vote, so at least one of the other two candidates will surely receive more.

In light of this, we propose a new approach, which adopts the philosophy that the winner of an election where some voters did not provide complete information should be selected in a way consistent with how voters *would have voted if they had complete information*. To accomplish this, we make the assumption that voters' true beliefs about the world (i.e. the beliefs they would express if they had complete, accurate, information about all the candidates, and adequate time to consider the implications of that information) are not distributed randomly, or adversarially, but in a way that resembles the views of at least some other voters. We can then apply an ordinary classification algorithm to learn a mapping between partial rankings of the candidates and complete rankings, using the ballots which have comparatively more information provided. The resulting model can be used to complete voters' ballots, so that a ranked ballot electoral system can pick a winner. For instance, suppose that the vast majority of voters who place the Communist Party first on their ballot, and ranked many candidates, put the Conservative Party relatively low in their ranking. It stands to reason then, that a voter who can tell us only that they like the Communist Party (it's their first pick), should be assumed to dislike the Conservative Party also, and that deciding the election on this basis could yield a choice more reflective of this voter's true viewpoint than assuming they have no preferences at all over the unranked candidates.

In present work [2], we consider primarily the case of "top-t" ballots, where voters can truthfully state their t most preferred alternatives, but are unsure of the relative rankings of the remaining candidates. In this domain, we proceed by converting the problem into a set of classification tasks. In the first task, a classifier must be learned that can accurately predict a voter's second preference from its first. In the second task, third preferences are predicted on the basis of first and second, and so on. We have proven that every deterministic classification algorithm (i.e. one which when trained on a given dataset always outputs the same model, and which outputs only models that are deterministic mappings) is equivalent to a ranked ballot voting rule, allowing us to analyze the proposed technique as a voting rule (for instance, we can ask questions like: how easy is it to vote strategically if logistic regression is used to complete ballots?). We have also provided a theoretical assessment of the ability of this system to recover the correct opinions of voters. We relate the expected error of the classifier (which is in turn related to the amount of available data), to the minimum margin of victory achieved by one candidate over another in the hypothetical "ground truth election" — an election where every voter knows and expresses a complete ranking over the candidates. When there is a larger margin of victory, the classifier needs less data to make the correct decision. When the margin is smaller, it needs more.

Beyond this initial theoretical work, we have implemented the proposed system using a powerful modular architecture that allows arbitrary combinations of classification algorithms and ranked ballot voting systems to be compared. We have evaluated the resulting system on the ballots from many real world elections, including the leadership elections of the Debian society between 2002 and the present, and the Irish national election ballots from the Dublin North, Dublin West, and Meath ridings, using the preflib.org data repository. Our experiments used the subset of ballots that had complete rankings (so that we had a ground truth to compare against), but ablated them in a way consistent with those in the actual electoral data [1]. Using only the ablated ballots, our system is shown to recover the correct winner in nearly 100% of problem instances, outperforming MMR (a state of the art algorithm), which sometimes picks a "safer" candidate (one more voters have provided information on), rather than the true winner, on the same datasets. We also show that the correlation between the overall rankings of candidates output by our system and the overall rankings of candidates in the ground truth election is very high (always more than 75%, and typically over 90%). This means that the technique does not just select the correct first place candidate, but also the selects the correct second place candidate, and third place candidate, and so on. In general, mistakes are made only under the conditions predicted by our theoretical assessment: when deciding the relative order of two very unpopular candidates, the system sometimes puts the wrong one first. Our system sometimes reverses the order of these two parties when predicting the complete order of candidates. However, we view these sorts of mistakes relatively minor. We have also performed some initial work assessing the resistance of the system to strategic voting. We show that, although in theory a strategic voter can leverage the classification algorithm to greatly expand their power, the system can be made resistant to strategic voting at a minimal cost to the classifier's accuracy.

3   Conclusion and Future Work

The system we describe has significant applications within AI, both to areas like crowdsourcing systems and human computation (where the opinions of human workers must be aggregated efficiently), and in areas like multi-agent coordination, cooperation, and resource allocation (where the opinions of strategic, computational, agents need to be aggregated). Our results showing that certain classification algorithms can be analyzed as voting systems is also of theoretical interest to the computational social choice community, where the intersection between learning and voting is a highly active topic.

In the future, we plan to expand our theoretical assessment of the system by expressing the axioms used to assess the quality of different voting systems as constraints on the behavior of classification algorithms, and connecting different voting systems to related classification algorithms (for instance, plurality to something like the OR classifier). We also plan a more detailed assessment of the manipulability of different classification algorithms in this context, and on whether the assumptions made by specific classification algorithms might offer implicit protection from strategic voting. Finally, we plan to consider whether specific classification algorithms combined with specific voting rules might be equivalent to simply applying another voting rule directly to the partial ballots (i.e. always making the same decision as another rule). For instance, using any classifier with the Plurality system is identical to just using Plurality on

---

[1] The aggregate views of voters providing complete ballots is strongly correlated with the aggregate views of the electorate as a whole, across all the elections considered.

top-t ballots directly. Tantalizingly, recent work suggests that combining pairs of voting rules can sometimes provide the best properties of both [4]. We also plan to explore additional applications of the system, including to problems in coordinating robot teams, resource allocation and matching. Human trials with the system could also answer questions about how voters' behaviors might change if they know that their ballot could be completed. [2]

References

1. Brandt, F., Conitzer, V., Endriss, U.: Computational social choice. In: Weiss, G. (ed.) Multiagent Systems, pp. 213–283. MIT Press (2012)
2. Doucette, J.A., Larson, K., Cohen, R.: Conventional machine learning for social choice. In: Proceedings of AAAI 2015 (to appear, 2015)
3. Lu, T., Boutilier, C.: Robust approximation and incremental elicitation in voting protocols. In: Proceedings of IJCAI 2011, vol. 22, pp. 287–293 (2011)
4. Narodytska, N., Walsh, T., Xia, L.: Combining voting rules together. In: Proceedings of ECAI 2012, pp. 612–617 (2012)
5. Procaccia, A.D., Shah, N., Zick, Y.: Voting rules as error-correcting codes. In: Proceedings of AAAI 2015, pp. 1142–1148 (2015)
6. Xia, L., Conitzer, V.: A maximum likelihood approach towards aggregating partial orders. In: Proceedings of IJCAI 2011, vol. 22, pp. 446–451 (2011)

---

[2] The author gratefully acknowledges the support and guidance of Professor Robin Cohen.

Novel Game Playing Strategies
Using Adaptive Data Structures

Spencer Polk[(✉)]

School of Computer Science, Carleton University, Ottawa, ON, Canada
andrewpolk@cmail.carleton.ca

Abstract. The problem of achieving intelligent game play has been a historically important, widely studied problem within Artificial Intelligence (AI). Although a substantial volume of research has been published, it remains a very active research area with new, innovative techniques introduced regularly. My thesis work operates on the hypothesis that the natural ranking mechanisms provided by the previously unrelated field of Adaptive Data Structures (ADSs) may be able to improve existing game playing strategies. Based on this reasoning, I have examined the applicability of ADSs in a wide range of areas within game playing, leading to the creation of two novel techniques, the Threat-ADS and History-ADS. I have found that ADSs can produce substantial improvements under a broad range of game models and configurations.

1 Introduction

Within the field of AI, a core problem is intelligently playing a game against a human player, such as the famous board games Chess and Go [1]. This problem has been deeply examined from the field's earliest days, with the earliest efforts to create a theoretical Chess playing engine dating from the 1950s, with many refinements and new techniques introduced over the decades [2,3]. Despite this substantial corpus of knowledge, however, game playing remains a very active research area, with many new innovations and applications introduced in recent years [4,5].

In the subfield of game playing, the majority of research has dealt only with two-player (2P) games, rather than multi-player (MP) games where the number of opponents the intelligent agent faces may be greater than one [6–8]. Techniques for MP games tend to be extensions of 2P techniques, and often have difficulty performing at the level of their 2P counterparts [7,9]. Given the disparity in performance, it is worthwhile to examine if new techniques may be constructed, that can take advantage of the qualities of MP games. Furthermore, such insight may be able to provide contributions to 2P game playing as well.

2 Adaptive Data Structures

Formerly unrelated to game playing, the field of ADSs deals with query optimization within a data structure over time [10,11]. The field of ADSs describes

© Springer International Publishing Switzerland 2015
D. Barbosa and E. Milios (Eds.): Canadian AI 2015, LNAI 9091, pp. 330–333, 2015.
DOI: 10.1007/978-3-319-18356-5_29

a variety of inexpensive techniques by which a data structure may be internally reorganized, in response to queries, so that it will converge over time to place the most frequently accessed elements at the head of the data structure [12]. List-based examples include moving the queried element to the front of the list (Move-to-Front), or moving it one step towards the front (Transposition), and many less intuitive varieties.

Given that in both MP and 2P games, there are many things that could be ranked to provide more intelligent game play, such as opponent threats, move effectiveness, valuable board positions, pieces, and so on, I hypothesized that the natural ranking mechanism provided by ADSs may be applicable in this environment. Particularly, such ranking techniques have many applications within the context of move ordering, to improve pruning in game trees.

3 Contributions

In the course of my thesis work, I have investigated the applicability of ADS-based techniques in a wide range of areas within game playing. Specifically, I have examined the following:

- Use of ADSs to rank opponents in a MP game
- Ordering moves through move history in 2P and MP games using ADSs
- Applicability of ADS-based techniques in MP games with varying numbers of opponents
- Effectiveness of ADSs at both initial board positions, and a wide range of reasonable mid-game board states
- ADS performance in game trees of different ply depths
- Relative strengths and weaknesses of a broad range of known ADS update mechanisms in game playing

To facilitate this exploration, I have developed two move ordering techniques, the Threat-ADS and History-ADS heuristics, which I have tested under a broad range of MP and 2P game models and configurations.

3.1 Threat-ADS

The first technique, the Threat-ADS, is developed specifically for MP environments, based on the recent Best-Reply Search (BRS) [8]. The BRS functions by grouping all opponents together into a single "super-opponent", and considering a slightly unrealistic perspective of the game where this super opponent will only get one minimizing move between each of the perspective player's moves, thus transforming the MP game into a 2P game [8]. The Threat-ADS operates by attaching an adaptive list to the BRS, which holds markers for each opponent, and is queried with an opponent's identity when he is found to pose the greatest threat. When the BRS groups opponent's moves together, they are gathered in the order specified by the ADS. Thus, over time, the Threat-ADS will adapt to relative opponent threats, and help achieve move ordering with a very low overhead. I first introduced the Threat-ADS in [13].

I tested the Threat-ADS' performance under a variety of MP games, including Chinese Checkers and Focus, used in the BRS' introductory paper, and a theoretical game of my own creation called the Virus Game. Experiments discovered that the Threat-ADS would produce a statistically significant improvement, under the Mann-Whitney non-parametric test, in terms of tree size in the vast majority of cases, generally at around a 10% reduction in node count with a ≥ 0.5 Effect Size. This improvement has remained consistent across all three games, with a variable number of opponents, at different ply depths and both initial and midgame states, with some specific exceptions [14]. Threat-ADS was found to perform well with a range of ergodic update mechanisms, with slight differences suggesting the Move-to-Front performs best in most cases [14].

3.2   History-ADS

Considering the success of a simple application of ADSs to rank opponents, it is reasonable to suspect that ADS-based techniques can rank move history as well, which is known to be useful for accomplishing move ordering, as demonstrated by the well-known Killer Moves and History heuristics [3]. Based on this, I developed another move ordering heuristic, called History-ADS. In this case, a Mini-Max or similar search is augmented with an adaptive list, which contains possible moves, and is queried with the identity of that move when it is found to produce a cut. Moves are then ordered based on this ADS, thus prioritizing those moves found to be effective earlier in the search.

I tested the History-ADS in the same range of MP games and configurations as the Threat-ADS, and found that, due to its larger list and thus more available information, it produced generally larger reductions in tree size, at over 50% in some cases. Furthermore, as move history may be employed in both MP and 2P environments, I also tested it in the 2P case, using the games Checkers, Othello, and the 2P variant of Focus. Again, large savings in terms of node count were discovered in all cases, that were statistically significant.

4   Future Work

My results to this point strongly support the hypothesis that ADSs can produce meaningful gains in game playing, in both 2P and MP cases. Building on the History-ADS, the next steps in my thesis work are the development of ADS-enhanced variants of the popular Killer Moves and History heuristics, mentioned earlier as part of its inspiration. While I have found that the History-ADS performs very well, it treats cuts at all levels of the tree equally, which is known to be a weakness these heuristics can overcome [3]. It is thus reasonable to believe that a more reasoned invocation of ADSs, compensating for this issue, could perform even better than the History-ADS heuristic.

Outside of move ordering, my results demonstrate the wide potential applications of ADSs to other areas of game playing. Specifically, I intend to investigate the use of an ADS to break near-ties in the Monte-Carlo UCT algorithm, as well as the use of an ADS to break ties in the MP Max-N algorithm, which does not specify a method by which this is done, despite potential changes to the semantic meaning of the tree [7].

Acknowledgments. I am thankful to my supervisor, Dr. John Oommen, for his continuing support and assistance in my thesis work.

References

1. Russell, S.J., Norvig, P.: Artificial Intelligence: A Modern Approach, 3rd edn. pp. 161–201. Prentice-Hall Inc., Upper Saddle River (2009)
2. Campbell, M.S., Marsland, T.A.: A comparison of minimax tree search algorithms. Artificial Intelligence, 347–367 (1983)
3. Schaeffer, J.: The history heuristic and alpha-beta search enhancements in practice. IEEE Transactions on Pattern Analysis and Machine Intelligence **11**, 1203–1212 (1989)
4. Gelly, S., Wang, Y.: Exploration exploitation in go: UCT for monte-carlo go. In: Proceedings of NIPS 2006, the 2006 Annual Conference on Neural Information Processing Systems (2006)
5. Szita, I., Chaslot, G., Spronck, P.: Monte-carlo tree search in settlers of catan. In: van den Herik, H.J., Spronck, P. (eds.) ACG 2009. LNCS, vol. 6048, pp. 21–32. Springer, Heidelberg (2010)
6. Luckhardt, C., Irani, K.: An algorithmic solution of n-person games. In: Proceedings of the AAAI 1986, pp. 158–162 (1986)
7. Sturtevant, N.: Multi-Player Games: Algorithms and Approaches. PhD thesis, University of California (2003)
8. Schadd, M.P.D., Winands, M.H.M.: Best Reply Search for multiplayer games. IEEE Transactions on Computational Intelligence and AI in Games **3**, 57–66 (2011)
9. Sturtevant, N., Bowling, M.: Robust game play against unknown opponents. In: Proceedings of AAMAS 2006, the 2006 International Joint Conference on Autonomous Agents and Multiagent Systems, pp. 713–719 (2006)
10. Gonnet, G.H., Munro, J.I., Suwanda, H.: Towards self-organizing linear search. In: Proceedings of FOCS 1979, the 1979 Annual Symposium on Foundations of Computer Science, pp. 169–171 (1979)
11. Hester, J.H., Hirschberg, D.S.: Self-organizing linear search. ACM Computing Surveys **17**, 285–311 (1985)
12. Albers, S., Westbrook, J.: Self-organizing data structures. In: Online Algorithms, pp. 13–51 (1998)
13. Polk, S., Oommen, B.J.: On applying adaptive data structures to multi-player game playing. In: Proceedings of AI 2013, the Thirty-Third SGAI Conference on Artificial Intelligence, pp. 125–138 (2013)
14. Polk, S., Oommen, B.J.: On enhancing recent multi-player game playing strategies using a spectrum of adaptive data structures. In: Proceedings of TAAI 2013, the 2013 Conference on Technologies and Applications of Artificial Intelligence (2013)

Auto-FAQ-Gen: Automatic Frequently Asked Questions Generation

Fatemeh Raazaghi[(✉)]

Faculty of Computer Science, University of New Brunswick, Fredericton, NB, Canada
f.razzaghi@unb.ca

Abstract. Using Frequently Asked Questions (FAQs) is a popular way of documenting the list of common questions on particular topics or specific contexts. Most FAQ pages on the Internet are static and can quickly become outdated. We propose to extend the existing work on question answering systems to generating FAQ lists. In conventional Question Answering (QA) systems, users are only allowed to express their queries in a natural language format. A new methodology is proposed to construct the questions by combining of question extraction and question generation methods. The proposed system accepts, extracts, or generates user questions in order to create, maintain, and improve the FAQs quality. In addition to present the basis of QA system, complimentary units will be added to conduct the FAQ list. The research proposed here will contribute to the field of Natural Language Processing, Text Mining, QA, particularly to provide high quality automatic FAQ generation and retrieval.

Keywords: Dynamic frequently asked questions · FAQ · Question Answering · Question Generation · Text mining

1 Introduction

The Question Answering (QA) technology is needed to fulfill the user requirements in the increasing amount of information on the web [3]. The QA systems can be classified into two main categories, open domain and restricted domain. Frequently Asked Question (FAQ) systems have seen been added to those two previously approved research studies [1][2]. FAQs satisfy most notably two goals: first, to provide users with an easy access to browse the key information to solve their problems, and second, to aid the party responsible for frequently answering the same queries from people interested in the same topic. Research in dynamic FAQ generation has the aim of creating lists of questions and answers for FAQ pages automatically or semi-automatically. The scope of the problems that arise in automatic FAQ generation is very broad: from extracting the questions to determining a repetition of a question, formulation of answers, knowledge base creation, and other challenges in automatic content generation.

The proposed Automatic Frequently Asked Question Generation (Auto-FAQ-Gen) system, creates a reliable and dynamic FAQ system. Auto-FAQ-Gen is

© Springer International Publishing Switzerland 2015
D. Barbosa and E. Milios (Eds.): Canadian AI 2015, LNAI 9091, pp. 334–337, 2015.
DOI: 10.1007/978-3-319-18356-5_30

built upon the QA and Question Generation (QG) paradigms to provide an architectural model, a process, system models, and the supporting tools to build FAQs that can adapt effectively in response to different contexts.

2 Challenges

In addition to the difficulties of creating QA and QG systems, developing an automatic FAQ generation system has its own problems:

- *Question Extraction and Question Generation in Question Analysis* What are the effects of QE and QG in Question Analysis? Based on our studies none of the existing QA systems architectures are able to generate questions or extract interrogative sentences [5]. This work will analyze the feasibility of extracting indirect questions or generating latent questions in dynamic FAQ generation.
- *Question Occurrence* How can one determine if a question is asked often? How can one choose relevant questions to add to FAQ lists? There is a lack of methodology to decide on the frequency threshold that qualifies a certain question will appear in an FAQ list or not.
- *Answer Generation* How to generate answers to FAQ questions? Research in the field of automatic QA has focused on factoid (simple) questions for many years [6], and complex types have been mostly ignored by researchers. Complex questions in FAQ lists often seek multiple different types of information simultaneously and do not presuppose that a single answer could meet all of their information needs.
- *Knowledge Base* How to answer Natural Language (NL) questions based on existing knowledge bases? How to create a knowledge base for FAQ generation system?

3 Proposed Work

I intend to take an approach to the problems outlined in Section 2. On the technical side, I will design a framework to automatically build a FAQ page including appropriate questions and related answers. I will elaborate on these components in the following sections:

Question Construction. The quality of an FAQ list depends on predicting the range of questions that a user might ask. Therefore, the questions that will finally be included in the FAQ list for Auto-FAQ-Gen could be generated from three different sources: it can be a question that a user defines explicitly as input to the system (NL format), the output of Question Extraction module, or the result of the Question Generation component. The primary task of this unit is producing a question. Users might use the question mark character in informal language in cases other than questions, a question may be stated in the declaration and indirect form using phrases e.g. "I was wondering..., Could you tell me..., I'd like to know", or rhetorical questions may not require to be associated with the answer segments [4]. After preprocessing the text, the QE step extracts direct and indirect questions using the Natural Language Parser, rule-based methods, and dictionary. In the QG, given the source text, after the data is cleaned and tokenized, the relevant sentences are passed to the Named Entity (NE) tagger and the Part Of Speech (POS) tagger. The given text can be in a form of a group of simple sentences, complex sentences, paragraphs, or a longer textual entity.

Input and Output Layer. This proposed system receives a written question as input, normally as a set of documents consisting of plain text, or a web page. The final output of the system is a ranked list of FAQs (implicit, explicit, or hidden) and their corresponding answers.

Question Processing. The question processing unit has two components, namely: (1)Question Analysis and (2)Question Interpretation. As the Question Analysis component parses the question string, it determines the expected answer type, extracts named entities and other terms, and creates a concise interpretation of the question consisting of its key phrases and a more specific answer type. Given this information, the Query generator from the Question Interpretation module builds several queries for document retrieval.

Answer Generator. The Answer Generator part contains two modules: Answer Extraction (AE) and Answer Selection (AS). The task of the AE and the AS components are to obtain the desired answer from the best-scored answer candidates and to present the proper formulation back to the user. The search results will pass through a pipeline of filters to produce the final list of ranked answers. A filter can drop unresponsive answers and rearrange the answers according to features such as confidence scores from the Answer Extractor and Ranker component.

Ranker. The Ranker unit contains a Question Ranker, Answer Ranker, and QA set Selector modules. These three modules will employs syntactic, semantic, and machine learning techniques for ranking.

4   Evaluation

This project plans to evaluate various independent components that are defined in Auto-FAQ-Gen independently and also using a theoretical framework to compare the general model with other existing QA systems. For instance, to comprehensive investigate the performance of the Question Generation module a precision and recall based evaluation technique will be used for the TREC-2007 data experiment.

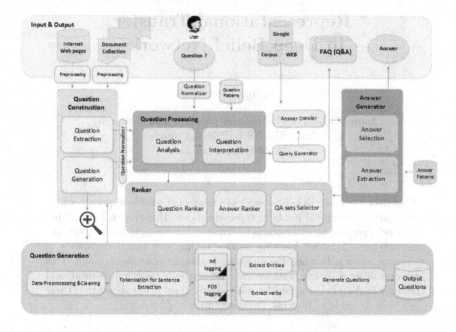

Fig. 1. Architecture of the Auto-FAQ-Gen: Modules and information flow between them

References

1. Hu, W.-C., Yu, D.-F., Jiau, H.C.: A faq finding process in open source project forums. In: 2010 Fifth International Conference on Software Engineering Advances (ICSEA), pp. 259–264. IEEE (2010)
2. Kothari, G., Negi, S., Faruquie, T.A., Chakaravarthy, V.T., Subramaniam, L.V.: Sms based interface for faq retrieval. In: Proceedings of the Joint Conference of the 47th Annual Meeting of the ACL and the 4th International Joint Conference on Natural Language Processing of the AFNLP, vol. 2, pp. 852–860. Association for Computational Linguistics (2009)
3. Mendes, A.C., Coheur, L.: When the answer comes into question in question-answering: survey and open issues. Natural Language Engineering **19**(1), 1–32 (2013)
4. Shrestha, L., McKeown, K.: Detection of question-answer pairs in email conversations. In: Proceedings of the 20th International Conference on Computational Linguistics, COLING 2004. Association for Computational Linguistics, Stroudsburg (2004)
5. Unger, C., Bühmann, L., Lehmann, J., Ngonga Ngomo, A.-C., Gerber, D., Cimiano, P.: Template-based question answering over rdf data. In: Proceedings of the 21st International Conference on World Wide Web, WWW 2012, pp. 639–648. ACM, New York (2012)
6. Verberne, S., Boves, L., Oostdijk, N., Coppen, P.: Data for question answering: the case of why. In: Proceedings of the 5th edition of the International Conference on Language Resources and Evaluation, LREC 2006, Genoa, Italy (2006)

Representational Transfer
in Deep Belief Networks

Xiang Jiang[(✉)]

Jodrey School of Computer Science, Acadia University,
Wolfville, NS B4P 2R6, Canada
120308j@acadiau.ca

Abstract. A Deep Belief Network is a machine learning approach which can learn hierarchical levels of representations. However, a Deep Belief Network requires large amounts of training examples to learn good representations. Transfer learning is able to improve the performance of learning, especially when the number of training examples is small. This paper studies different transfer learning methods using representational transfer in deep belief networks, and experimental result shows that these methods are able to improve the performance of learning.

1 Introduction

A Deep Belief Network is a type of machine learning method which can learn hierarchical levels of features or representations [1]. Deep learning can be categorized into discriminative deep learning architectures, such as convolutional neural networks [2], generative deep learning architectures, such as deep belief networks [3], hybrid deep learning architectures, such as a model pretrained using deep belief network and fine-tuned using backpropagation [4], and hierarchical deep learning architectures, such as hierarchical multitask learning and hierarchical deep models [5–7]. Deep learning has been very successful and has advanced the state-of-the-art of many machine learning applications, because they can automatically create hierarchical abstract features. However, deep learning architectures require a large amount of training examples to learn good representations. Transfer learning is a type of machine learning method which can use the knowledge of a previously learned task to inductively bias the learning of future tasks [8]. Transfer learning is a natural process in human learning, and it can improve the performance of learning when only a small number of examples are available for the new task. Representational transfer starts learning from direct or indirect assignment of representation taken from the model of a previous task.

2 Background

The most popular representational transfer method in deep learning architectures is layer-wise model adaptation. This method first builds a model for the source task, then starts with the representation of the source task and retrains

© Springer International Publishing Switzerland 2015

D. Barbosa and E. Milios (Eds.): Canadian AI 2015, LNAI 9091, pp. 338–342, 2015.
DOI: 10.1007/978-3-319-18356-5_31

the model for the target task layer by layer. This method can learn good mid-level representations from the source task to help the learning of the target task. Layer-wise model adaptation has been successfully applied to handwritten character recognition [9,10], video and image classification [11,12], and speech recognition [13].

3 Theory

There are at least three reasons why transfer learning can benefit from deep learning architectures: 1. *Shared internal representation:* Deep learning can learn shared internal representation in an unsupervised fashion from the examples of a number of different tasks, which enables the learner to generate useful features of the task domain and the context of the problem [14]. 2. *Hierarchical levels of representation:* Deep learning builds hierarchical levels of representation where each layer generate features from representation in the layer below it. Even for two different but related tasks, it is very probably that these tasks can share some lower levels of representations. This suggests that if we fix the lower levels of representations and retrain the higher levels of representations, we would be able to improve learning with a relatively small number of training data [10]. 3. *Learning from unlabeled data*: Since the layer-wise training in deep learning architectures is unsupervised, it enables us to leverage a small number of labeled data with a large amount of unlabeled data [15].

This paper studies different configurations of representational transfer in both backpropagation neural networks and Deep Belief Networks, and we propose that the transfer of representations learned in an unsupervised fashion is the best method of representational transfer.

4 Experiments

4.1 Objectives

The objectives of these experiments are: (1). To study whether representational transfer in a Deep Belief Network will improve classification accuracy. (2). To study whether representational transfer in a Deep Belief Network is better than representational transfer in backpropagation neural networks. (3). To study the effects of backpropagation for the source task in the process of representational transfer.

4.2 Experimental Configuration

The dataset used in the experiments is the MNIST database of handwritten digits where each example is a 28x28 normalized grey scale image [2]. 70% of the examples in the target task are used for training and the rest 30% are used for validation. An independent test set of 1000 examples was used for each digit. The source task is to classify one set of handwritten digits, and the target task is to classify a different set of handwritten digits. The number of examples of each digit ranges from 10 to 110. Our approach uses a four layer network with 784 inputs, two hidden layers and a softmax output layer for classification. The software for these experiments is a modified version of Theano [16].

4.3 Experiment 1 and 2: Single Task Learning

This is a control experiment to help evaluate the effectiveness of transfer learning.

Experiment 1 Build a classifier for the target task using backpropagation alone. **Experiment 2** Build a classifier for the target task using stacked Restricted Boltzmann Machines (RBMs) followed by backpropagation that updates the RBM weights.

4.4 Experiment 3: Representational Transfer in Backpropagation Neural Networks

This experiment has two steps: (1) Build a classifier for the source task using backpropagation; (2) Starting from the learned representation of the source task, build a classifier for the target task using backpropagation without examples from the source task. This experiment can provide some baseline results to compare with representational transfer in deep belief networks.

4.5 Experiment 4 and 5: Representational Transfer in Deep Belief Networks

The purpose of these experiments is to examine the performance of representational transfer in deep belief networks. There are two different configurations for representational transfer.

Experiment 4 has three steps: (1) Learn a generated representation for the source task using stacked RBMs; (2) Starting from that representation, learn a generated representation for the target task using stacked RBMs; (3) Finally starting from the generated representation, build a classifier for the target task using backpropagation with no use of source task examples.

Experiment 5 has four steps: (1) Learn a generated representation for the source task using stacked RBMs; (2) Starting from that learned representation of the source task, build a classifier for the source task using backpropagation for the softmax layer or all layers; (3) Using the existing representation learn a generated representation for the target task using stacked RBMs; (4) Finally starting from the existing representation, build a classifier for the target task using backpropagation with no use of source task examples.

The difference between experiment 4 and experiment 5 is the use of backpropagation after unsupervised learning of the source task. The comparison of these two experiments can provide insights into whether backpropagation learning of the source task will affect the performance of transfer learning.

4.6 Experimental Results

The classification accuracy is shown in Figure 1.

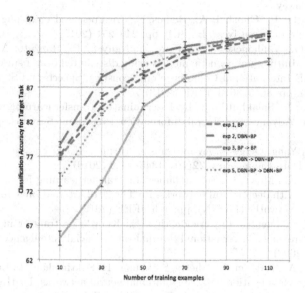

Fig. 1. Classification accuracy

The best model is achieved in experiment 4, which learns the source task in an unsupervised fashion, then, starting from the representation to learn a generated representation for the target task using stacked RBMs, finally builds a classifier for the target task using backproagation.

When comparing the classification accuracy in experiment 4 with that of experiment 3 and 5, we can infer that the process of backpropagation tend to make the target task get stuck in poor local minimal, and the unsupervised learning of the source task can develop better features for transfer.

5 Conclusions and Future Work

This paper studied different transfer learning methods using representational transfer in deep belief networks. Experimental results show that representational transfer in deep belief networks can improve the performance of supervised learning. The future work is to apply representational transfer on some larger datasets.

References

1. Bengio, Y.: Learning deep architectures for ai. Foundations and trends® in Machine Learning **2**(1), 1–127 (2009)
2. LeCun, Y., Bottou, L., Bengio, Y., Haffner, P.: Gradient-based learning applied to document recognition. Proceedings of the IEEE **86**(11), 2278–2324 (1998)

3. Hinton, G., Osindero, S., Teh, Y.-W.: A fast learning algorithm for deep belief nets. Neural Computation **18**(7), 1527–1554 (2006)
4. Deng, L.: Three classes of deep learning architectures and their applications: A tutorial survey
5. Swietojanski, P., Ghoshal, A., Renals, S.: Unsupervised cross-lingual knowledge transfer in dnn-based lvcsr. In: SLT, pp. 246–251 (2012)
6. Heigold, G., Vanhoucke, V., Senior, A., Nguyen, P., Ranzato, M., Devin, M., Dean, J.: Multilingual acoustic models using distributed deep neural networks. In: 2013 IEEE International Conference on Acoustics, Speech and Signal Processing (ICASSP), pp. 8619–8623. IEEE (2013)
7. Srivastava, N., Salakhutdinov, R.: Discriminative transfer learning with tree-based priors. In: Advances in Neural Information Processing Systems, pp. 2094–2102 (2013)
8. Pan, S.J., Yang, Q.: A survey on transfer learning. IEEE Transactions on Knowledge and Data Engineering **22**(10), 1345–1359 (2010)
9. Ciresan, D.C., Meier, U., Schmidhuber, J.: Transfer learning for latin and chinese characters with deep neural networks. In: The 2012 International Joint Conference on Neural Networks (IJCNN), pp. 1–6. IEEE (2012)
10. Gutstein, S., Fuentes, O., Freudenthal, E.: Knowledge transfer in deep convolutional neural nets. International Journal on Artificial Intelligence Tools **17**(03), 555–567 (2008)
11. Karpathy, A., Toderici, G., Shetty, S., Leung, T., Sukthankar, R., Li, F.-F.: Large-scale video classification with convolutional neural networks. In: IEEE Conference on Computer Vision and Pattern Recognition, CVPR (2014)
12. Oquab, M., Bottou, L., Laptev, I., Sivic, J., et al.: Learning and transferring mid-level image representations using convolutional neural networks (2013)
13. Huang, J.-T., Li, J., Yu, D., Deng, L., Gong, Y.: Cross-language knowledge transfer using multilingual deep neural network with shared hidden layers. In: 2013 IEEE International Conference on Acoustics, Speech and Signal Processing (ICASSP), pp. 7304–7308. IEEE (2013)
14. Baxter, J.: Learning internal representations. In: Proceedings of the Eighth Annual Conference on Computational Learning Theory, pp. 311–320. ACM (1995)
15. Raina, R., Battle, A., Lee, H., Packer, B., Ng, A.Y.: Self-taught learning: transfer learning from unlabeled data. In: Proceedings of the 24th International Conference on Machine Learning, pp. 759–766. ACM (2007)
16. Bergstra, J., Breuleux, O., Bastien, F., Lamblin, P., Pascanu, R., Desjardins, G., Turian, J., Warde-Farley, D., Bengio, Y.: Theano: a CPU and GPU math expression compiler. In: Proceedings of the Python for Scientific Computing Conference (SciPy). Oral Presentation (June 2010)

Unsupervised Multi-modal Learning

Mohammed Shameer Iqbal[(✉)]

Jodrey School of Computer Science, Acadia University,
Wolfville, NS B4P 2R6, Canada
111271i@acadiau.ca

Abstract. We present an unsupervised deep belief network that can learn from multiple channels and is capable of dealing with missing information. The network learns transferable features to accommodate the addition of a new channel using a combination of unsupervised learning and a simple back-fitting.

Keywords: Unsupervised learning · Associative memory · Restricted boltzmann machines · Multi-modal data · Sensory-motor learning

1 Introduction

Humans receive information through multiple channels and in multiple forms for a given task. As a human we learn to deal with different data channels by learning them in an associative manner which allows us to perform tasks even when only partial information is available. To emulate this associated learning, we develop a system that is capable of learning from multiple channels and responding to partial information. The learned representations can be transferred to a new sensory or modality channel. We develop a deep belief network that can learn multi-modal data in an associative manner so that network can reconstruct partial information such as an associated image given an audio signal. The associative representations are then transferred over to a motor channel without having to retrain the whole model by using a combination of unsupervised learning and simple back-fitting.

2 Background

Multi-modal systems that are capable of learning multiple sensory channels can be developed as one whole system rather than being developed as separate expert systems for each modality. By placing several deep belief network together to create a common shared representation, the whole system may produce better results than individual deep-belief networks [9].

A multi-modal approach has been adopted in deep learning by other researchers. Srivastava et al. [8] developed a system that learns images and text simultaneously and Ngiam et al.[7] developed a system capable of associating video and audio.

© Springer International Publishing Switzerland 2015
D. Barbosa and E. Milios (Eds.): Canadian AI 2015, LNAI 9091, pp. 343–346, 2015.
DOI: 10.1007/978-3-319-18356-5_32

They use back propagation to fine tune weights between two modalities, however this method is not practical for 3 or more modalities. By avoiding supervised techniques like back-propagation our system is able to associate more than two modalities with each other in a manner that progressively builds knowledge of the learners environment.

3   Theory

A deep belief network contains stacked layers of Restricted Boltzmann Machines (RBM) capable of learning multiple levels of abstraction [2][4]. Initially, we use two deep belief networks that learns handwritten images and representations of audio signals respectively. Each contain two stacked layers of RBMs. We bring these two system together by a shared layer of RBMs creating one whole deep belief network that associates one channel with another. The image channel learns hand-written images of digits 0-9 while the audio channel learns audio representation of spoken digits 0-9. These two channels are associated at top by the associative layer.

The weights connecting the associative layer and the two channels are then sharpened using the back-fitting algorithm [3]. The back-fitting algorithm takes the bidirectional weights from RBM and turns into two unidirectional set of weights, referred to as recognition weights going upwards and generative weights going downwards. Only the recognition weights are changed and generative weights remain the same as RBM weights, which allows the deep belief network to retain its ability to reconstruct the image and audio signal. Hence we can regenerate one channel given the other. Back-fitting process is local to the associative layer and the RBM layer immediately below it while the weights in the layers below remain unaltered. This plays a vital role when we add more channels, as the number of times back propagation applied increases in a quadratic fashion with respect to the number of channels.

After learning the associated image and audio representations, a third channel, which consists of 10 softmax units indicating the category, is learned as shown in Figure 1 [5]. The new modality is connected to the top level associative layer and trained using the same back-fitting algorithm.

4   Experiments

The objective is to test the ability of our deep belief network to learn multiple channels and reconstruct one channel given the other channel. We use a dataset of images and audio recordings collected from 20 male students. Each user was asked to write the digits 0 through 9 and then asked to speak them respectively, 10 times. The images are re-sized to 28-by-28 pixels and the audio is transformed to the frequency spectrum using STFT (Short-Time Fourier Transform) which produces 680 frequency bins [1]. We train the system with data from 18 users and tested it with the remaining 2 users' data. We repeat this process by rotating the dataset ten times. A portion of the training data (2 users) is used as a validation

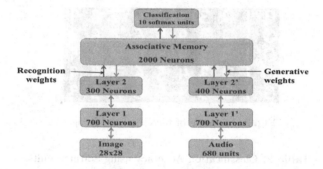

Fig. 1. System with added modality

set to prevent the network from over-fitting. After the layer-wise training of RBMs are done, the top associative layer weights are fine-tuned by using back-fitting algorithm.

4.1 Reconstructing One Channel from the Other

Our deep belief network is capable of regenerating the input on each channel as it is an associated generative network. We use this feature to reconstruct one channel given the other. The STFT transformation of audio signals is non-reversible and hence we cannot produce an actual audio signal after reconstruction, however its frequency domain representation can be evaluated for error. At the image channel, we can see the reconstructed image given audio signal as input. Since our system is unsupervised, we use a deep belief network developed by Hinton et al.[6] as an oracle to classify the reconstructed images to save time. Example images produced by the network are shown in Figure 2. The reconstructed images are given to Hinton's network to classify them. Table 1 shows the classification accuracy of each digit over 10 runs. Hinton's DBN classified only 3.67% of the reconstructed "7"s and 67.4% overall, which is the result of differences in pre-processing of MNIST and our dataset.

Table 1. Classification Accuracy from Hinton DBN

Digit	0	1	2	3	4	5	6	7	8	9	Average
Accuracy (%)	68.89	77.78	83.78	92.33	67.56	77.56	96.00	3.67	82.78	23.67	67.40

4.2 Learning a New Modality

We add a new channel to the existing deep belief network. The channel consists of 10 binary softmax units indicating the category of the digits. These are trained using the back-fitting algorithm as described in Section 3. Since we do not use back-propagation the generative features of the existing network are preserved. We have effectively transferred the associated representation from audio and image channels to this new channel which is able to classify the images and audio signals. The averaged results of 10 runs are tabulated in table 2.

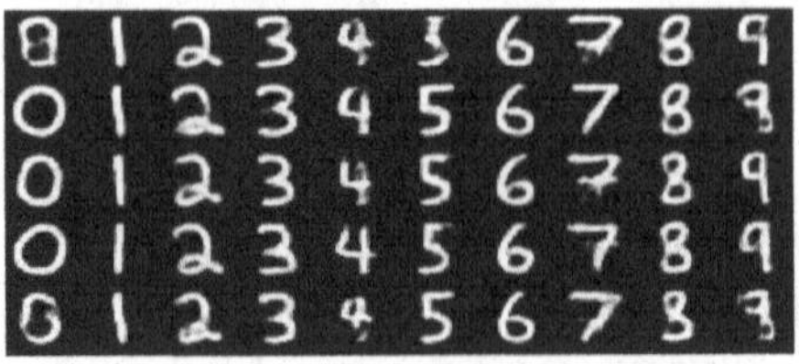

Fig. 2. Example of reconstructed Images

Table 2. Classification Accuracy using Softmax units

Input	0	1	2	3	4	5	6	7	8	9	Average
Image (%)	97	94	95	93	94	98	97	100	92	97	96
Audio (%)	86	76	91	84	83	82	99	88	98	77	86

5   Conclusion and Future Works

We have developed a system that can learn information from multiple channels and respond accurately to partial information. We used an associative layer to combine multiple channels using back-fitting which allows to easily add new channels. We found we were able to transfer representation from the associative layer to the new channel. In the future we will test adding different channels such as one that simulates a sensory-motor channel.

References

1. Allen, J.: Short-term spectral analysis, and modification by discrete fourier transform. IEEE Transactions on Acoustics Speech and Signal Processing **25**(3), 235–238 (1977)
2. Bengio, Y.: Learning deep architectures for AI. Foundations and Trends in Machine Learning **2**(1), 1–127 (2009)
3. Hinton, G., Osindero, S., Teh, Y.-W.: A fast learning algorithm for deep belief nets. Neural Computation **18**(7), 1527–1554 (2006)
4. Hinton, G.E.: Learning multiple layers of representation. Trends in Cognitive Sciences **11**(10), 428–434 (2007)
5. Hinton, G.E.: A practical guide to training restricted boltzmann machines, Technical report (2010)
6. Geoffrey, E.: Hinton and Ruslan R Salakhutdinov. Reducing the dimensionality of data with neural networks. Science **313**(5786), 504–507 (2006)
7. Ngiam, J., Khosla, A., Kim, M., Nam, J., Lee, H., Ng, A.Y.: Multimodal deep learning. In: Proceedings of the 28th International Conference on Machine Learning, ICML 2011, pp. 689–696 (2011)
8. Srivastava, N., Salakhutdinov, R.: Multimodal learning with deep boltzmann machines. Journal of Machine Learning Research **15**, 2949–2980 (2014)
9. Wang, T.: Classification via reconstruction using a multi-channel deep learning architecture. Master's thesis, Acadia University (2014)

On the Usage of Discourse Relations Across Texts with Different Readability Levels

Elnaz Davoodi[(✉)]

Department of Computer Science and Software Engineering,
Concordia University, Montréal, QC, Canada
e_davoo@encs.concordia.ca

1 Introduction

In a coherent text, text spans are not understood in isolation but in relation with each other through discourse relations, such as CAUSE, CONDITION, ELABORATION, etc. Discourse analysis involves modeling the coherence relations between text segments which allows readers to interpret and understand the communicative purpose of text's constitutive segments. Many natural language processing applications such as text summarization, question answering, text simplification, etc. can benefit from discourse analysis. In the proposed research project, we plan to use discourse analysis in the context of text simplification in order to enhance a text's readability level.

2 Background

Discourse Analysis: Most of the research in discourse analysis is based on two main annotation frameworks. The first, Rhetorical Structure Theory (RST) [8], assumes a hierarchical representation of texts consisting of a number of discourse units, called EDUs (Elementary Discourse Units) are related to each other through discourse relations that hold between them. In this framework, the recognition of relations do not rely on morphological or syntactic signals but rests on functional and semantic judgement. There are a number of RST-annotated corpora such as the RST corpus [1], Taboada's review corpus [14] and Instructional corpus [13]. In the original RST framework [8], 78 discourse relations are defined; however, the relations list is defined as being open.

Recently, a lexically-grounded framework called D-LTAG [15] was introduced which identifies discourse relations using discourse markers. In this annotation framework, relations are categorized as being *explicit* or *implicit*. An explicit relation is one where a discourse marker (e.g. *but, however*, etc.) is present in the text; otherwise it is implicit. As opposed to the relations inventory in the RST framework, in D-LTAG, relations are categorized into 3 levels of granularity. The Penn Discourse Tree Bank (PDTB) [10] and the Biomedical Discourse Relation Bank (BioDRB) [11] are two of the copora annotated using this framework.

I would like to thank my supervisor Dr. Leila Kosseim for the guidance, encouragement and advice she has provided throughout.

© Springer International Publishing Switzerland 2015

D. Barbosa and E. Milios (Eds.): Canadian AI 2015, LNAI 9091, pp. 347–351, 2015.
DOI: 10.1007/978-3-319-18356-5_33

Automatic tools called discourse parsers have been designed in order to identify discourse relations and their arguments automatically. Different discourse parsers use different approaches to segment the text, identify the arguments of a relation and label the relations. SPADE (Sentence-level PArsing for DiscoursE) [12], HILDA (HIgh-Level Discourse Analyzer) [6], the Feng & Hirst parser [5] and the PDTB-style End-to-End discourse parser [7] are four of the most well-known discourse parsers. In the proposed research, we plan to use the End-to-End discourse parser which is based on D-LTAG.

Text Complexity: Text complexity is a feature that can be reduced to make texts more accessible to readers. Text simplification techniques typically involve reducing the complexity of a text at the lexical (e.g. [4]) or syntactic level (e.g. [2]), while preserving its meaning. However, to our knowledge, very little work has focused on discourse-level simplification. Existing attempts are mostly based on a small set of rules extracted from small resources. The Simple English Wikipedia [3] released in 2011 was the first large scale parallel corpus of texts with different levels of complexity which can now be harvested for large scale analysis of discourse properties across texts with different complexity levels.

3   Research Questions

Discourse-level properties of a text such as the use of discourse relations, choice and/or position of discourse markers, the realisation of discourse relations as well as explicit or implicit relations, etc. can play a significant role in a text's readability level. The main goal of this research is to build a computational model to recommend appropriate discourse-level properties in order to increase a text's readability. Our main research questions will be:

1. What are the differences in the usage and realisation of discourse relations across texts with different readability levels?
2. What linguistic features can be used to determine discourse-level choices across readability levels?
3. How can a tool will be developed to recommend appropriate discourse-level choices to increase text readability?

4   Proposed Research Methodology

Given the availability of textual data and efficient discourse parser, our research we will follow a data-driven approach. We plan to use the End-To-End discourse parser developed based on D-LTAG and annotate parallel texts at different readability levels. Our proposed methodology consists of two main steps:

Corpus Analysis: The corpus analysis will involve collecting and annotating corpora, investigating possible discourse-level changes that can occur across readability levels and investigating which features seem to influence these changes. The parallel corpora that we have started to investigate are the Simple English Wikipedia and LiteracyWork parallel news articles[1]. The Simple English Wikipedia consists of two parts: (1) 60K aligned articles and (2) 167K aligned sentences. We have already investigated some discourse-level changes across simple and regular aligned sentences within this corpus. Four phenomenon have been identified:

1. Change in the choice and/or position of a discourse marker.
2. Change in the expression of the discourse relation from implicit to explicit or vice versa.
3. Change in the overall usage of discourse relations.
4. Removal of relation arguments.

The discovery of these four phenomenon will be valuable for building our recommendation model. As the readability level of a text is changed, the meaning and the communicative goal of a text should be preserved. Hence, it is forseeable that discourse relations are maintained. In order to verify this, we have performed an analysis of the sentence-aligned part of the Simple English Wikipedia corpus [3] to see if discourse relations are generally preserved across readability levels. We have used the End-To-End discourse parser [7] to identify the explicit discourse relations[2]. The results demonstrate that discourse relations seem to be preserved across readability levels. However, the usage of discourse markers is different across readability levels. In particular, we observed that the relative frequency of discourse markers is higher in more complex texts. Additionally, our analysis revealed that the distribution of discourse markers to convey specific relations is different across readability levels. These results seem to indicate that although the same logical and semantic information is conveyed in both simple and regular versions of a text; how they are signalled is different.

To complete our corpus analysis, we plan to focus on changes in specific relations and extract the exact changes along with the discourse-level features which cause these changes. Peterson and Ostendorf ([9]) used 21 features (mainly structural and syntactic features) in order to classify texts based on their complexity levels as complex or simple. However, we plan to use a wider range of discourse-level features and use Petersen and Ostendorf's features as baseline.

Model Development: We will develop a computational model from our corpus analysis to recommend appropriate discourse-level changes in order to modify a text's readability level.

---

[1] http://literacynet.org/cnnsf/index_cnnsf.html
[2] We only considered explicit discourse relations as the accuracy of the parser in detecting implicit relation is not very high (as low as 24.54%)

Once the corpus analysis and feature extraction is performed, we plan to develop a model in order to recommend discourse-level changes to enhance text readability. The changes we will consider include:

1. Change in the choice and position of discourse markers.
2. Change in the relation (e.g. from CONTRAST to CONCESSION).
3. Change in the expression of discourse relations from explicit to implicit or vice versa.

Our proposed methodology for this step is to use a supervised machine-learning approach trained using the corpus used for corpus analysis. Each instance in the corpus will be expressed as a vector in the feature space model and each relation will be modeled as a triplet structure: < *Choice of discourse marker, Position of discourse marker, Relation label*>. Then, each class label, will be a triplet structure which reflects the relation representation in the simpler counterpart.

5   Conclusions and Future Work

The main goal of this research is to use discourse analysis to improve textual readability. To acheive this goal, we proposed a research methodology consisting two main steps: a corpus analysis and a model development. We have already investigated some properties of the Simple English Wikipedia corpus across two levels of readability. Our preliminary work shows that explicit discourse relations are preserved across readability levels, while their representations using discourse markers are changed. Further in this research, we plan to investigate other characteristics of such corpora, extract discourse-level features and build a computational model to recommend discourse-level changes. We plan to use an intrinsic evaluation in order to evaluate the performance of our recommendation model.

Acknowledgments. The author would like to thanks anonymous referees for their useful comments. This work was financially supported by NSERC.

References

1. Carlson, L., Okurowski, M.E., Marcu, D.: RST discourse treebank. Linguistic Data Consortium, Catalog Number-LDC2002T07 (2002)
2. Chandrasekar, R., Srinivas, B.: Automatic induction of rules for text simplification. Knowledge-Based Systems **10**(3), 183–190 (1997)
3. Coster, W., Kauchak, D.: Simple English Wikipedia: A new text simplification task. In: Proceedings of NAACL-HLT, pp. 665–669. Portland (2011)
4. Devlin, S., Tait, J.: The use of a psycholinguistic database in the simplification of text for aphasic readers. In: Linguistic Databases, pp. 161–173 (1998)
5. Feng, V.W., Hirst, G.: Text-level discourse parsing with rich linguistic features. In: Proceedings of ACL, pp. 60–68. Jeju, Korea (2012)
6. Hernault, H., Prendinger, H., Duverle, D., Ishizuka, M.: Hilda: A discourse parser using support vector machine classification. Dialogue & Discourse **1**(3) (2010)

7. Lin, Z., Ng, H.T., Kan, M.Y.: A PDTB-styled end-to-end discourse parser. Natural Language Engineering **20**(2), 151–184 (2014)
8. Mann, W.C., Thompson, S.A.: Rhetorical structure theory: A framework for the analysis of texts. Tech. rep., IPRA Papers in Pragmatics 1 (1987)
9. Petersen, S.E., Ostendorf, M.: Text simplification for language learners: a corpus analysis. In: Proceeding of the SLaTE, pp. 69–72, Farmington, PA, USA (2007)
10. Prasad, R., Dinesh, N., Lee, A., Miltsakaki, E., Robaldo, L., Joshi, A., Webber, B.L.: The Penn discourse treebank 2.0. In: Proceedings of LREC, Marrakesh (2008)
11. Prasad, R., McRoy, S., Frid, N., Joshi, A., Yu, H.: The Biomedical Discourse Relation Bank. BMC Bioinformatics **12**, 188 (2011)
12. Soricut, R., Marcu, D.: Sentence level discourse parsing using syntactic and lexical information. In: Proceedings of NAACL-HLT, pp. 149–156, Edmonton (2003)
13. Subba, R., Di Eugenio, B.: An Effective Discourse Parser that uses Rich Linguistic Information. In: Proceedings of NAACL-HLT, pp. 566–574. Colorado (2009)
14. Taboada, M., Anthony, C., Voll, K.: Methods for creating semantic orientation dictionaries. In: Proceedings of the LREC, pp. 427–432, Genova, Italy (2006)
15. Webber, B.: D-LTAG: Extending lexicalized tag to discourse. Cognitive Science **28**(5), 751–779 (2004)

A Framework for Discovering Bursty Events and Their Relationships from Online News Articles

Pooria Madani[✉]

Faculty of Computer Science, University of New Brunswick, Fredericton, NB, Canada
s.p.madani@gmail.com

Abstract. The main goal of this research work is to develop a framework with a set of methods for discovery bursty events and their relationships from streams of online news articles. Our aim is to build a fully functional system that indexes news text documents and their corresponding terms occurrences according to the timestamp (temporal indexing) in order to discover bursty terms. The discovery of a bursty event can then be done using the discovered bursty terms which are significantly smaller in size compared to the original feature-set. Furthermore, the discovered bursty events are compared against each other in order to discover any potential relational link between any of two. It is the assumption of this work that the bursty events and their relations in time can provide useful information to firms and individuals whose decision-makings tasks are significantly affected by news events.

1 Introduction

With advances in technology and the world wide web, access to vital information became easier and faster. As a result, analyzing such information and acting on them is more important than any time in the history of civilizations. Today, a typical investor has to read a huge volume of news articles and base his decision on them by connecting the dots and predicting future of stock market. A CEO of a big food corporation has to be on top of the ongoing news in the world since different events can effect the supply chain of the company and result in a saviour market lost. Consequently, with today's availability of information, only those who analyze these information quicker are able to see the big picture and succeed in the competitions.

Event extraction and tracking, a subfield of Topic Modelling and Topic Tracking (TDT) in temporal text mining, studies models of event detection and extraction from stream of unstructured text data. Events play important role in our daily life and every organization utilizes news events differently. For example, scientist who wants to study all the earthquakes that occurred in the United States since 1980 can use event detection and extraction system towards finding

© Springer International Publishing Switzerland 2015
D. Barbosa and E. Milios (Eds.): Canadian AI 2015, LNAI 9091, pp. 352–355, 2015.
DOI: 10.1007/978-3-319-18356-5_34

such events and its related information from historical news articles (retrospective event extraction). Or it may be important for an hedge fund in Wall Street to detect events as they occur that may affect their investments.

The goal of this research work is to construct a framework for bursty event discovery system from news articles. Due to the importance of bursty events, it is our objective goal to build a system capable of discovery bursty events and find their possible relation with old bursty events that have occurred in past.

Our contributions are providing a general framework for temporal document indexing, developing a novel procedure in discovering bursty events, and discovering link between the bursty events in course of time and visualizing them as a graph.

2 Related Works

Kleinberg [4] proposed use of infinite-state automaton to model the arrival times of documents in a streams in order to identify bursts that have high intensity over limited durations of time. The states of the probabilistic automaton correspond to the frequencies of individual words, while the state transitions capture the burst, which correspond to a significant change in word frequency. However, this model assumes documents in the stream are all about one single topic which renders this model useless for applications tracking multiple topics.

Chang et al. [3] applied spectral analysis using discrete (DFT) to categorize features for different event characteristics (e.g., important or not, and periodic or aperiodic events). DFT converts the signals from the time domain into the frequency domain, such that a burst in the time domain corresponds to a spike in the frequency domain. However, DFT cannot identify the period of a bursty event. Therefore, they employed Gaussian mixture models to identify feature bursts and their associated periods.

Snowsill et al. [5] presented an online approach for detecting events in news streams based on statistical significant tests of n-gram word frequency within a time frame. An incremental suffix tree data structure was applied to reduce the time and space constraints required for online detection.

Story Link Detection(SLD) is the final task in a TDT system by finding link between two group of selected stories. Unfortunately, SLD has not been studied extensively by the research community due to the unclarity of its applications [1]. Normally SLD is done by comparing returned value of a *similarity* function, as a result the main researches in this field is done on comparing different similarity function computations.

3 Proposed Framework

Figure 1 shows an overview of our proposed framework. Since we are considering the timestamp of news articles as one of the major feature in this work, it is important to index the news articles according to their timestamp. Therefore, the first step in this framework is the temporal indexing. Our proposed temporal indexing module is responsible for document tokenization, computing values such as *Document Term Frequency(TF)*, and storing the relational links between occurred words in a document.

Not all the tokenized words are important and may not carry vital information for the purpose of bursty event discovery. In order to improve the stability and accuracy of the clustering algorithms, it is important that only a subset of features get selected for bursty event discovery purposes. The idea behind bursty feature identification is to generate a probabilistic model of $P_g(n_{i,j})$ for each feature f_i in a window w_j by counting the documents that contain the feature denoted as $n_{i,j}$. This model proposed by Fung et al. [2], estimate an unknown hyper-geometric distribution by performing the estimation on binomial distribution. The probability distribution will model the success probability of occurrence of $n_{i,j}$ of feature f_i in a window w_j given the number of the documents in the window.

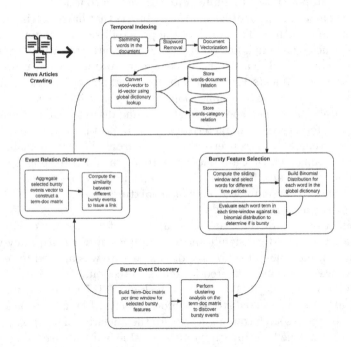

Fig. 1. Overview of The Proposed Event Relation Discovery Framework

Feature-pivot clustering, as term suggests, is done by clustering the feature set instead of the document set. documents can belong to more than one cluster as its forming features are taking apart in different clusters. For example, a document that discusses government shutdown and its effect on stock markets can very well contribute to both events of *government shutdown*, and *collapse in the stock market*. Such documents help to create link between different event clusters and helps to summarize and explain some of these events. Therefore, it is reasonable to consider that some news articles may discuss more than one event and by clustering features instead of documents it is possible to discover such interrelated events. In this study, we adopt feature-pivot clustering as our main approach on event discovery and extraction. Once the bursty features are selected, they will be clustered using a dynamic clustering algorithms in order to form bursty events. *Silhouettes Score* will be used to automatically verify the quality of the form clustered, and clusters with lower score are discarded as noise.

Once the bursty events are discovered (i.e. nodes), they needs to be compared to previous discovered nodes back in time in order to draw any conclusion about their relation. Cosine similarity function is been proposed to be used in order to discover any potential relationship. However, more complex methods could be used in order to accomplish the relation discovery task. Discovering relation between different real world events, is a hard problem and often is executed by domain experts. Therefore, it is reasonable to expand the framework to incorporate a knowledge-base in order to execute the relation discovery task by conducting semantic analysis. This system requires large number of news articles in order to be able to detect bursty features. As a result, validating such system with a large unlabeled dataset is a hard task. One of the challenges that this research is faced with is validation methods that are going to be used for the discovered bursty events and their relationships.

References

1. Allan, J.: Introduction to topic detection and tracking. In: Allan, J. (ed.) Topic Detection and Tracking. The Information Retrieval Series, vol. 12. Springer, US (2002)
2. Fung, G.P.C., Yu, J.X., Yu, P.S., Lu, H.: Parameter free bursty events detection in text streams. In: Proceedings of the 31st International Conference on Very Large Data Bases, pp. 181–192 (2005)
3. He, Q., Chang, K., Lim, E.-P., Zhang, J.: Bursty feature representation for clustering text streams. In: SDM, pp. 491–496. SIAM (2007)
4. Kleinberg, J.: Bursty and hierarchical structure in streams. Data Mining and Knowledge Discovery 7(4), 373–397 (2003)
5. Snowsill, T., Nicart, F., Stefani, M., De Bie, T.: Finding surprising patterns in textual data streams, pp. 405–410. IEEE (2010)

Behavioural Game Theory: Predictive Models and Mechanisms

James R. Wright[✉]

University of British Columbia, Vancouver, BC, Canada
jrwright@cs.ubc.ca

1 Introduction

Classical economic models proceed from strong rationality assumptions which are known to be inaccurate (as no human is perfectly rational), but which are thought to reasonably approximate aggregate human behaviour. However, there is now a wealth of experimental evidence that shows that human agents frequently deviate from these models' predictions in a predictable, systematic way. Using this data, there is now an opportunity to model and predict human economic behaviour more accurately than ever before. More accurate predictions will enable the design of more effective multiagent mechanisms and policies, allowing for more efficient coordination of effort and allocation of resources.

Prediction (as distinct from description or explanation) is valuable in any setting where counterfactuals need to be evaluated, as where the impact of a policy needs to be determined before it is enacted. In this work, I am most interested in approaches that make explicit predictions about which actions a player will adopt, and that are grounded in human behaviour.

In a multiagent setting, perhaps the most standard game-theoretic assumption is that all participants will adopt Nash equilibrium strategies. However, experimental evidence shows that Nash equilibrium often fails to describe human strategic behaviour [e.g., Goeree and Holt, 2001]—even among professional game theorists [Becker et al., 205].

The relatively new field of *behavioural game theory* extends game-theoretic models to account for deviations from the standard models of behaviour, and proposes new models to account for human behaviour by taking account of human limitations [Camerer, 2003]. Researchers have developed many models of how humans behave in strategic situations based on experimental data. But this multitude of models presents a problem: which model should be used?

My thesis is that human behaviour can be predicted effectively in a wide range of settings by a single model that synthesizes known deviations from economic rationality. In particular, I claim that such a model can predict human behaviour better than the standard economic models. Economic mechanisms are currently designed under behavioural assumptions (i.e., full rationality) that are known to be unrealistic. A mechanism designed on the basis of a more accurate model of behaviour will be more able to achieve its goal, whether that goal is social welfare, revenue, or any other aim.

© Springer International Publishing Switzerland 2015
D. Barbosa and E. Milios (Eds.): Canadian AI 2015, LNAI 9091, pp. 356–359, 2015.
DOI: 10.1007/978-3-319-18356-5_35

2   Progress to Date

In my completed work I analyzed and evaluated several behavioural models for simultaneous-move games, eventually identifying a specific class of models (iterative models) as the state of the art. I then proposed and evaluated an extension that improves the prediction performance of any iterative model by better incorporating the behaviour of nonstrategic agents.

2.1   Model Comparisons [Wright and Leyton-Brown, 2010]

In my initial project, I explored the question of which of the quantal response equilibrium [McKelvey and Palfrey, 1995], level-k [Costa-Gomes et al., 2001; Nagel, 1995], cognitive hierarchy [Camerer et al., 2004], and quantal level-k [Stahl and Wilson, 1994] behavioural models was best suited to predicting out-of-sample human play of normal-form games. I also evaluated the standard game theoretic solution concept, Nash equilibrium.

Using a large set of experimental data drawn from the literature, I identified a single "best" model, quantal level-k, which performed best or nearly-best on each source dataset, plus a combined dataset. This is a striking result, as one might reasonably expect different models to perform well on different datasets.

2.2   Bayesian Parameter Analysis [Wright and Leyton-Brown, 2012]

In this work, I used a Bayesian approach to better understand the entire parameter space of two behavioural models: quantal level-k, the best-performing model identified in my previous work (Section 2.1); and cognitive hierarchy.

The parameter analysis identified several anomalies in the parameter distributions for quantal level-k, suggesting that a simpler model could be preferable. I identified a simpler, more predictive family of models based in part on the cognitive hierarchy concept. Based on a further parameter analysis of this family of models, I derived a three-parameter model, QCH, that predicts better than the five-parameter quantal level-k.

2.3   Level-0 Meta-Models [Wright and Leyton-Brown, 2014]

Iterative models such as QCH (Section 2.2) predict that agents reason iteratively about their opponents, building up from a specification of nonstrategic behaviour called level-0. The modeller is in principle free to choose any description of level-0 behaviour that makes sense for the given setting; however, in practice almost all existing work specifies this behaviour as a uniform distribution over actions. In most games it is not plausible that even nonstrategic agents would choose an action uniformly at random, nor that other agents would expect them to do so. In this work I considered "meta-models" of level-0 behaviour: models of the way in which level-0 agents construct a probability distribution over actions, given an arbitrary game. A linear weighting of features that can be computed from any normal form game achieved excellent performance across the board, yielding a combined model that unambiguously outperforms the previous state of the art.

3   Proposed Research

My proposed research will build upon my existing work on prediction in one-shot games, and extend it in two main directions: first, by leveraging machine learning techniques to improve the quality of the predictive models; second, by studying the implications of more accurate models of human behaviour for designing mechanisms and protocols.

3.1   Feature Discovery for Predicting Human Behaviour

Thus far, game properties that people might find *salient*—and hence be favoured by nonstrategic agents–are discovered primarily through introspection about particular examples, by asking oneself "How might I reason about playing this specific game?" *Deep learning* (see, e.g., [Bengio, 2009]) is a recent machine learning paradigm that has shown success in a wide range of (mostly signal processing) domains for automatically determining problem features. To my knowledge, deep learning has never before been applied in a game-theoretic context. I plan to study the adaptation of these techniques to the behavioural game theory domain to discover new representations of salient game characteristics.

I will begin work on this project by seeking to build highly predictive deep models. However, another important task will be integrating the discovered properties and insights into the framework of explicitly cognitive, lower-dimensional models, which I expect to be much easier to fit and use in applications.

3.2   Endogenous Levels

Most iterative models, including QCH, take the distribution of levels as a parameter. This implicitly assumes that the proportion of agents playing at a given level k will be identical regardless of the setting. This is unlikely to be true; rather, agents should be willing to perform more counterspeculation when it is easier or may plausibly yield greater rewards, and less otherwise.

In this phase, I will investigate ways of making the choice of level *endogenous* to the QCH model—that is, having the properties of the game or setting itself determine the distribution of levels, rather than having the distribution of levels be a parameter that must explicitly be learned.

3.3   Theoretical Implications for Mechanism Design

Standard mechanism design makes implicit assumptions about how agents will behave in response to incentives; typically that agents will play a Nash equilibrium. In this work, I will consider the implications for mechanism design of moving to a more accurate behavioural model. For example, what objectives are implementable according to an accurate model of behaviour, compared to those that can be implemented in equilibrium? What approximation guarantees are possible in terms of the properties of the agent population?

References

Becker, T., Carter, M., Naeve, J.: Experts playing the traveler's dilemma. Diskussionspapiere aus dem Institut fr Volkswirtschaftslehre der Universitt Hohenheim 252/2005, Department of Economics, University of Hohenheim, Germany (2005)

Bengio, Y.: Learning deep architectures for AI. Foundations and Trends in Machine Learning **2**(1), 1–127 (2009)

Camerer, C., Ho, T., Chong, J.: A cognitive hierarchy model of games. QJE **119**(3), 861–898 (2004)

Camerer, C.F.: Behavioral Game Theory: Experiments in Strategic Interaction. Princeton University Press (2003)

Costa-Gomes, M., Crawford, V., Broseta, B.: Cognition and behavior in normal-form games: An experimental study. Econometrica **69**(5), 1193–1235 (2001)

Goeree, J.K., Holt, C.A.: Ten little treasures of game theory and ten intuitive contradictions. AER **91**(5), 1402–1422 (2001)

McKelvey, R., Palfrey, T.: Quantal response equilibria for normal form games. GEB **10**(1), 6–38 (1995)

Nagel, R.: Unraveling in guessing games: An experimental study. AER **85**(5), 1313–1326 (1995)

Stahl, D., Wilson, P.: Experimental evidence on players' models of other players. JEBO **25**(3), 309–327 (1994)

Wright, J.R., Leyton-Brown, K.: Beyond equilibrium: predicting human behavior in normal-form games. In: Twenty-Fourth Conference of the Association for the Advancement of Artificial Intelligence, AAAI 2010, pp. 901–907 (2010)

Wright, J.R., Leyton-Brown, K.: Behavioral game-theoretic models: a Bayesian framework for parameter analysis. In: Proceedings of the 11th International Conference on Autonomous Agents and Multiagent Systems, AAMAS 2012, vol. 2, pp. 921–928 (2012)

Wright, J.R., Leyton-Brown, K.: Level-0 meta-models for predicting human behavior in games. In: Proceedings of the Fifteenth ACM Conference on Economics and Computation, pp. 857–874. ACM (2014)

Author Index

Printed in the United States
By Bookmasters

Printed in the United States
By Bookmasters